Microwave Technology

Holger Heuermann

Microwave Technology

Field Simulation, Non-linear Circuit Technology, Components and Subsystems, Plasma Technology, Antennas and Propagation

Springer

Holger Heuermann
IMP
FH Aachen
Aachen, Germany

ISBN 978-3-658-45685-6 ISBN 978-3-658-45686-3 (eBook)
https://doi.org/10.1007/978-3-658-45686-3

Translation from the German language edition: "Mikrowellentechnik" by Holger Heuermann, © Springer Fachmedien Wiesbaden GmbH, ein Teil von Springer Nature 2020. Published by Springer Fachmedien Wiesbaden. All rights reserved.

This Springer imprint is published by the registered company Springer Fachmedien Wiesbaden GmbH, part of Springer Nature.
The registered company address is: Abraham-Lincoln-Str. 46, 65189 Wiesbaden, Germany

Foreword

Towards the close of the previous century, the ubiquitous presence of microwave products within personal attire was virtually nonexistent; conversely, contemporary society finds itself heavily reliant upon such commodities. The primary objective of this tome is to furnish nascent engineers with comprehensive instruction while concurrently acquainting seasoned practitioners with contemporary advancements in the field.

The exposition of microwave technology herein commences with an elucidation of modern electromagnetic simulation methodologies and the rudiments of non-linear switching techniques. The discourse is pragmatically oriented, eschewing excessive abstraction in favor of meticulous yet accessible elucidation, striving for completeness while endeavoring to maintain simplicity. To this end, the narrative, where feasible, operates at the level of Kirchhoff's laws, opting for their relative simplicity over the more intricate domain of Maxwell's equations, as favored by some. Foundational to this discussion is the introduction to mixed-frequency measurement and S-parameter technology, underpinning a novel technological framework, notably more streamlined than the extant X-parameter technology, and successfully leveraged by the IMP Institute, under my stewardship, in the realization of various innovations.

Furthermore, the text acquaints the reader with the requisite components and subsystems pertinent to the single and double-digit GHz spectrum, serving applications spanning radio, navigation, localization, measurement, medical science, heating, and fusion, among others, with an emphasis on non-linear phenomena. While certain chapters, such as those addressing power amplification, draw heavily from established literature, others, like the discourse on oscillators, adopt a novel pedagogical approach, incorporating the latest research findings.

A notable contribution originating from the IMP is the development of GHz plasma technology, boasting manifold applications arising from several groundbreaking innovations vis-à-vis conventional plasma torches. Detailed exposition is provided on applications ranging from luminaires to spark ignition systems, surgical implements, 3D printing apparatus, and decoupled plasma, the latter finding utility in highly efficient culinary devices. Additionally, the narrative elucidates the trajectory towards fusion power sources, culminating with discussions on antennas and propagation phenomena, thereby offering a comprehensive treatment of the subject matter.

This volume draws upon the syllabi of the courses "Microwave Technology" and "Antennas and Propagation" offered as part of the Master's program in Electrical Engineering at the FH Aachen University of Applied Sciences. Supplementary examination materials, available online, afford students a practical means to engage with the material presented herein. Furthermore, this text complements my preceding work, "Hochfrequenztechnik" (Radio Frequency Technology), which addresses undergraduate coursework and, in conjunction with the present volume, furnishes graduates with the requisite proficiency to navigate the developmental landscape of GHz components and systems. Notably, an English translation of "Radio Frequency Technology" is slated for publication in 2025.

Indeed, a myriad of motivations underpin the substantial effort invested in crafting such a tome. For the author, paramount among these is the aspiration to inspire international students to pursue their Master's studies at our esteemed university, thereby mitigating the dearth of skilled professionals in our locale. Given the absence of tuition fees in Germany, the allure of academic pursuit is inherently heightened, rendering it an attractive proposition from the outset. Moreover, the development of English-language instructional materials holds promise for facilitating guest lectures overseas post-retirement, perhaps serving as an interim measure during sabbaticals or research sojourns.

Acknowledgments

I would like to express my greatest thanks to my three wonderful daughters Helena, Clara and Emelie, to whom I dedicate this book and to whom I was not able to give nearly as much time as I would have liked, but to whom I could not do justice due to my ambitious work goals (including this book project, to make the world a better place through research). Of course, this also applies to my beloved wife, who was the main reason why we can be so proud of our children today.

I would also like to thank the FH Aachen, starting with the rectorate member Prof. Ritz as well as the administration, the Dean's Office and all my colleagues in the department for their great support and the excellent environment.

I would also like to thank all current and former IMP employees who, like me, enjoy working on the exciting and complex research topics in GHz technology and who also hope to make the world a cleaner and safer place.

Last but not least, I would also like to thank the many students whose motivation is a great driving force for my teaching and whose many questions and comments have turned the chapters in this book into what you, dear readers, now have before you.

Aachen Holger Heuermann
July 31, 2024

Contents

Chapter 1
Waveguides

1.1 Introduction

The history of waveguides traces back to 1893 with the first proposal from J.J. Thomson and to the seminal work of J.C. Bose, whose scientific article "On the Determination of the Wave-Length of Electric Radiation by Diffraction Grating" in 1897 laid foundational principles for their development. Bose's pioneering research marked the earliest exploration into the propagation of electromagnetic waves through metallic tubes, laying the groundwork for subsequent advancements in waveguide theory and application. His contributions paved the way for the modern understanding and utilization of waveguides across a myriad of fields, from telecommunications to microwave engineering.

Until 1970, so-called waveguides were the most commonly used conductor systems in high-frequency and microwave technology. All predominantly military applications and radio relay systems used waveguides. Waveguides were the most widely used conductor system in the GHz range. However, with the advent of coaxial cables and other technologies, waveguide systems began to diversify, leading to new applications and an expanded range of possibilities. In this chapter, the important propagation mode in the most commonly used waveguide is graphically derived in detail, all waveguide sizes are introduced, various waveguide shapes are presented and the associated conductor modes as well as waveguide components are visualized in detail using electromagnetic simulations.

Definition: Waveguide: Tubular metallic guide with a specific cross-sectional shape, hollow inside or dielectric filled

Waveguides conduct electromagnetic waves in a similar way to light – electromagnetic waves. Waveguides are used in applications mainly in the GHz range. Waveguides differ in many respects many respects from TEM transmission lines (Fig. 1.1).

	TEM transmission lines	**Waveguides**
Cross-section geometry:	Multiple contiguous area	Single contiguous area

Nowadays, waveguides are used in particular for systems with very high performance – such as radar systems, microwave ovens – are used. In the consumer sector, waveguides are used as part of the antennas of low-noise converters (LNC).

H. Heuermann, *Microwave Technology*,
https://doi.org/10.1007/978-3-658-45686-3_1

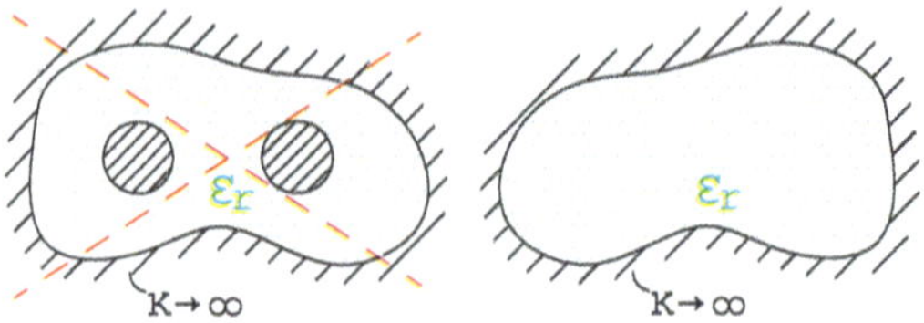

Fig. 1.1 Cross-sections of a multi-conductor (left) and a waveguide (right)

It is very important to understand waveguides in order to find interference sources in electronic high-frequency circuits: Each shielded housing corresponds to a waveguide with a shorting plate at the input and output. These usually metallic housings have natural resonances at frequencies that can be calculated using the wavelengths $n \cdot \lambda/2$ (n: integer number). In these frequency ranges, the electromagnetic energy is coupled into the entire circuit via the waveguide mode. This means that an oscillation often occurs in an active circuit at these frequencies via this feedback (via the waveguide mode).

Characteristics of the Waveguides

- No electrical forward and return conductor
- No transmission of direct current and voltage

Technical Forms of Implementation

- Rectangular waveguides (mainly)
- Round waveguide (e.g. antennas, rotary joints)

Advantages Waveguide compared to coaxial conductor

- Very low-loss (up to lossless!)
- Very high power compatibility
- High-pass behavior with strong filter effect

Disadvantages Waveguide compared to coaxial conductor

- Rigid cable guide
- Expensive manufacturing costs
- Limited bandwidth use
- Multimode problems

Applications

- Transmission systems with high performance such as radar or directional radio

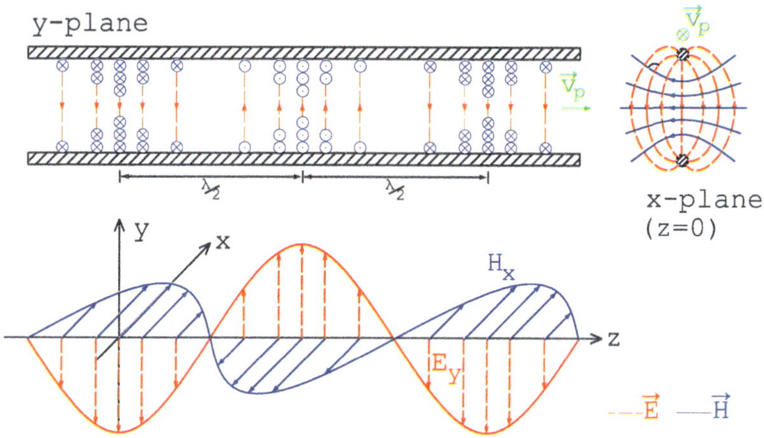

Fig. 1.2 TEM wave along a two-wire line with propagation in z-direction

- Microwave ovens, GHz measurement technology, antenna technology (here also in the consumer range up to 77 GHz)
- 77 GHz-Radar in the automotive sector (here also within circuit boards)

Before discussing the theory of waveguides, a very brief summary of the TEM waveguides with the numerous TEM waveguides with the numerous representatives such as microstrip, coaxial and two-wire lines (Fig. 1.2).

1.1.1 Brief Summary of the TEM Wave

In a TEM transmission line, a wave that has no field component in the direction of propagation through two-wire conductors. The simple relationship applies to the so-called "right hand rule":

Two wire conductor „$\vec{E}$ turns to $\vec{H}$"
In a two-wire line, two transverse field components field components (here E_y, H_x) occur. The propagation is similar to a homogeneous plane wave (HPW) in free space with linear polarization.

1.1.2 Field Theory for Waveguides

In numerous publications (e.g. [94, 116]), the exact field-theoretical derivations and solutions are presented in analytical form. In short form, the analysis is as follows:

Procedure: Given: Maxwell's equations and cross section

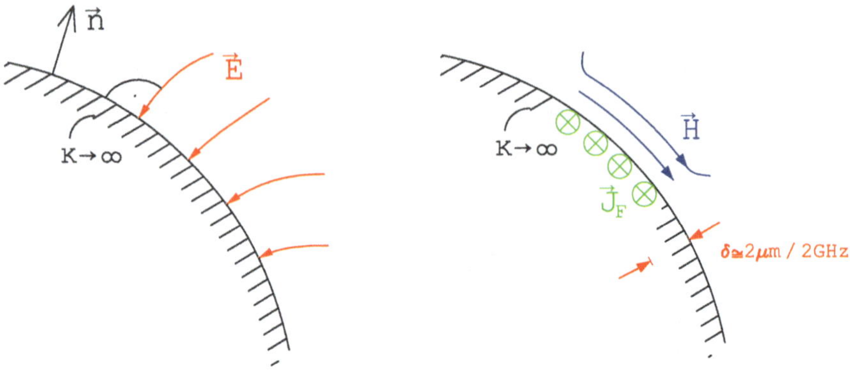

Fig. 1.3 Left part: Typical E-field on a metallic surface; right part: H-field lines and via a metallic surface and associated current density J

with boundary conditions (e.g. $\vec{E} = 0$)

⇒ Approach: Eigenvalue problem

Results: Infinite number of discrete field types (not TEM)

a) Wave types, propagation modes ($\lambda_C \geq \lambda$, transport of energy)

b) Attenuation types, no propagation ($\lambda_C \leq \lambda$, no transport of energy, but storage of energy)

The wave propagation only takes place above a certain cut-off frequency f_c (transverse dimensions $\geq \lambda/2$).

This estimation can also be used to estimate the danger of possible waveguide mode propagation in housings.

In the following, the fields within a rectangular waveguide will be introduced using a graphical derivation.

1.2 Graphical Derivation of the H_{10}-Wave

1.2.1 Boundary Conditions on the Ideal Conductor

The boundary conditions for the electric field are visualized in Fig. 1.3. The electric fields are perpendicular to the surface of a metallic plane. **E-Field** vertical $\vec{n} \times \vec{E} = 0$ The electric fields are perpendicular to the surface of a metallic plane.

The 2D sectional view is shown (as in Fig. 1.3) of an extended metallic surface. In very approximation, the field lines have the same behavior for the metallic materials used in practice. The magnetic fields form parallel to the surface of a metallic plane.

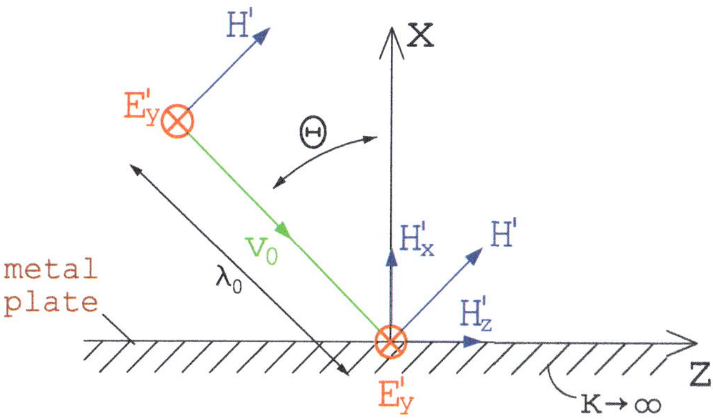

Fig. 1.4 Field components and associated angles of an incident HPW

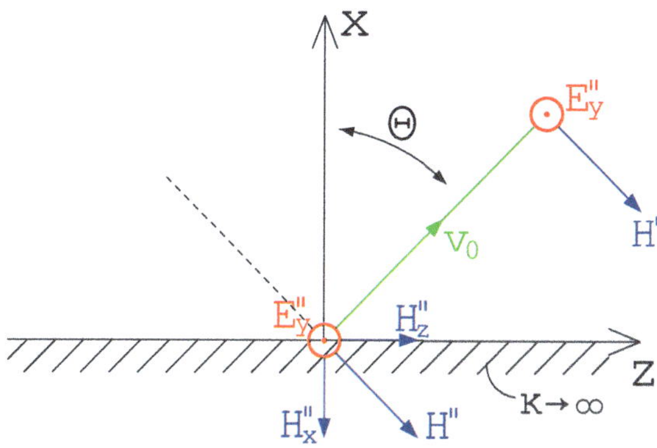

[scale=0.9]

Fig. 1.5 Field components and associated angles of a reflected HPW

1.2.2 Reflection of a HPW on a Metallic Plate

The incident homogeneous plane wave (HPW) on a perfectly conducting plane is shown in Fig. 1.4 at time t_0.

Incident HPW

This incident wave is totally reflected. This must the boundary condition that the E-field is zero on the metallic surface must be fulfilled. The corresponding reflected wave at time t_0 shows the Fig. 1.5.

Reflected HPW

The following applies to the field components shown in the geometric zero point:

$$-E'_y = E''_y, \qquad H'_x = -H''_x \qquad \text{and} \qquad H'_z = H''_z \ . \tag{1.1}$$

1.2.3 Construction of the H_{10}-Wave from Multiple Reflections

In the following, the so-called H_{10}-wave is constructed in its basic form as a transverse electric wave in a rectangular waveguide. The following Fig. 1.6 shows an incident homogeneous plane wave (HPW) on a metallic plate in the upper part. This electromagnetical wave is reflected. The green arrow indicates the direction of incidence. V_0 should stand for the velocity of the incident wave. For the medium air, this is at the speed of light.

The middle part of the Fig. 1.6 shows the reflected HPW on a metallic plate. The direction of the green arrow indicates the direction of the electromagnetic wave.

Superimposing the two waves results in an image of the H_{10} wave (lower part in the Fig. 1.6). The superposition of the magnetic fields results in closed magnetic fields. At the height $x = a$, both the magnetic fields and the electric fields are zero. The same conditions apply as at the height $x = 0$.

In the plane x=a hold E=0 .
⇒ At this level, a metal wall can be installed!

If the implementation of a substance does not change the electromagnetic fields, it is also permissible. Between these two plates, an H_{10}-wave propagates.

No boundary conditions are violated and therefore the field solution is also a correct solution. The Fig. 1.6 can be used to calculate the waveguide wavelength $\frac{\lambda_h}{2}$ can be seen.

1.2.4 Field Images of the H_{10}-Wave in the Rectangular Waveguide

The H_{10}-wave is the technically used wave in the rectangular guide. The following images illustrate the field of the H_{10}-wave from different perspectives (Fig. 1.7).

In all illustrations, the electric field lines are shown in red and the magnetic field line in blue. The wavelength in the waveguide is denoted by λ_h (Fig. 1.8).
Fig. 1.9 shows the H_{10}-wave, which was created using a numerical field simulation (these are explained in the next chapter).

The intensities are shown according to the rainbow colors and the lengths of the arrows. The red areas are therefore the zones with a high energy density.
On the right-hand gate, you can clearly see from the protruding arrows that the starting point of the arrows represents the energy point.

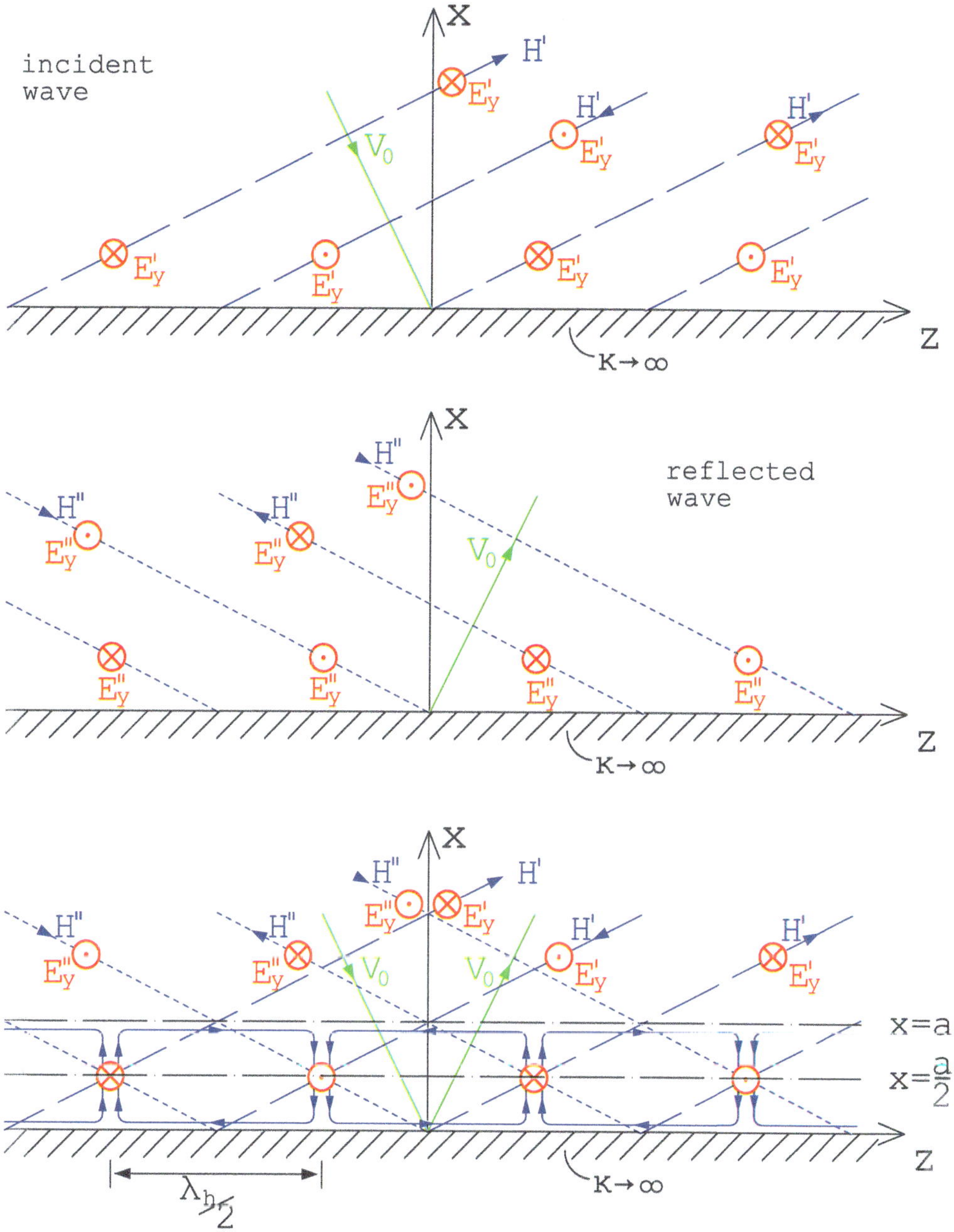

Fig. 1.6 Incident and reflected wave and their superposition to construct the of the H_{10}-wave (plane x=a is later the wall of the waveguide)

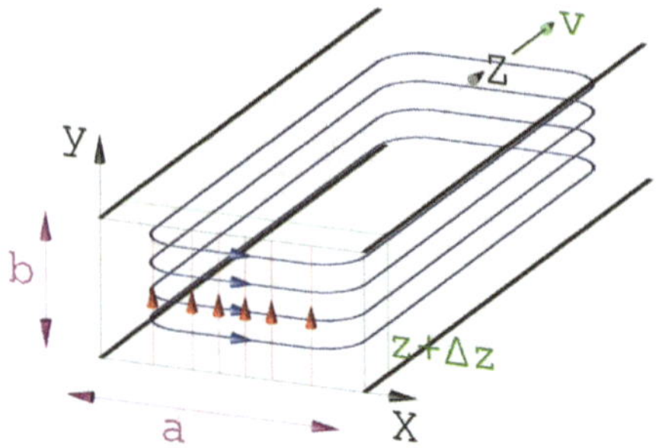

Fig. 1.7 Representation of the H_{10}-wave as a 3D image

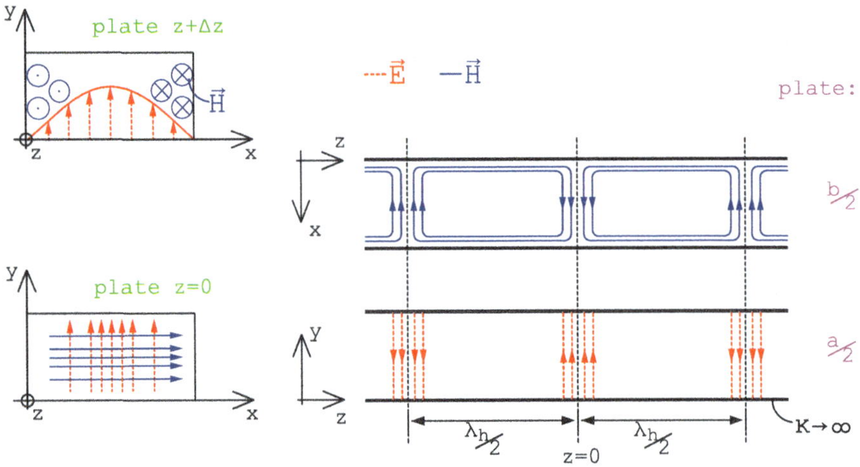

Fig. 1.8 Sectional views of the H_{10}-wave in the waveguide

The pure E-field is shown in Fig. 1.10.

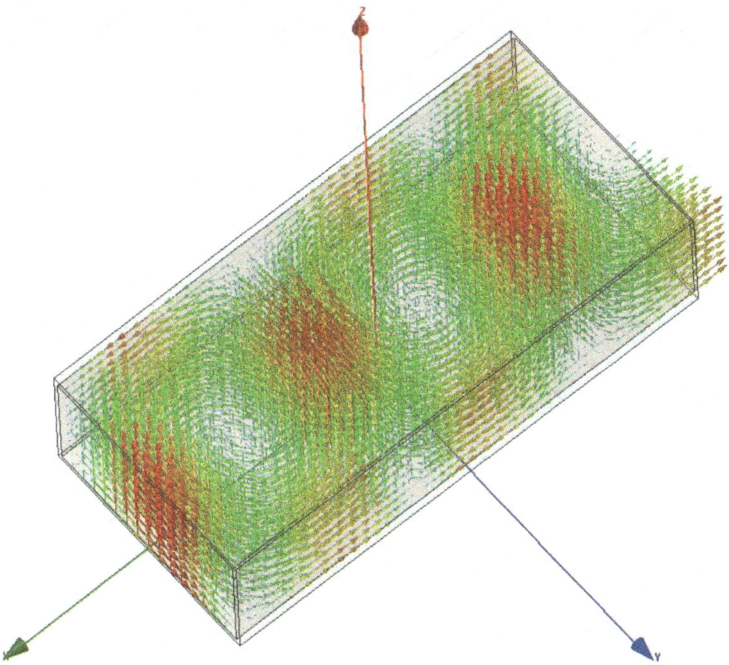

Fig. 1.9 3D representation of the electric and magnetic field of the H_{10}-wave in the waveguide (created using the simulation program HFSS)

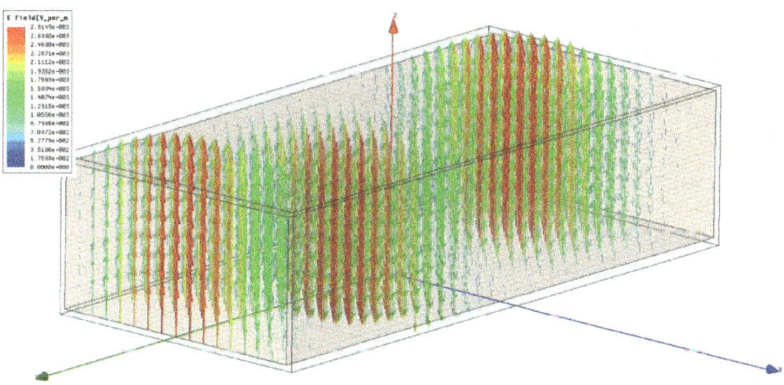

Fig. 1.10 3D representation of the electric field of the H_{10}-wave in the waveguide (from HFSS)

1.3 Waveguide Wavelength

From an enlarged image of the H_{10}-wave λ_h the wavelength in the waveguide is to be calculated as a function of the geometry and the free-space wavelength λ_0. Waveguides generally have air as filling and therefore $\varepsilon_r = 1$ and therefore $\lambda = \lambda_0$. The following further correlations can be taken from Fig. 1.11:

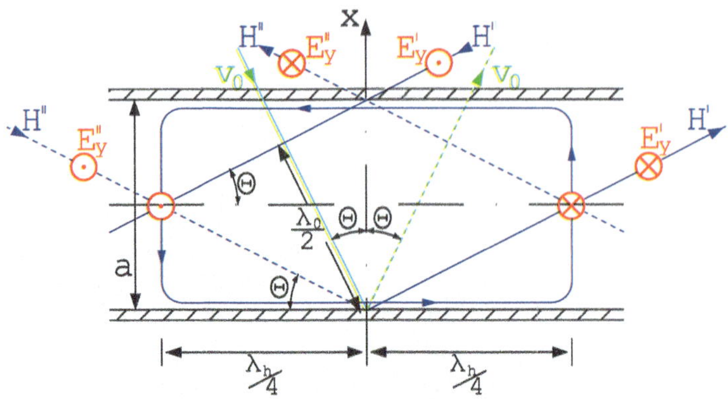

Fig. 1.11 Enlarged H_{10}-wave between two metallic plates

$$\cos\Theta = \frac{\lambda_0}{2\cdot a} \qquad \text{as well as} \qquad \cot\Theta = \frac{\lambda_h/4}{a/2} = \frac{\lambda_h}{2\cdot a}, \tag{1.2}$$

$$\Rightarrow \frac{\lambda_0}{\lambda_h} = \sin\Theta\,. \tag{1.3}$$

With

$$\sin\Theta = \sqrt{1-\cos^2\Theta} \qquad \text{resp.} \qquad \sin\Theta = \sqrt{1-\left(\frac{\lambda_0}{2a}\right)^2} \tag{1.4}$$

holds

$$\lambda_h = \frac{\lambda_0}{\sqrt{1-\left(\frac{\lambda_0}{2a}\right)^2}}\,. \tag{1.5}$$

With $a \geq \lambda_0/2$, the waveguide wavelength is greater than the free-space wavelength ($\lambda_h > \lambda_0$). Usually, the wavelength in a dielectric-filled cable or on a substrate is shorter than the free-space wavelength ($\lambda_{Substrat} < \lambda_0$).

Borderline Cases

In the following, the two limiting cases for the angles of incidence Θ (horizontal and vertical) will be considered..

$\Theta = 0^\circ \Rightarrow$ No propagation in z-direction
$\Theta = 90^\circ \Rightarrow$ Propagation similar to HPW

Cut-Off Wavelength

It has already been shown that the following applies: $\lambda_h > \lambda_0$. On closer consideration of Eq. (1.5), it can be seen that at a height of height of $a = \lambda_0/2$ the waveguide wavelength approaches goes to infinity. Therefore, the wavelength

$$\lambda_C = 2a \tag{1.6}$$

is called cut-off wavelength. If this relationship is used in Eq. (1.5), the result is the waveguide wavelength:

$$\lambda_h = \frac{\lambda_0}{\sqrt{1 - \left(\frac{\lambda_0}{\lambda_C}\right)^2}} . \tag{1.7}$$

1.4 Phase and Group Velocity

In the Basic RF books, e.g. [38] the phase constant or phase coefficient is introduced in detail. It applies to the waveguide:

$$\beta_h = \frac{2\pi}{\lambda_h} = \frac{2\pi}{\lambda_0} \cdot \sqrt{1 - \left(\frac{\lambda_0}{\lambda_C}\right)^2} . \tag{1.8}$$

The phase constant is proportional to the transmission phase ($\sim \sphericalangle S_{21}$). The following applies to a lossless line $S_{21} = e^{-j\beta l}$.

In [38] the phase velocity is also introduced in general and in application for TEM lines is introduced. Applied to the waveguide applies:

$$v_p = \frac{\Delta z}{\Delta t} = \frac{\omega}{\beta_h} = f\lambda_h = \frac{f \cdot \lambda_0}{\sqrt{1 - \left(\frac{\lambda_0}{\lambda_C}\right)^2}} = \frac{v_0}{\sqrt{1 - \left(\frac{\lambda_0}{\lambda_C}\right)^2}} = \frac{v_0}{\sin\Theta} . \tag{1.9}$$

The following applies $\lambda_0 > \lambda_C$. If you insert this relationship into Eq. (1.9), you get the result that the phase velocity is greater than the speed of light: $v_p > c_0$.

Propagation of Electromagnetic Energy

The electromagnetic energy propagates at the group velocity v_g.

The mathematical relationship can be seen in Fig. 1.12.

$$v_g = v_0 \sin\Theta = v_0 \cdot \sqrt{1 - \left(\frac{\lambda_0}{\lambda_C}\right)^2} \tag{1.10}$$

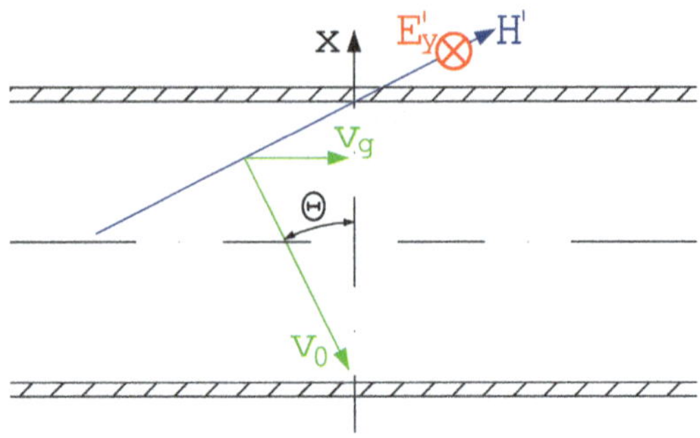

Fig. 1.12 The group velocity v_g is less than the speed of light

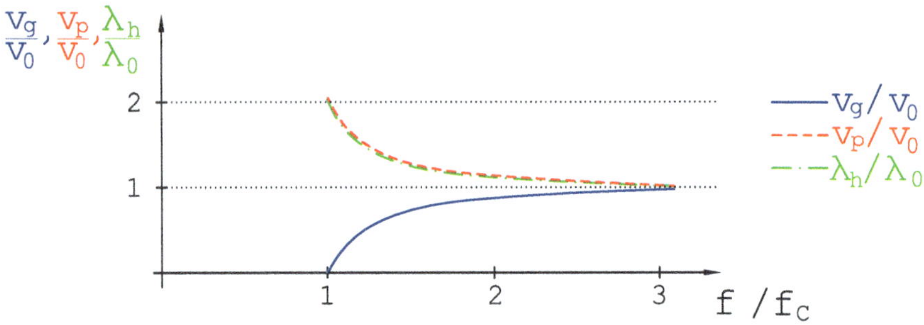

Fig. 1.13 Frequency dependencies of the group and phase velocity as well as the waveguide wavelength (normaly for $c_0 = v_0$)

The multiplication of group velocity and phase velocity results in the square of the speed of light:

$$v_g \cdot v_p = v_0^2 \,. \tag{1.11}$$

Frequency Dependencies of the Waveguide

The frequency dependencies (also referred to as dispersion) of the group and phase velocity as well as the waveguide wavelength described mathematically above are shown in Fig. 1.13. Energy transport starts at the cut-off frequency f_C and above $4 \cdot f_C$ the conditions are similar to those in free space.

Fig. 1.13 shows that the waveguide is very dispersive. This property is very disruptive for broadband transmission technology, but can be helpful for special analog signal shaping. These strong dispersions are the reason why waveguides cannot be used for the transmission of modulated broadband signals. This fundamental problem also reduces the use of THz systems, as they at least require antennas (with feed lines) based on high conductors.

1.5 General Field Wave Impedance and Field Wave Impedances of the H_{10}-wave

For the propagation in the technical waveguide is the characteristic impedance of the vacuum

$$Z_0 = \frac{|\vec{\mathbf{E}}|}{|\vec{\mathbf{H}}|} \tag{1.12}$$

is a very good approximation of that in air and is used in the following with the alternative designation as field wave impedance. The following applies to the field wave impedance $Z_F = Z_0$ of free space:

$$Z_F = \sqrt{\frac{\mu_0}{\varepsilon_0}} \simeq 120\pi\,\Omega\,. \tag{1.13}$$

The general characteristic impedance in the waveguide depends on whether it is a TE- or H-type wave or whether it is a TM- or E-type wave.

A derivation for the following equations can be found in [94] under the keyword wave impedance and in [116].

H-Typ: There is no E-field component in the direction of propagation z.

In this case applies:

$$E_y = -Z_{FH} \cdot H_x \quad \text{and} \quad E_x = Z_{FH} \cdot H_y \qquad \text{with} \tag{1.14}$$

$$Z_{FH_{10}} = \frac{Z_F}{\sqrt{1-\left(\frac{f_c}{f}\right)^2}} = f \cdot \mu_0 \cdot \lambda_h\,. \tag{1.15}$$

E-Typ: There is no H-field component in direction of propagation z.

In this case:

$$E_x = Z_{FE} \cdot H_y \quad \text{and} \quad E_y = -Z_{FE} \cdot H_x \qquad \text{with} \tag{1.16}$$

$$Z_{FE_{10}} = Z_F \sqrt{1-\left(\frac{f_c}{f}\right)^2} \tag{1.17}$$

1.6 Propagation Coefficient and Characteristic Impedance of the H_{10}-Wave Over the Frequency

As a rule, the waveguide is filled with air and therefore $\varepsilon_r = 1$ applies as a very good approximation. The propagation coefficient γ_h can be divided into an attenuation coefficient and a phase coefficient.

$$\gamma_h = \alpha_h + j\beta_h \tag{1.18}$$

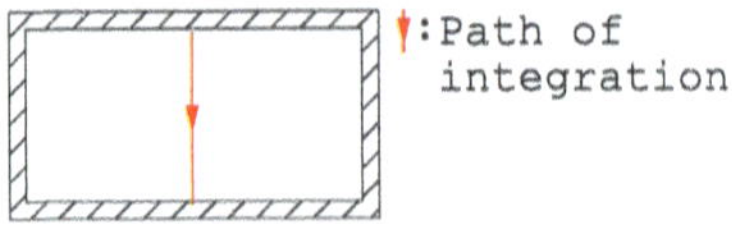

Fig. 1.14 Integration path for the H_{10}-wave

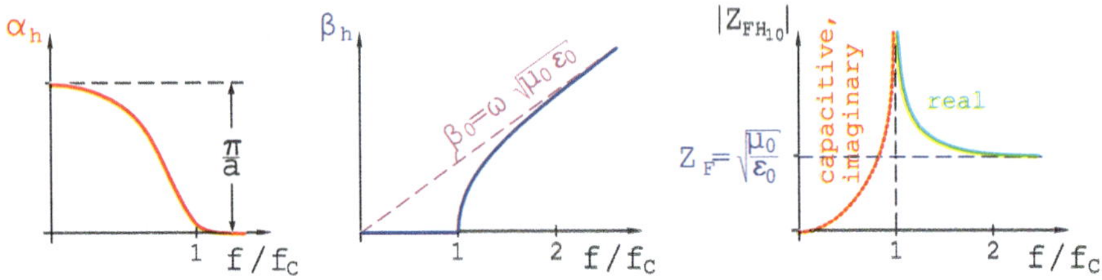

Fig. 1.15 Frequency-dependent propagation coefficient and field impedance of the H_{10}-wave

To introduce or define the field wave impedance in the waveguide, an integration path must be defined in the waveguide. The usual integration path for the H_{10}-wave is shown in Fig. 1.14.

This integration path is required in practice in the field simulation. If a different path is selected, the waveguide wave a different field characteristic impedance[1].

For $\lambda_0 > \lambda_C$ or $f_0 < f_C$ holds:

$\gamma_h = \alpha_h$, purely real $\quad Z_{FH_{10}}$: imaginary

For $\lambda_0 \leq \lambda_C$ holds:

$\gamma_h = \alpha_h + j\beta_h \simeq j\beta_h$: purely imaginary $\quad Z_{FH_{10}}$: purely real

The diagram of Fig. 1.15 shows the frequency-dependent quantities for the wave propagation of the H_{10} wave.

1.7 Current Distribution in Rectangular Waveguide

The current distribution in the metallic walls is illustrated in the Fig. 1.16. The image was split into two parts for this purpose:

a) Magnetic field lines in the waveguide (top view in section),
b) Currents in the side walls (development: underside and side walls).

Furthermore, Fig. 1.16 shows the areas in which probes can be optimally implemented and decoupling elements can be inserted.

Ranges for probes: $\alpha' \parallel$ Current paths, selected as:

- Few field disturbance,

[1] Details on the choice of integration paths can be found in the next chapter on page 49 (Fig. 2.22).

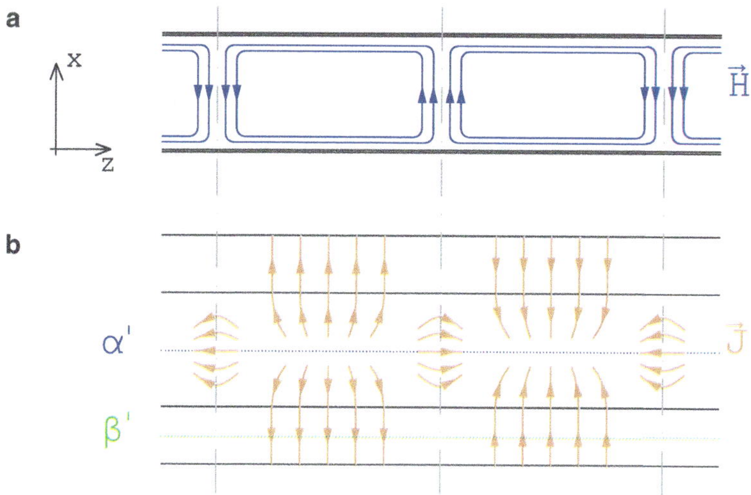

Fig. 1.16 Current distribution for the H_{10}-wave in the side walls (rear, bottom, front) of the rectangular waveguide (the fourth side wall is missing)

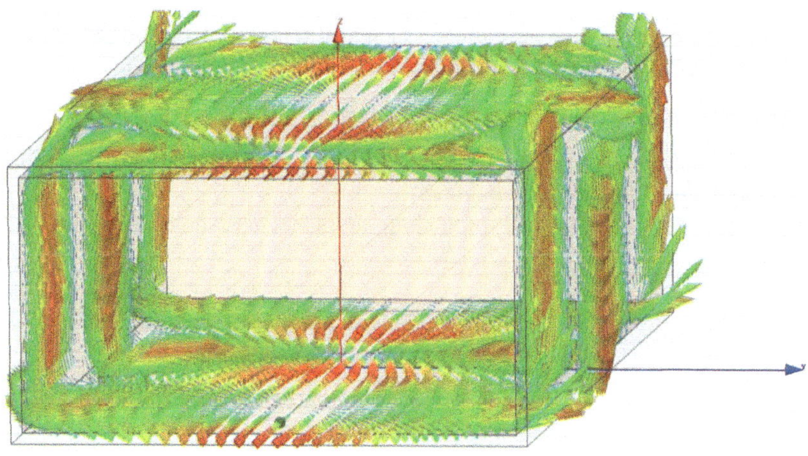

Fig. 1.17 Current distribution of the H_{10}-wave in 3D from HFSS

- Little decoupling.

Areas for strong decoupling: $\beta' \perp$ current paths, selected there:

- Strong E-field in the decoupling gap.

In Fig. 1.17 a result of a field simulation visualizes the current distribution.

Here, too, there are slight distortions of the field image, as the vector arrows start at the point of the calculated value and are not centered on this value. However, it is easy to see that the currents are highest in the middle (area of the integration line) and flow in the z-direction.

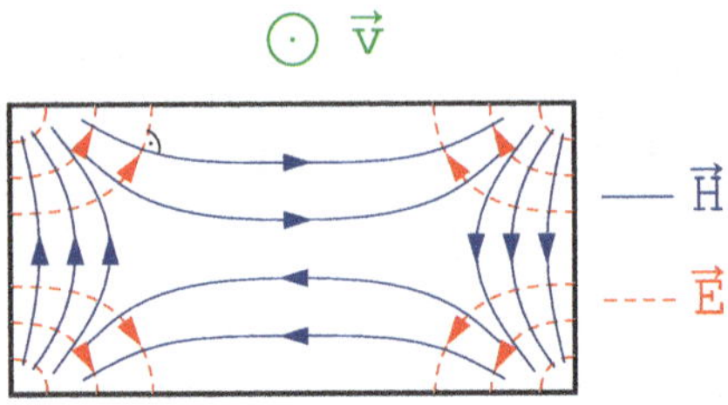

Fig. 1.18 Example of the field image of an H_{11}-wave in the x-y section

1.8 Higher Modes in the Rectangular Waveguide

Higher order modes in rectangular waveguides refer to electromagnetic wave patterns that exhibit more complex spatial variations within the waveguide cross-section. These modes arise due to multiple reflections and interactions between the wave and the waveguide walls, leading to additional possible configurations beyond the fundamental mode. Understanding higher order modes is crucial in microwave engineering as they affect the propagation characteristics and performance of devices utilizing rectangular waveguides, such as filters, couplers, and antennas.

In rectangular waveguides, the following modes can be propagated.

1. **H_{mn}-waves** (also called TE_{mn})

   m: Number of half-periods of the transverse electric field transverse electric field along the large inner side (a)
   n: Number of half-periods of the longitudinal electric field (b)

2. **E_{mn}-waves** (also called TM_{mn})

   m: Number of half-periods of the longitudinal magn. field (a)
   n: Number of half-periods of the longitudinal magn. field (b)

In general, m and n are integers and positive: m, n = 0, 1, 2 ... However, there must be at least one half-period must be present.

Example of the H_{11}-wave

Fig. 1.18 shows the image for the E- and H-fields of the H_{11} wave as the first example of a higher mode.

Cut-Off-Frequencies

The lower cut-off frequencies of the modes in the rectangular waveguide depend on the cross-sectional ratio $\frac{b}{a}$, Fig. 1.19.

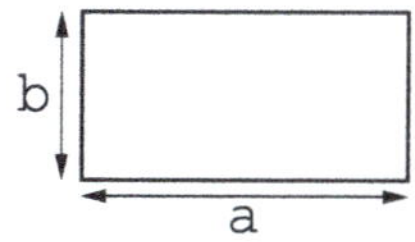

Fig. 1.19 Ratio of the cross-sections (standard $a = 2 \cdot b$)

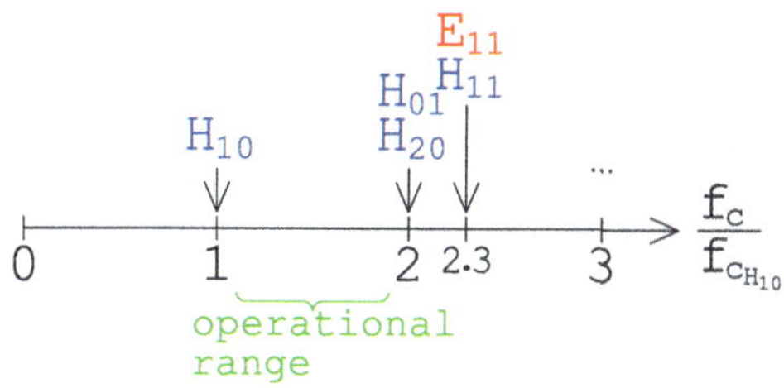

Fig. 1.20 Lower cut-off frequencies of the H_{11}-wave and the higher modes in the rectangular waveguide for the cross-sectional ratio $\frac{b}{a} = \frac{1}{2}$

The lower cut-off frequencies in the rectangular waveguide are shown in Fig. 1.20 for the first five possible modes. It can be seen that for this waveguide there is a relatively large working range in which only the H_{10} mode is capable of propagation.

In addition to the H_{10} wave, Fig. 1.21 shows other higher modes and their electric and magnetic fields as well as the surface current densities.

The cut-off wavelength of the different modes is calculated for the standard rectangular waveguide from

$$\lambda_{cmn} = \frac{2ab}{\sqrt{m^2 b^2 + n^2 a^2}} . \tag{1.19}$$

The so-called cut-off frequencies of the various modes are calculated as follows:

$$f_{cmn} = c_0 / 2 \cdot \sqrt{(m/a)^2 + (n/b)^2} . \tag{1.20}$$

The corresponding propagation coefficients are obtained from

$$\gamma_{cmn} = \frac{2\pi}{\lambda_0} \cdot \sqrt{\left(\frac{\lambda_0}{\lambda_{cmn}}\right)^2 - 1} . \tag{1.21}$$

The field wave resistances for H waves are calculated using the following correlations:

$$Z_{FH_{mn}} = -\frac{E_y}{H_x} = \frac{E_x}{H_y} , \tag{1.22}$$

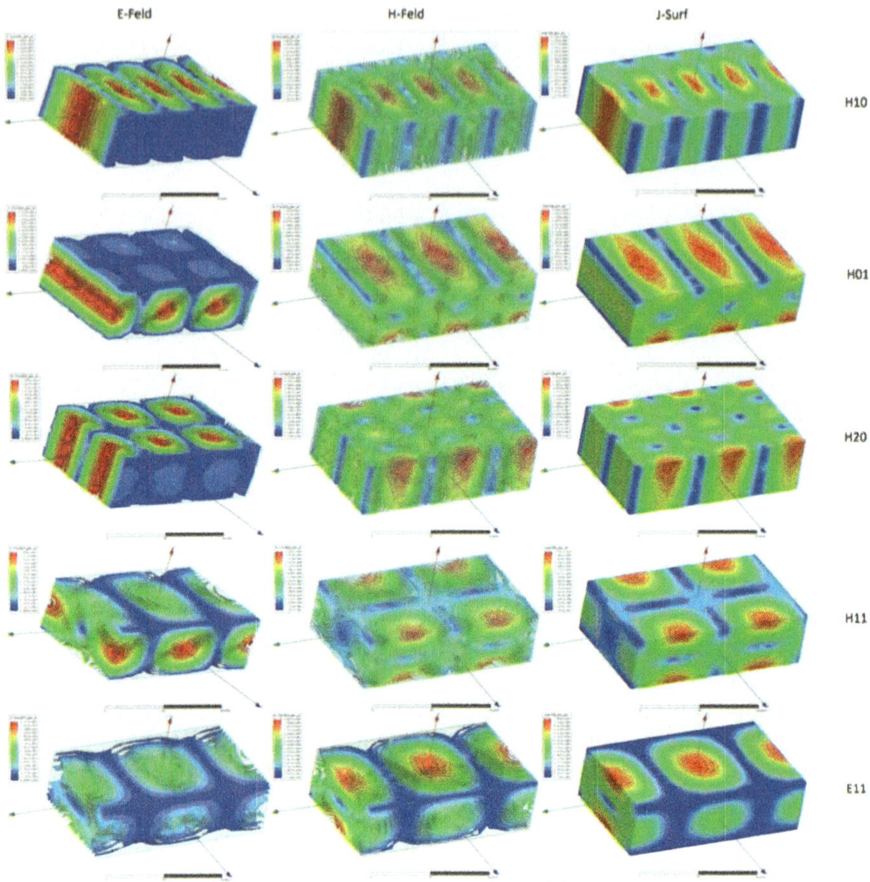

Fig. 1.21 Representation of different electromagnetic fields for different modes in the standard rectangular waveguide

$$Z_{FH_{mn}} = \frac{Z_F}{\sqrt{1 - \left(\frac{\lambda_0}{\lambda_{cmn}}\right)^2}}. \tag{1.23}$$

The field wave resistances for E-waves can be calculated in a similar way:

$$Z_{FE_{mn}} = Z_F \cdot \sqrt{1 - \left(\frac{\lambda_0}{\lambda_{cmn}}\right)^2}. \tag{1.24}$$

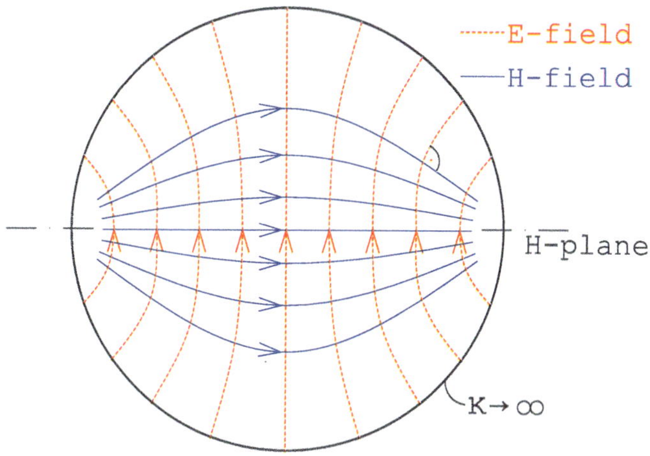

Fig. 1.22 Fundamental wave mode in the circular waveguide: H_{11}-mode

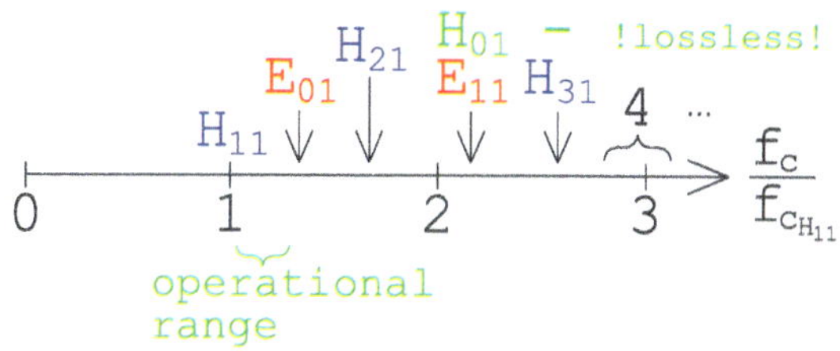

Fig. 1.23 Cut-off frequencies of the higher modes in the circular waveguide

1.9 Round Waveguide and Coaxial Conductor Modes

Round waveguides are cylindrical structures commonly used to guide electromagnetic waves, particularly in microwave and millimeter-wave applications. Their circular cross-section allows for the propagation of fundamental modes with lower losses compared to rectangular waveguides. Round waveguides find extensive application in radar systems, satellite communications, and medical imaging devices due to their efficiency in transmitting high-frequency signals with minimal attenuation.

The first waveguide mode capable of propagation (basic mode) in the round waveguide and in the coaxial conductor is the H_{11} mode.

As an estimate, the propagation wavelength can be calculated using the diameter:

$$2 \cdot r \simeq \frac{\lambda_0}{2} \,. \tag{1.25}$$

Fig. 1.22 shows the field images for the H_{11} wave in the circular waveguide (Fig. 1.23).

Higher Modes in the Circular Waveguide

Higher modes in circular waveguides denote electromagnetic wave configurations beyond the fundamental mode, characterized by more complex spatial variations within the waveguide. These modes emerge due to reflections and interactions between the waves and the circular waveguide boundaries, influencing the propagation characteristics and energy distribution. While higher modes are typically minimized in practical applications to maintain signal integrity, they find relevance in certain contexts such as mode conversion devices and specialized antenna designs requiring tailored radiation patterns.
The corresponding propagation coefficients of the circular waveguide modes (here lossless, i. h. $\alpha = 0$) are obtained with the diameter D from:

$$\lambda_{cmn} = \frac{\pi \cdot D}{x_{mn}} . \tag{1.26}$$

The following values apply for the H-modes:

$$x_{11} = 1.841 \qquad x_{21} = 3.054 \qquad x_{01} = 3.832 .$$

The following values apply for the E-modes:

$$x_{01} = 2.405 \qquad x_{11} = 3.832 .$$

Lossless Mode H_{01}

With the H_{01} wave in the circular waveguide and in the coaxial conductor, the electromagnetic fields decay towards the edge (metallic boundary). Therefore, almost no currents (ideal for f$\rightarrow \infty$GHz in the circular waveguide) are induced in the metallic conductors. As a result, the lines have almost no losses (Fig. 1.24 and 1.25).

The lossless mode of the circular waveguide represents a fundamental electromagnetic wave configuration with minimal energy dissipation during propagation. Older research works extensively explored the theoretical properties and practical implementations of this mode in the context of microwave engineering. Potential applications of the lossless mode include high-performance communication systems, radar arrays, and particle accelerators where efficient transmission of electromagnetic waves with minimal signal degradation is essential for reliable operation.

Researchers have investigated various techniques to harness the benefits of the lossless mode, such as mode converters and impedance matching strategies, to optimize its utilization in diverse engineering applications. Despite advancements in modern waveguide technologies, the exploration of the lossless mode continues to be relevant for enhancing the efficiency and reliability of electromagnetic systems in telecommunications, aerospace, and scientific research.

The following estimation applies to the H_{01}-wave in the coaxial conductor

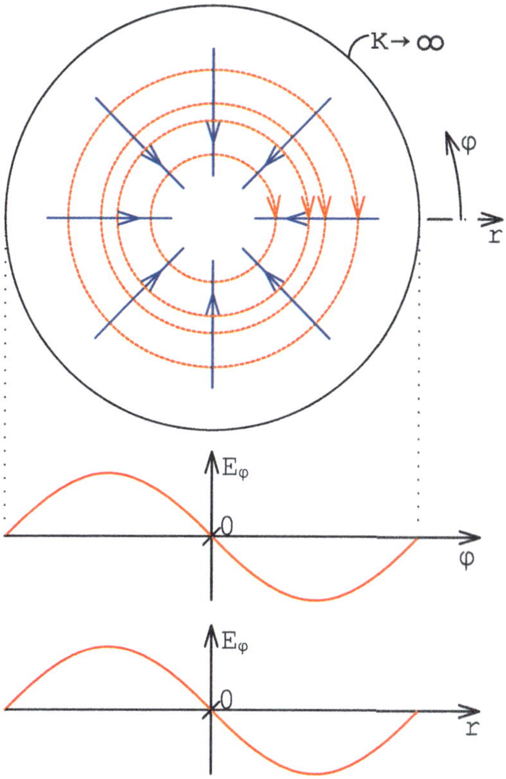

Fig. 1.24 Lossless mode in the circular waveguide

$$\pi \cdot (r_a + r_i) \simeq \lambda_0 \; . \tag{1.27}$$

The basic mode of the waveguide wave in the coaxial conductor is the H_{11}-mode starts first. Its cutoff frequency is estimated as follows and outer diameter (d+D) is estimated as follows:

$$f_{c_{H_{11}}} \simeq \frac{1}{\pi \left(\frac{D+d}{2}\right) \sqrt{\mu \, \varepsilon}} \, , \tag{1.28}$$

or with the speed of light c_0 via

$$f_{c_{H_{11}}} \simeq \frac{c_0}{\pi \left(\frac{D+d}{2}\right) \sqrt{\mu_r \, \varepsilon_r}} \, , \tag{1.29}$$

whereby in practice the following often applies: $\mu_r = 1$.

1.10 Waveguide to Coaxial Transition with Adapter

Magnetron sources in the kW range are often used as powerful sources. If their output power and/or the reflected signals are to be measured, directional couplers (in waveguide technology) and waveguide to coaxial adapters are used.

A coaxial waveguide adaptor is often used for the convertion of coaxial modes to waveguide mode or vice versa. Typically achievable matching values of these low-loss transitions are $|S_{11}| < -20\,\text{dB}$. Fig. 1.26 shows such a transition in section. An image from a field simulation is shown in Fig. 1.27. Only the magnitudes of the electric field are shown here.

1.11 Transmission Resonator

Resonators in various technologies are described in detail in a number of basic books (such as [38]). Energy is stored in a resonator in the same way as in a capacitor. As waveguides are very low-loss conductors and can also handle a lot of power, they are ideal for developing resonators.

Fig. 1.28 shows the S-parameter results of a field simulation of a transmission resonator designed in the X-band.

The results show the desired narrowband transmission behavior of a transmission resonator with very good matching at 10 GHz

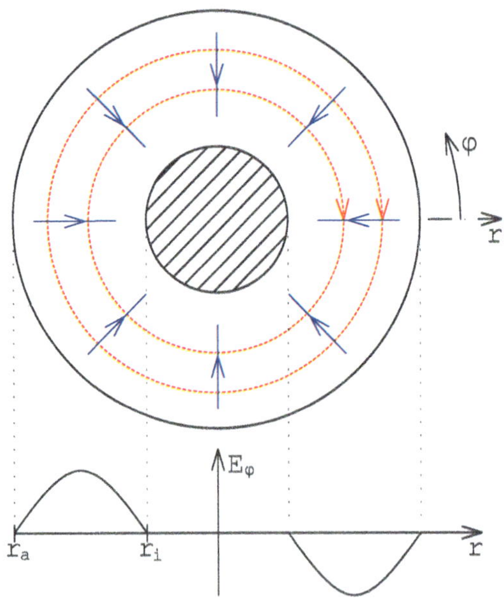

Fig. 1.25 H_{01}-mode in coaxial conductor

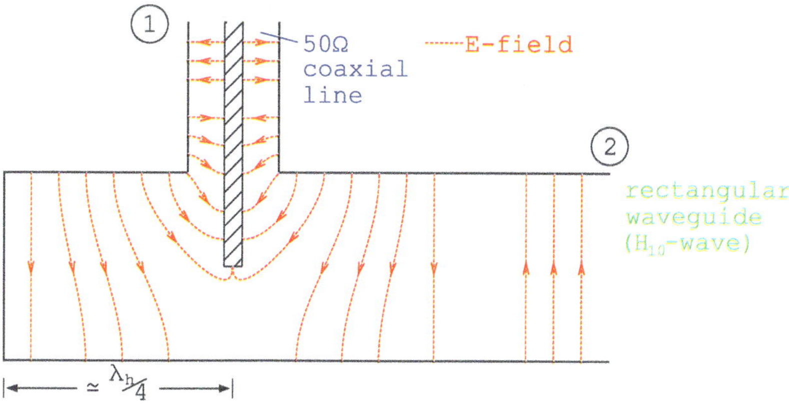

Fig. 1.26 Coaxial waveguide transition with E-field pattern

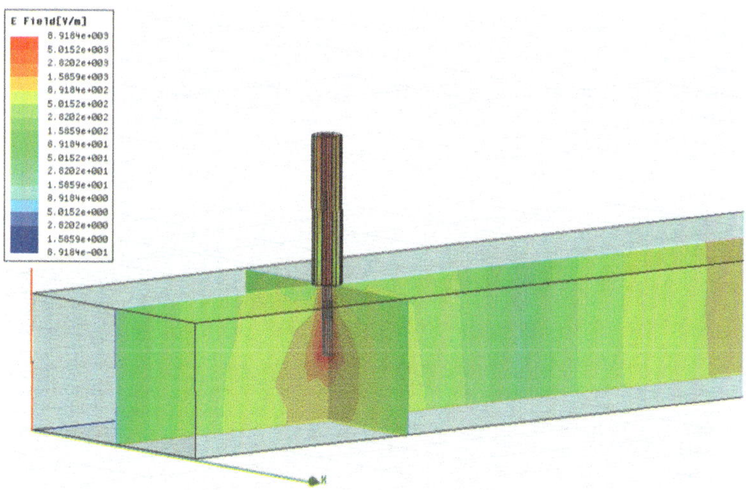

Fig. 1.27 3D representation of a coaxial waveguide transition with E-field images

An image of the associated E-field from this field simulation is shown in Fig. 1.29. Here, the field strengths in the resonator are so high that the fields of the coupled-in and coupled-out H_{10}-wave are barely recognizable.

The corresponding H-field is illustrated in Fig. 1.30. Here, too, the field strengths in the resonator are so high that the fields of the coupled-in and coupled-out H_{10}-wave are barely recognizable.

The resonator consists of a very high-impedance waveguide with an electrical length of $\lambda/2$. The high-impedance design ensures low ohmic losses as the current densities are kept low. The coupling and decoupling of the resonator is realized in

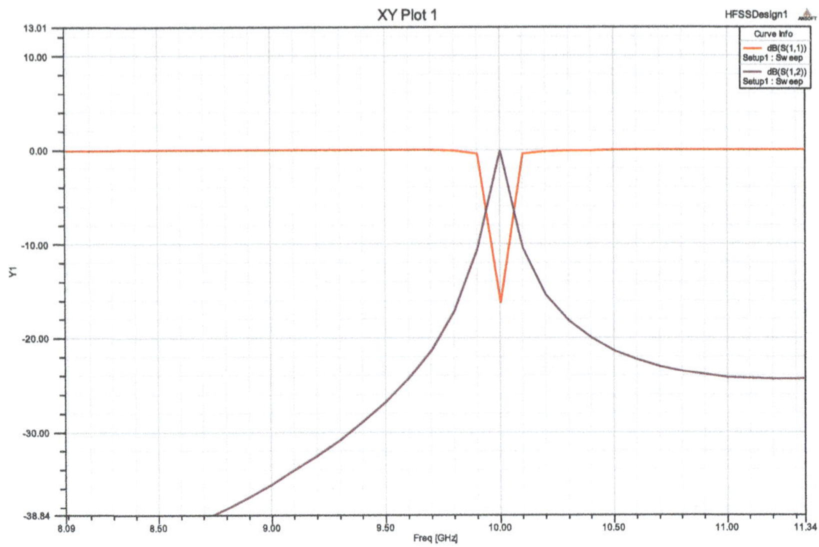

Fig. 1.28 Reflection and transmission scattering parameters of a simulated transmission resonator for 10 GHz

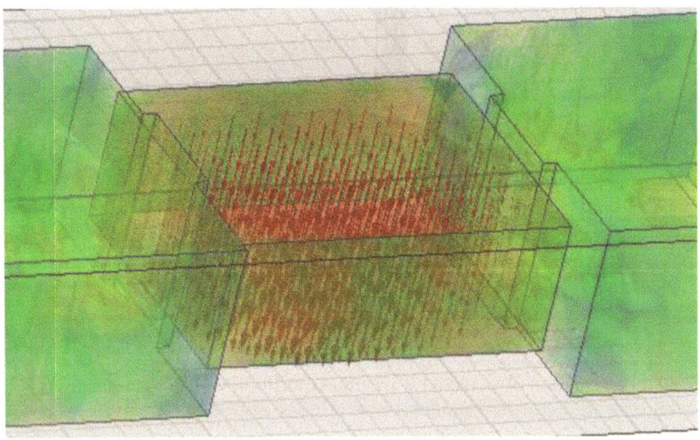

Fig. 1.29 3D representation of a transmission resonator (embedded in X-band waveguides) with E-field images

the circuit is modeled by small series capacitors. The hardware realization is a very short waveguide section waveguide section below the cut-off frequency[2].

[2] Calculation examples can be found in the numerous examples in the microwave exams. These exams are freely accessible with the sample solutions, [35].

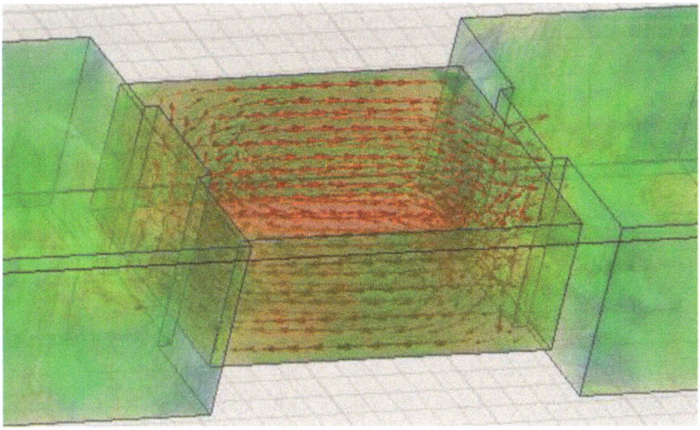

Fig. 1.30 3D field image representation of a transmission resonator with H-field images

1.12 Frequency Bands for Rectangular Waveguides

Waveguides are standardized in various standards. The following table shows the most important standards for rectangular waveguides, their designations and the associated dimensions.

Frequency range (GHz)	Name of Band	Standards DIN47302 IEC153	EIA	Wdith a (mm)	(inch)
0.77-1.14	UHF/L	R 9	WR 975	247.65	9.750
1.12-1.70	L	R 14	WR 650	165.10	6.500
1.70-2.60	LA	R 22	WR 430	109.22	4.300
2.20-3.30	LS	R 26	WR 340	86.36	3.400
2.60-3.95	S	R 32	WR 284	72.14	2.840
3.22-4.90	A	R 40	WR 229	58.17	2.290
3.95-5-85	G	R 48	WR 187	47.55	1.872
4.64-7.05	C	R 58	WR 159	40.39	1.590
5.85-8.20	J	R 70	WR 137	34.85	1.372
7.05-10.0	H	R 84	WR 112	28.50	1.122
8.20-12.4	X	R 100	WR 90	22.86	0.900
10.0-15.0	M	R 120	WR 75	19.05	0.750
12.4-18.0	P	R 140	WR 62	15.80	0.622
15.0-22.0	N	R 180	WR 51	12.95	0.510
18.0-26.5	K	R 220	WR 42	10.67	0.420
21.7-33.0		R 260	WR 34	8.64	0.340
25.5-40.0	R	R 320	WR 28	7.11	0.280
33.0-50.0	Q		WR 22	5.69	0.224
40.0-60.0	U		WR 19	4.78	0.188
50.0-75.0	V		WR 15	3.76	0.148
75 - 110	W		WR 10	2.54	0.100
90 - 140			WR 8	2.032	0.080
110 - 170			WR 6	1.651	0.065
140 - 220			WR 5	1.295	0.051
170 - 260			WR 4	1.092	0.043
220 - 325			WR 3	0.864	0.034
260 - 400			WR 2.8	0.711	0.028
325 - 500			WR 2.2	0.559	0.022
400 - 600			WR 1.9	0.483	0.019
500 - 750			WR 1.5	0.381	0.015
600 - 900			WR 1.2	0.305	0.012
750 -1100			WR 1.0	0.254	0.010

The WRxxx designation of the waveguide is derived from the width of the waveguide in inches (1 inch = 25.4 mm). A WR 28 waveguide is therefore 28 % of an inch = 7.11 mm wide. The abbreviation WR stands for Waveguide Rectangular.

This table is based on a height/width ratio (b/a) of 1:2. The lower recommended transmission frequencies are 1.26 times higher than the critical lower cut-off frequencies f_c. The upper transmission frequencies are 1.48 times the lower recommended transmission frequencies. The average factor of 1.86 of the upper transmission frequencies to the respective critical lower cut-off frequency ensures monomode propagation (value <2) of the H_{10} wave.

Chapter 2
Fundamentals of Field Simulation

In recent decades, a large number of software solutions have been developed that enable microwave engineers to calculate electromagnetic field problems with good and very good quality and thus make good predictions.

Field simulation is widely employed in microwave applications to analyze and optimize the performance of components and systems. By numerically solving Maxwell's equations, field simulation tools enable engineers to visualize electromagnetic fields, study wave propagation, and assess device behavior under various operating conditions. This approach aids in the design of antennas, filters, matching networks for amplifiers, plasma applications and other microwave devices, allowing for precise tuning and enhancement of performance characteristics prior to physical prototyping and testing.

This chapter presents two aspects of the practical calculation of electromagnetic fields using commercially available software:

1. Basics for calculating fields over long distances,
2. Fundamentals for the precise numerical analysis of electromagnetic fields in 3D volumes.

2.1 Free Space Simulation

Free space simulations take into account propagation effects including the calculation of the complex transfer function of direct radiation and interactions caused by buildings, wind turbines and mountains, among others.

The simulation software for the calculation of electromagnetic fields is based on closed-form solutions for antennas and objects located in the 2D or 3D far field.

2D observations are merely sections (horizontal planes) of a 3D problem.

The following Fig. 2.1 shows an object (for example a cube-shaped house) as well as one transmitting and two receiving antennas and the possible propagation paths between the transmitter and the receivers.

H. Heuermann, *Microwave Technology*,
https://doi.org/10.1007/978-3-658-45686-3_2

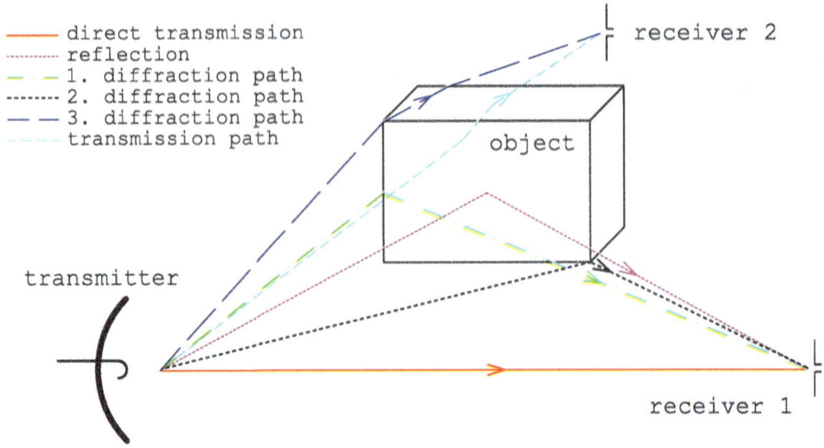

Fig. 2.1 Object in a free space simulation with the signal paths supported by simulators

2.1.1 Transmitter and the Receivers

In these free-space simulations, the transmitter and receiver must be created as shown below.

The transmitter is defined by

a) Signal form and power,
b) Antenna, position and pose.

The receiver is only defined by the antenna and its position and pose.
Antennas play a central role. Precise knowledge of the technical properties of the antennas is necessary.

Basics of Antennas

- Normally linearly polarized,
- Diverse constructions (⇒ Directional characteristic),
- Defined by dimensions or antenna gain,
- File input of the directional characteristic is possible.

Antenna Types Used

- Isotropic omnidirectional antenna, dipoles, aperture antenna,
- Horn radiators, patch antennas.

The example of the Hertzian dipole is used to explain the technical background of a free-space simulation.

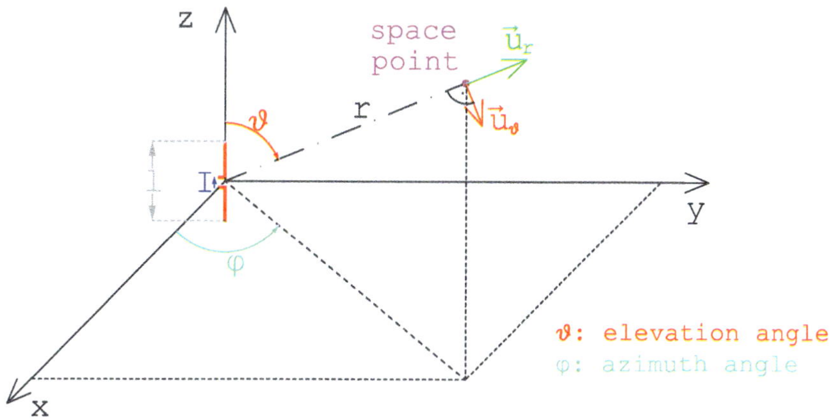

Fig. 2.2 Representation of the Hertzian dipole in the center of the coordinate system and the corresponding coordinates to the center point

Hertzian Dipol

The Hertzian dipole[1] (named after Heinrich Hertz) consists of two short, thin wires that are fed by a concentrated current source.
The following Fig. 2.2 shows the Hertzian dipole at the origin and the coordinates for a receptor point for which the electromagnetic field can be precisely calculated from closed analytical solutions.

The main beam direction (= x-axis) is defined by the angle values: $\varphi = 0°$; $\vartheta = 90°$.

The isotropic spherical radiator often used for receivers has the same detection values for all φ- and ϑ-values, i.e. it is independent of direction. However, a dipole only radiates for all φ-values with the same intensity. The radiation varies greatly over the ϑ-angle values.
It results exactly for „Free space conditions":

$$\vec{E}(r,\vartheta) = \mathrm{j}Z_0 \frac{Il}{2\lambda_0 r} e^{-\mathrm{j}k_0 r} \left\{ \left[1 + \frac{1}{\mathrm{j}k_0 r} + \frac{1}{(\mathrm{j}k_0 r)^2}\right] \sin\vartheta \cdot \vec{u}_\vartheta + 2\left[\frac{1}{\mathrm{j}k_0 r} + \frac{1}{(\mathrm{j}k_0 r)^2}\right] \cos\vartheta \cdot \vec{u}_r \right\}, \tag{2.1}$$

$$\vec{H}(r,\vartheta) = \mathrm{j}\frac{Il}{2\lambda_0 r} e^{-\mathrm{j}k_0 r} \left[1 + \frac{1}{\mathrm{j}k_0 r}\right] \sin\vartheta \cdot \vec{u}_\varphi . \tag{2.2}$$

The electromagnetic wave propagates with the propagation constant $\gamma_0 = \mathrm{j}k_0$. In antenna technology, the wavenumber $k_0 = (2\pi)/\lambda_0$ is preferred.
The following applies for $\varepsilon_r = 1$ for the phase constant: $\beta = k_0$.

[1] This antenna is also known as an elementary radiator or dipole.

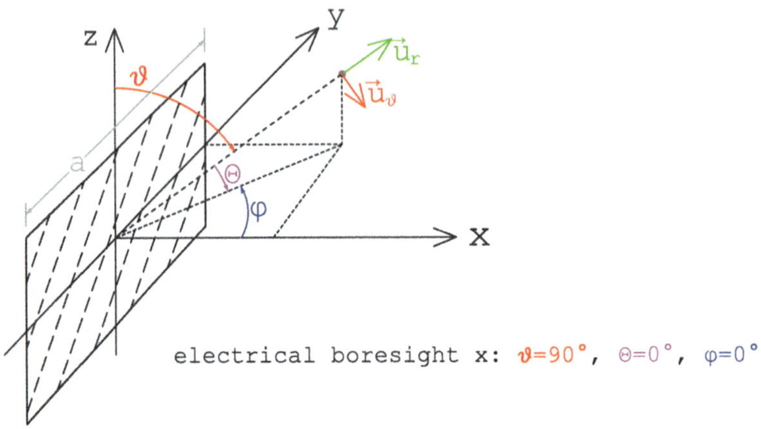

Fig. 2.3 Representation of a surface with known field assignment for calculating the reflection from objects

All the information contained in Eqs. (2.1) and (2.2) is entered into the software for the transmitter and receiver as well as the antennas involved via the points shown above. These equations are used to calculate the fields of the waves incident on the object and for the direct path of Fig. 2.1!

2.1.2 Reflection on Objects

The electric field on a rectangular surface is given. This was calculated for the surface of the object using Eq. (2.1), for example.

There are two applications for the following calculations in antenna technology and free-space simulation (Fig. 2.3).

Sources: a) Field of a transmitting antenna ⇒ Object reflection
b) Internal source ⇒ Horn radiator or general aperture antenna

With the so-called directional characteristics $\mathrm{W}(\sin\varphi)$, $\mathrm{W}(\sin\Theta)$ the exact solution to this problem applies:

$$\vec{E}(r,\Theta=0,\varphi)=\frac{\mathrm{j}(1+\cos\varphi)}{2\lambda_0}\left[\vec{u}_\varphi W_y(\sin\varphi)-\vec{u}_\vartheta W_z(\sin\varphi)\right]\frac{e^{-\mathrm{j}k_0 r}}{r}, \tag{2.3}$$

$$\vec{E}(r,\Theta,\varphi=0)=\frac{\mathrm{j}(1+\cos\Theta)}{2\lambda_0}\left[\vec{u}_\varphi W_y(\sin\Theta)-\vec{u}_\vartheta W_z(\sin\Theta)\right]\frac{e^{-\mathrm{j}k_0 r}}{r} \tag{2.4}$$

with

$$W_y(\sin\varphi)=\frac{1}{\lambda_0}\int\int E_y(y,z)\cdot e^{\mathrm{j}k_0 y\sin\varphi}\cdot dy\cdot dz\ldots. \tag{2.5}$$

For the sake of clarity, Eq. (2.5) has not been printed in full[2].

For a uniform assignment E_0 of E_y over y and z applies:

$$W(\sin\varphi) = \frac{a \cdot E_0}{\lambda_0} \cdot \frac{\sin(\sin(\varphi) \cdot k_0 \cdot a/2)}{\sin(\varphi) \cdot k_0 \cdot a/2}, \tag{2.6}$$

$$W(\sin\varphi)|_{\sin\varphi \simeq \varphi} \approx \frac{a \cdot E_0}{\lambda_0} \cdot \mathrm{si}(\varphi \cdot k_0 \cdot (a/2)) \tag{2.7}$$

with the 3 dB limit $\equiv$ half-width.

$\Delta\varphi = 0.886 \cdot \frac{\lambda_0}{a} \qquad \Rightarrow \qquad \Delta\varphi \simeq 51^\circ \cdot \frac{\lambda_0}{a}$.

Eq. (2.7) contains the si-function for the directional characteristic. This is shown in Fig. 2.4.

For oblique incidence of the electromagnetic wave with the angle φ_0 according to Fig. 2.5, Eq. (2.8).

$$W(\sin\varphi) = \frac{a \cdot E_0}{\lambda_0} \cdot \frac{\sin((\sin\varphi - \sin\varphi_0)k_0 \cdot (a/2))}{(\sin\varphi - \sin\varphi_0)k_0 \cdot (a/2)} \tag{2.8}$$

In Eq. (2.8) there is a maximum for $\varphi = \varphi_0$. However, the following applies to the E-field: $\vec{E}(r, \Theta, \varphi) \sim (1 + \cos\varphi)$. This factor influences the maximum of the reflected wave.

[2] The complete solution can be found in the derivation in the antenna section of this book.

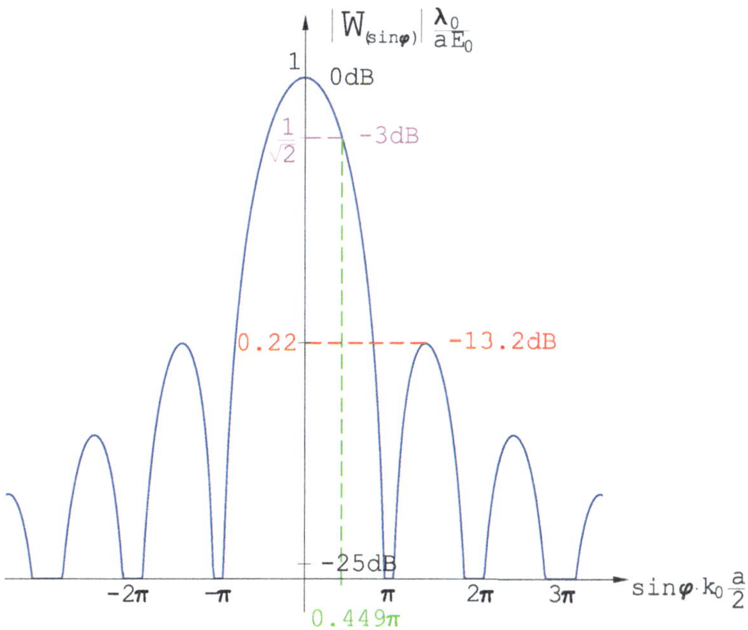

Fig. 2.4 Directional characteristic of a uniformly illuminated surface

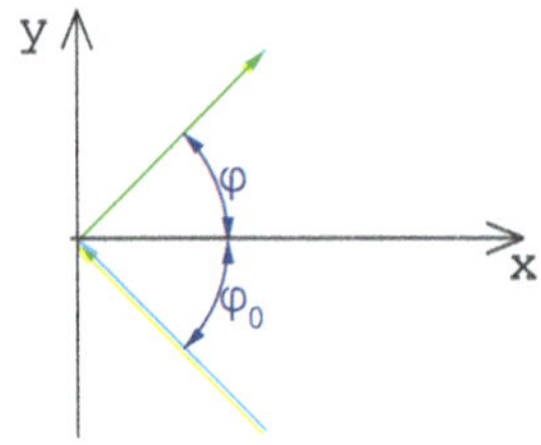

Fig. 2.5 Representation of the angles for the oblique incidence (often $\varphi_0 \neq \varphi$)

Only for a surface with the extension $a \to \infty$ holds:

Angle of incidence = angle of reflection.

2.1.3 Diffraction and Penetration of Objects

Diffraction

In further, the phenomena of diffraction of an electromagnetic wave discussed first.

The solutions for diffraction are widely used for

a) Right-angled edges and
b) "Knife edge".

Phenomenon for a):

The homogeneous plane wave (HPW) in Fig. 2.6 is incident on the edge. Currents are induced on both sides of the edge with finite conductivity. These currents generate new fields (similar to the reflections on surfaces). Constructive and destruc-

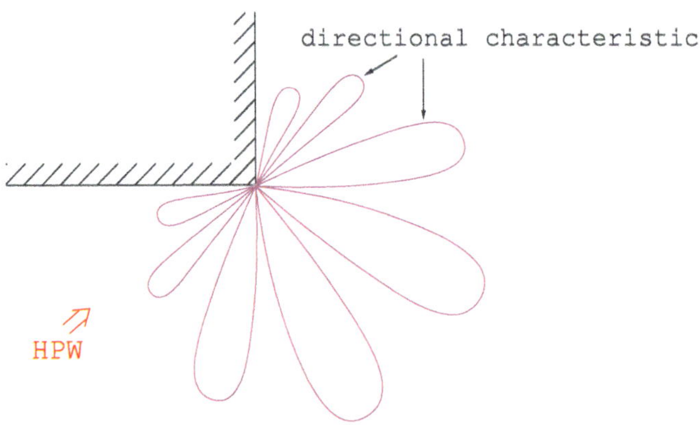

Fig. 2.6 Diffraction at right-angled edge

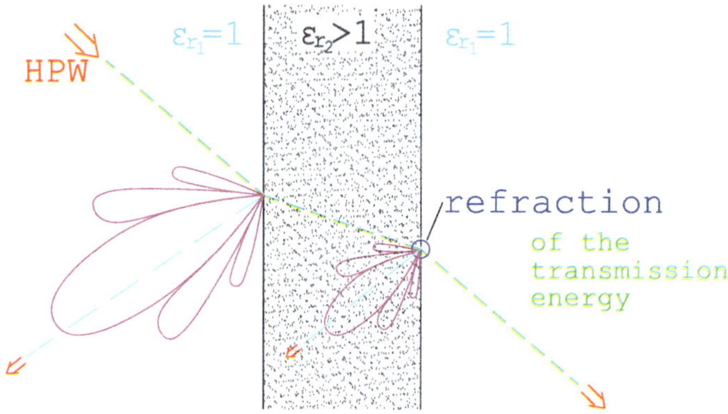

Fig. 2.7 Penetration (green dashed line) of electromagnetic energy through a wall with simultaneous reflection of signal components

tive superpositions arise and thus a more complex diffraction behavior, as shown in Fig. 2.6.

Diffraction effects are generally much smaller than reflection and transmission effects, but are often not negligible.

Penetration

The parts that are not reflected by an object (wall) penetrate, are attenuated if necessary, parts are reflected again before exiting and the rest exits. This residual energy penetrates the object. This is illustrated in Fig. 2.7.

For objects with small dielectric material constants, the reflections are very low and the penetration component predominates.

The polarization of the electric field to the surface is discussed below. Two cases are distinguished and visualized in Fig. 2.8.

Case 1: $\vec{E}_e \parallel$ Incident plane (image plane)
Case 2: $\vec{E}_e \perp$ Incident plane (image plane)

Fig. 2.8 shows that in case 1 only the E-field and in case 2 only the H-field are changed in direction.

For Infinitely Extended Surfaces the Following Applies

Reflection law: $\Theta_e = \Theta_r$
Refraction law: $\mathrm{k}_1 \cdot \sin\Theta_e = \mathrm{k}_2 \cdot \sin\Theta_t$
with the wave numbers $\mathrm{k}_i = \frac{2\pi}{\lambda_i} = \frac{2\pi}{\sqrt{\varepsilon_{ri}} \cdot \lambda_0}$

From this follows

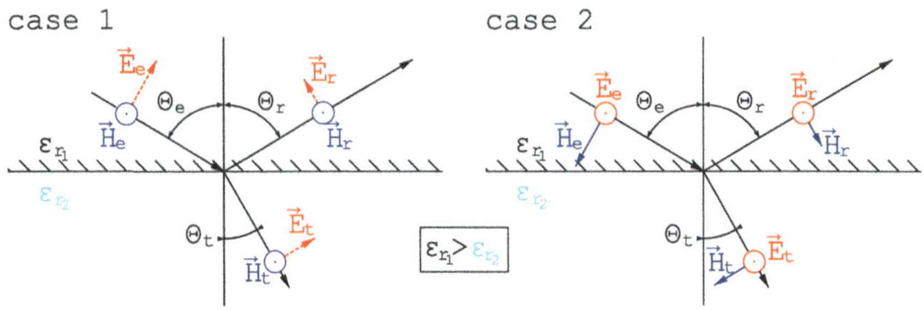

Fig. 2.8 Representation of the penetration for different polarizations of the E-field

$$\Rightarrow \sqrt{\varepsilon_{r2}} \cdot \sin\Theta_e = \sqrt{\varepsilon_{r1}} \cdot \sin\Theta_t \,. \tag{2.9}$$

The following statement can be taken from Eq. (2.9):

$\hookrightarrow$ Large angular differences in the transmission paths

2.1.4 Procedures and Products

Procedures

The following methods are implemented in both commercial and scientific software solutions:

a) PO: Physical Optic,
b) GTD: Geometrical Theory of Diffraction,
c) UTD: Uniform geometrical Theory of Diffraction.

These three processes have the following characteristics:

to a) Reflection and refraction law only
- No diffraction
- Wrong reflection angles
- No side lobes

to b) – Wrong reflection angles
assumes infinitely extended surfaces.

to c) Can perform everything shown here.

Products

Many facilities for analyzing dispersion in open spaces have their own in-house software. Many commercial and freely available software solutions are now also available. Some commercial manufacturers offer slimmed-down versions for students.

Commercial software:	Wireless Insite from Remcom Radioplan
Commercial and free software:	Radiowave Propagation Simulator (RPS) from Radioplan GmbH
Free software:	Cloud RF: www.cloudrf.com

The free ClouRF software is based on the GTD method and contains complex terrain profiles of the entire world. However, buildings are not implemented and cannot be taken into account.
Tip: When working with the software, it is important to always allow more reflections than diffractions, as the energy behind the first diffraction drops significantly more than the energy behind the first reflection.

Examples

Fig. 2.9 shows very clearly the calculation paths between the the transmitter, marked in green in the center, and the receivers.

There is only one direct path and possibly a second path via the reflection on the metal wall. In addition to the paths, the energy at the receivers is also shown in color coding. The constructive and destructive angle ranges, which overlap, can be clearly seen.

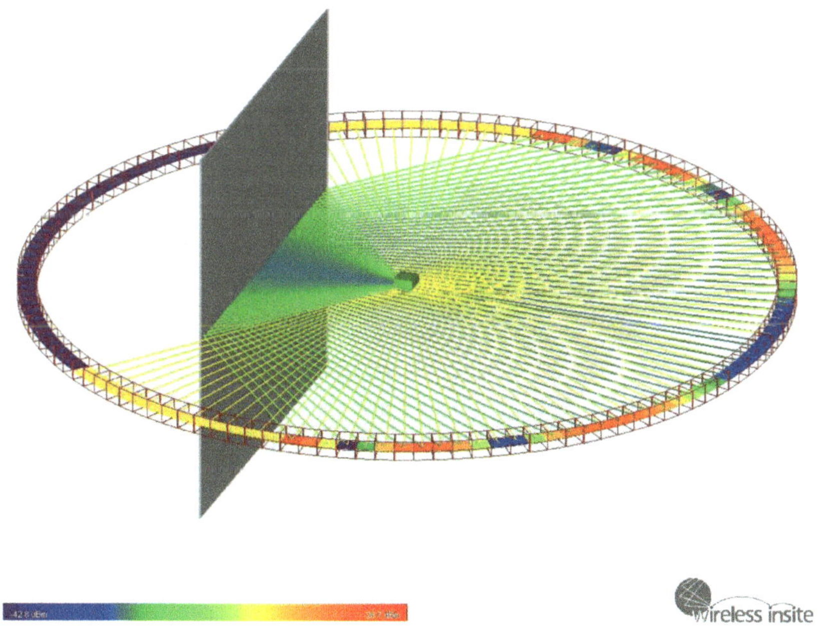

Fig. 2.9 Propagation calculation of a dipole in front of a metal wall

Figs. 2.10 and 2.11 show the practical use of such free-space simulation software. In this case, its influence on the night landing system was calculated before the construction of large maintenance hangars.

The left part of the Fig. 2.10 shows the transmitter, which on the one hand directly radiates the approach area and on the other also illuminates the maintenance hangars. The signals reflected by the maintenance hangars (Fig. 2.11) may interfere with the main signal, which is then distorted.

In the last example, only the energy received in one part of the city is shown. The transmitter is again marked in green. It is positioned in the area with the strongest received power (Fig. 2.12).

2.2 Numerical Field Simulation

Field simulations are based on numerical solutions of Maxwell's equations in the frequency domain (FD) or in the time domain (TD).

The following applies in the frequency range for time-harmonic fields:

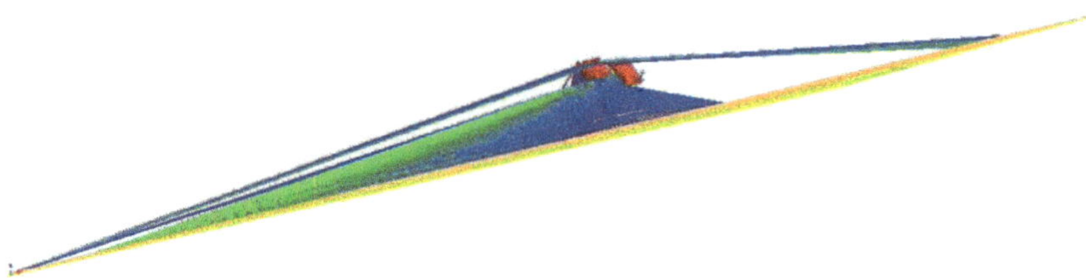

Fig. 2.10 Calculation of the influence of the new A380 maintenance hangars on the ILS landing system: general view

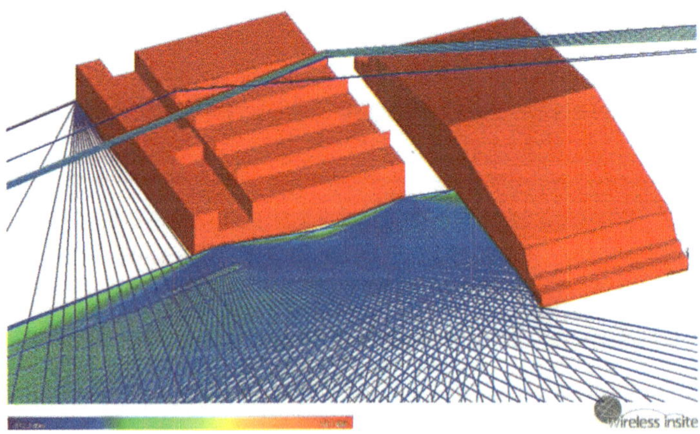

Fig. 2.11 Calculation of the influence of the new A380 maintenance hangars on the ILS landing system: hangar view

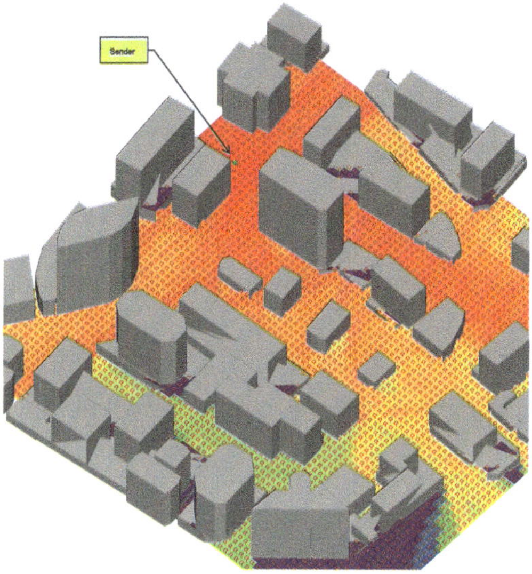

Fig. 2.12 Calculation of the received energy and the shaded areas

$$\text{rot}\vec{H} = \vec{J} + \text{j}\omega\vec{D}, \tag{2.10}$$

$$\text{rot}\vec{E} = \quad - \text{j}\omega\vec{B}. \tag{2.11}$$

$\vec{J}$: Current density
$\vec{D}$: Electrical displacement field $\quad \vec{D} = \varepsilon \cdot \vec{E}$
$\vec{B}$: Magnetic induction $\quad \vec{B} = \mu \cdot \vec{H}$

There are two ways of solving the Maxwell's equations:

1. The general solution of Maxwell's equations **with** inner sources,
2. The special solution of Maxwell's equations **without** inner sources.

2.2.1 General Solution of Maxwell's Equations

The initially presented general solution of Maxwell's equations contains inner sources.

Related example: Antenna problem

The solution method uses a) the scalar potential Φ and
b) the vector potential $\vec{A}$.

An inner source is impressed as current density $\vec{J}$ at the receptor point $\vec{r}$'.

The vector potential is calculated as the first solution step:

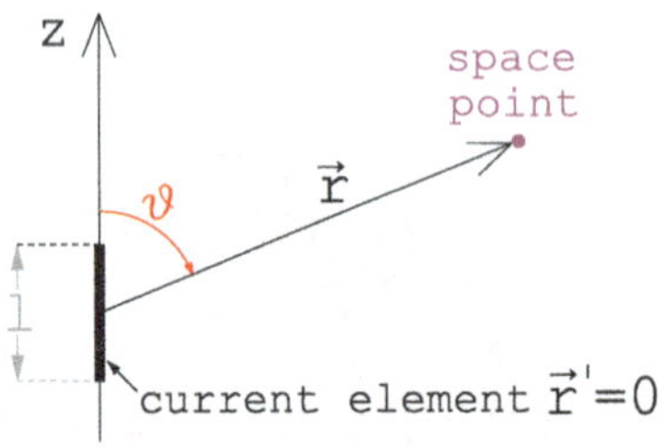

Fig. 2.13 Graphical representation of the Hertzian dipole solution approach

$$\vec{A}(\vec{r}) = \frac{\mu}{4\pi} \iiint \frac{\vec{J}(\vec{r}') \cdot e^{-jk|\vec{r}-\vec{r}'|}}{|\vec{r}-\vec{r}'|} \, dV' . \tag{2.12}$$

The scalar potential that may be required can be calculated in the second step using the so-called Lorentz calibration:

$$\Phi = \frac{j}{\omega\mu\varepsilon} \operatorname{div}\vec{A} . \tag{2.13}$$

In the last step, the electromagnetic fields are calculated:

$$\vec{B} = \operatorname{rot}\vec{A} , \tag{2.14}$$

$$\vec{E} = \frac{1}{j\omega\varepsilon\mu} \left(\operatorname{rot}\vec{B} - \mu\vec{J} \right) . \tag{2.15}$$

Example: Hertzian Dipole Solution
Fig. 2.13 illustrates the Hertzian dipole and a receptor point.

The following equations show the calculation steps for calculating the vector potential.

$$\vec{J} \cdot dV' = I \cdot \vec{e}_z \cdot ds' \tag{2.16}$$

$$\vec{A}(\vec{r}) = \frac{\mu_0}{4\pi} \cdot \vec{e}_z \int_0^l \frac{I \cdot e^{-jkr}}{r} \cdot ds' \tag{2.17}$$

$$\vec{A}(\vec{r}) = \frac{\mu_0}{4\pi} I \cdot l \cdot \frac{e^{-jkr}}{r} \cdot \vec{e}_z \tag{2.18}$$

In the last step, the calculation of Eqs. (2.14) and (2.15) of the Hertzian fields can be calculated from the above results. The corresponding results have already been shown in Eq. (2.1) and (2.2).

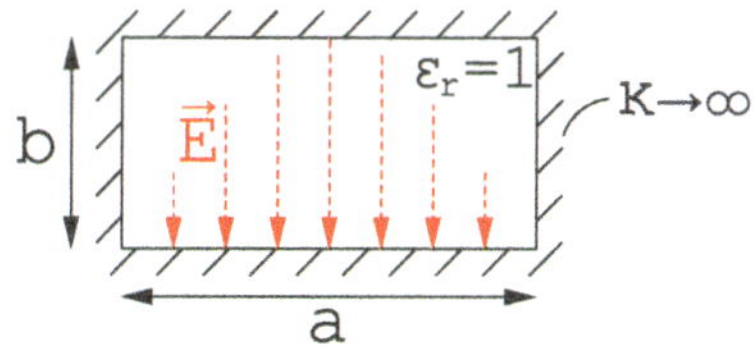

Fig. 2.14 2D waveguide (practice: here κ must be infinite)

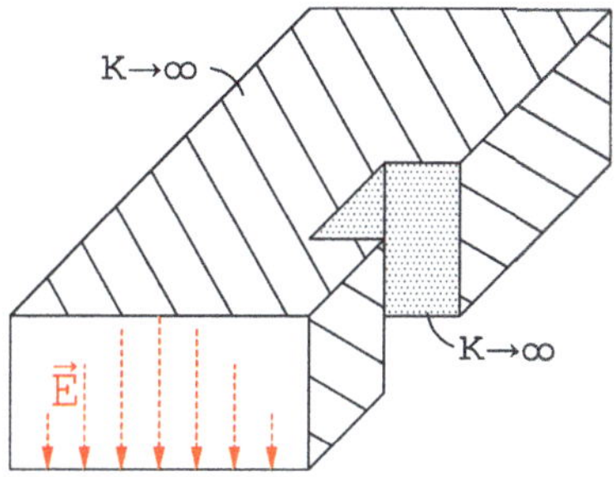

Fig. 2.15 3D waveguide (practice: here κ can also be finite)

2.2.2 Solutions of Maxwell's Equations for Source-Free Geometries

Most field simulation problems are source-free. Many can be categorized under the different line types and the different passive circuits.

Here are two examples for the calculation of waveguides.

Example 1: 2D Waveguide

Here, only the material constant and the geometries must be given (Fig. 2.14).

This 2D problem can only be used to calculate the possible waveguide modes (parameters and fields).

Example 2: 3D Waveguide

Substance constant, geometries and boundary conditions at the gates (2D solutions) must be given (Fig. 2.15).

This 3D problem can be used to obtain a wide range of information. In addition to field images the S-parameters of the n-port can also be calculated.

Maxwell's equations are solved using the homogeneous Helmholtz equations via the scalar operator

$$\Delta \vec{H} = \text{grad div} \vec{H} - \text{rot rot} \vec{H} \tag{2.19}$$

and the two equations

$$\Delta \vec{H} + k^2 \vec{H} = 0, \tag{2.20}$$

$$\Delta \vec{E} + k^2 \vec{E} = 0. \tag{2.21}$$

These two Eqs. (2.20) and (2.21) correspond to elliptical differential equations and can even be simplified for a division into TE and TM waves with the help of the scalar auxiliary variables Ψ_E and Ψ_H.

$$\Delta\Psi_E + k^2\,\Psi_E = 0 \tag{2.22}$$

$$\Delta\Psi_H + k^2\,\Psi_H = 0 \tag{2.23}$$

The five field components for the E and H waves are calculated from Ψ_E and Ψ_H respectively.

$$\text{TM- resp. E-Typ:}\quad \mathrm{E}_x = \frac{1}{j\omega\varepsilon}\,\frac{\delta^2\Psi_E}{\delta x\delta z} \quad \cdots \quad H_z = 0$$

$$\text{TE- resp. H-Typ:}\quad \mathrm{H}_x = -\frac{\delta\Psi_H}{\delta y} \quad \cdots \quad E_z = 0$$

The complete results can be found, for example in [94].

The finite difference method is presented as the first numerical method.

The finite element method is then explained in detail from a theoretical point of view and with regard to handling and use.

2.2.3 Finite Difference Method (FD)

Means the method of the Finite Differences (FD) (or also known as the finite difference method) can be used to solve ordinary and partial differential equations. This method is completely derived below for a special 2D problem as an example.

Preliminary Remarks on the FD Method

- Simplest numerical method and fully 3D-capable.
- Only differential equations can be solved numerically.
- Differential quotients are approximated by difference quotients.

2.2.3.1 Euler's Method

Euler's method works with one of the following three difference quotients.

$$\textit{on the right}: \quad \left.\frac{df}{dx}\right|_{x_0} = \lim_{h\to 0}\frac{f(x_0+h)-f(x_0)}{h} \tag{2.24}$$

$$\textit{on the left}: \quad \left.\frac{df}{dx}\right|_{x_0} = \lim_{h\to 0}\frac{f(x_0)-f(x_0-h)}{h} \tag{2.25}$$

$$\textit{central}: \quad \left.\frac{df}{dx}\right|_{x_0} = \lim_{h\to 0}\frac{f(x_0+h)-f(x_0-h)}{2h} \tag{2.26}$$

Eq. (2.26) is usually the best choice (error goes with $h^2 \to 0$).

The double derivative is formed from the difference of Eqs. (2.24) and (2.25) divided by h and is as follows:

$$\left.\frac{d^2 f}{dx^2}\right|_{x_0} = \lim_{h \to 0} \frac{f(x_0 - h) - 2f(x_0) + f(x_0 + h)}{h^2} . \tag{2.27}$$

Other numerical methods are as follows:

- Taylor series method (derivation from Eq. (2.27))
- Picard's method
- Method from Runge-Kutta

In the following, a complete mathematical derivation is presented using an example that starts general and becomes more and more specific, and is compared with an analytical solution at the end.

2.2.3.2 Example for General Potential Function $u_{(x,y)}$

Example 1

Given: $u_{xx} + u_{yy} = f$ in G, $u|_{\partial G} = \varphi \mathrel{\hat{=}}$ BC[3]

So it applies in a different notation:

$$\frac{d^2 u}{dx^2} + \frac{d^2 u}{dy^2} = f_{(x,y)} . \tag{2.28}$$

Given is a potential function that applies to the domain G. This area has a boundary whose potentials (green points in the Fig. 2.16) are known.

The mathematical function ($u_{(x,y)}$) for this 2D region is known. Based on this known function and the known values (potentials) at the boundary, the potentials in the interior are to be calculated numerically.

The potentials should be arranged in the x-direction with the distance h and in the y-direction with the distance k.

[3] BC: Boundary Conditions.

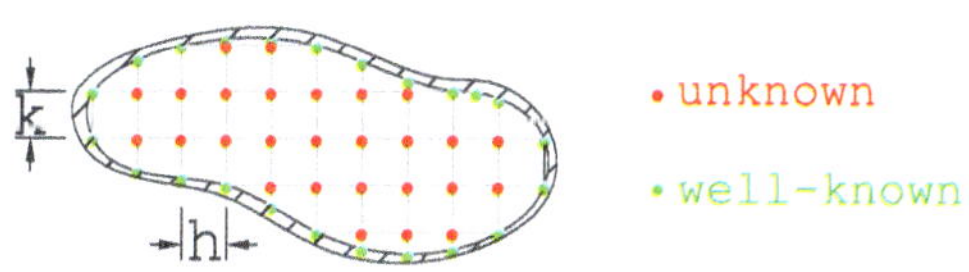

Fig. 2.16 Example of a potential function with known values on the edge

Example 1.0

Place a "mesh" over the area G and obtain the grid points (meshing), (x_i, y_j).
u_{ij} is introduced as a shortened notation for $u_{(x_i,y_j)}$.
The following applies at the edges: $u_{ij} = u_{(x_i,y_j)} = \varphi_{(x_i,y_j)}$.
The following therefore applies for Eq. (2.27):

$$u_{xx(x_i,y_j)} \approx \frac{1}{h^2}\left(u_{i-1,j} - 2\,u_{ij} + u_{i+1,j}\right) \tag{2.29}$$

$$u_{yy(x_i,y_j)} \approx \frac{1}{k^2}\left(u_{i,j-1} - 2\,u_{ij} + u_{i,j+1}\right) \tag{2.30}$$

in the interior of G with
h: Step size in x-direction and k: Step size in y-direction as well as
i: running index in x-direction and *j*: running index in y-direction.

Example 1.1

Opposite: Quadratic Structure (Fig. 2.17)
The step size is chosen to be the same in both directions: $h = k = \frac{1}{n}$.
The steps are therefore calculated as follows: $x_i = i \cdot h$; $y_j = j \cdot h$.
The range can be defined as follows: $0 \leq i, j \leq n$.
Thus (1+n) is the number of grid pionts in each direction.
The total number of grid points is calculated from $(1+n)^2$.
The known Boundary Conditions (BC) are as follows:

$$u_{(x_i,0)} = u_{i0} = \varphi_{1(x_i)} \quad \text{and} \quad u_{(x_i,1)} = u_{in} = \varphi_{3(x_i)} \tag{2.31}$$

$$u_{(0,y_j)} = u_{0j} = \varphi_{4(y_j)} \quad \text{and} \quad u_{(1,y_j)} = u_{nj} = \varphi_{2(y_j)} \tag{2.32}$$

Furthermore, for a clearer representation $f_{ij} = f_{(x_i,y_j)}$.
⇒ In the interior $(1 \leq i,\ j \leq n-1)$ should apply:

$$(u_{i-1,j} - 2\,u_{ij} + u_{i+1,j}) + (u_{i,j-1} - 2\,u_{ij} + u_{i,j+1}) = h^2 \cdot f_{ij}\,.$$

⇒ $(n-1)^2$ equations for $u_{11} \ldots u_{n-1,n-1}$ unknowns ⇒ correct!
The equation for $i = 1, j = 1$ is:

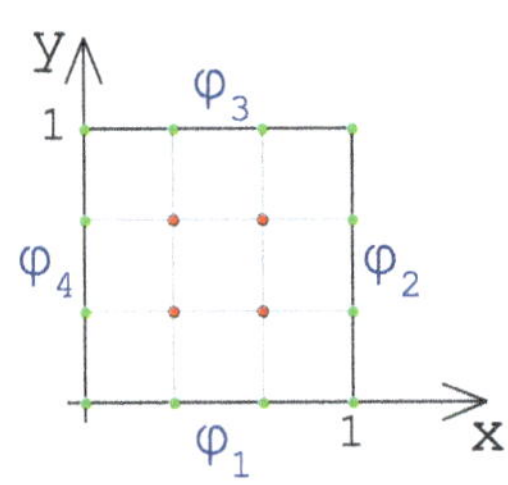

Fig. 2.17 Example of a square

$$u_{01} - 2u_{11} + u_{21} + u_{10} - 2u_{11} + u_{12} = h^2 \cdot f_{11}.$$

This can be re-sorted:

Unknown side		Known side
$-4\,u_{11} + u_{21} + u_{12}$	$=$	$h^2\,f_{11} - u_{01} - u_{10}$
	or	
$4\,u_{11} - u_{21} - u_{12}$	$=$	$u_{01} + u_{10} - h^2\,f_{11}$

This is the first row of the following matrix equation.

In matrix notation, you get a total of 16 unknowns for e.g. n=5 $\Rightarrow \hat{=}\, \vec{U}$:

$$\underbrace{\left(\begin{array}{cccc|cccc|c}
4 & -1 & 0 & 0 & -1 & 0 & 0 & 0 & \\
-1 & 4 & -1 & 0 & 0 & -1 & 0 & 0 & \\
0 & -1 & 4 & -1 & 0 & 0 & -1 & 0 & \\
0 & 0 & -1 & 4 & 0 & 0 & 0 & -1 & \\
\hline
-1 & 0 & 0 & 0 & 4 & -1 & 0 & 0 & \ddots \\
 & \ddots & & & -1 & 4 & -1 & 0 & \\
 & & \ddots & & 0 & -1 & 4 & -1 & \\
 & & & & 0 & 0 & -1 & 4 & \\
\hline
 & & & & \ddots & & & & \ddots
\end{array}\right)}_{=[A]}
\underbrace{\begin{pmatrix} u_{11} \\ u_{12} \\ u_{13} \\ u_{14} \\ \hline u_{21} \\ u_{22} \\ u_{23} \\ u_{24} \\ \hline u_{31} \\ \vdots \end{pmatrix}}_{=\vec{U}}
=
\underbrace{\begin{pmatrix} u_{01} + u_{10} \\ u_{02} \\ u_{03} \\ u_{04} + u_{15} \\ \hline -u_{20} \\ 0 \\ 0 \\ u_{25} \\ \hline \vdots \end{pmatrix}}_{=\vec{RB}}
- h^2
\underbrace{\begin{pmatrix} f_{11} \\ f_{12} \\ f_{13} \\ f_{14} \\ \hline f_{21} \\ f_{22} \\ f_{23} \\ f_{24} \\ \hline f_{31} \\ \vdots \end{pmatrix}}_{=\vec{f}}$$

$$[A] \cdot \vec{U} = \underbrace{\vec{RB} - \vec{f}}_{=\vec{C}} \tag{2.33}$$

$$\Rightarrow \vec{U} = [A]^{-1}\vec{C} \tag{2.34}$$

Example 1.1.1

The following properties should continue to apply.

Given.: $f_{x_i,y_j} = 0$; $\varphi_{1(x)} = x^2$ $\varphi_{2(y)} = 1 - y^2$ $\varphi_{3(x)} = x^2 - 1$ $\varphi_{4(y)} = -y^2$

The following calculation applies for the first boundary condition and for $n = 4$:

$$u_{01} + u_{10} = \varphi_{1(x_1)} + \varphi_{4(y_1)} = 0.25^2 - 0.25^2 = 0\,. \tag{2.35}$$

The exact solution to this equation is: $u_{x,y} = x^2 - y^2$.

For n=4 provides the numerical solution:

$$\left(\begin{array}{ccc|ccc|ccc} 4 & -1 & 0 & -1 & 0 & 0 & 0 & 0 & 0 \\ -1 & 4 & -1 & 0 & -1 & 0 & 0 & 0 & 0 \\ 0 & -1 & 4 & 0 & 0 & -1 & 0 & 0 & 0 \\ \hline -1 & 0 & 0 & 4 & -1 & 0 & -1 & 0 & 0 \\ 0 & -1 & 0 & -1 & 4 & -1 & 0 & -1 & 0 \\ 0 & 0 & -1 & 0 & -1 & 4 & 0 & 0 & -1 \\ \hline 0 & 0 & 0 & -1 & 0 & 0 & 4 & -1 & 0 \\ 0 & 0 & 0 & 0 & -1 & 0 & -1 & 4 & -1 \\ 0 & 0 & 0 & 0 & 0 & -1 & 0 & -1 & 4 \end{array}\right) \left(\begin{array}{c} u_{11} \\ u_{12} \\ u_{13} \\ \hline u_{21} \\ u_{22} \\ u_{23} \\ \hline u_{31} \\ u_{32} \\ u_{33} \end{array}\right) = \left(\begin{array}{c} 0 \\ -0.25 \\ -1.56 \\ \hline 0.25 \\ 0 \\ -0.75 \\ \hline 1.56 \\ 0.75 \\ 0 \end{array}\right)$$

The maximum error between u_{ij} and u_{x_i,y_j} is only $4.6 \cdot 10^{-10}$!

Applied to the equations of the field simulation, the equations are significantly more complex. Nevertheless, the basic procedure is the same.

2.2.4 *Finite Element (FE) Method*

The Finite Element method (FE), also called Finite Element Method (FEM), is a numerical solution method for solving partial differential equations.

FE software plays a pivotal role in modern microwave technology and provides, offering engineers and researchers powerful tools for analyzing and designing complex electromagnetic structures. Originating in the mid-20th century as a numerical technique for solving partial differential equations, the finite element method found early application in structural and mechanical engineering. However, with the growing demand for advanced microwave devices and systems, the method was adapted and refined to address electromagnetic phenomena.

In the field of microwave engineering, FE software enables the precise modeling and simulation of electromagnetic fields in various components, such as waveguides, antennas, filters, passive components, IC-design, and single or multilayer microwave circuits. This computational approach provides insights into electromagnetic behavior and facilitates the optimization of device performance and the exploration of innovative design concepts. Over the years, the FE software has evolved to meet the increasing complexity and demands of microwave engineering, incorporating advanced algorithms and computational techniques to enhance simulation accuracy and efficiency. FE software is now an indispensable tool for engineers and researchers in the pursuit of cutting-edge microwave technologies and applications.

Preliminary Remark

FE method: This is a special technique for the direct numerical solution of "variational tasks".

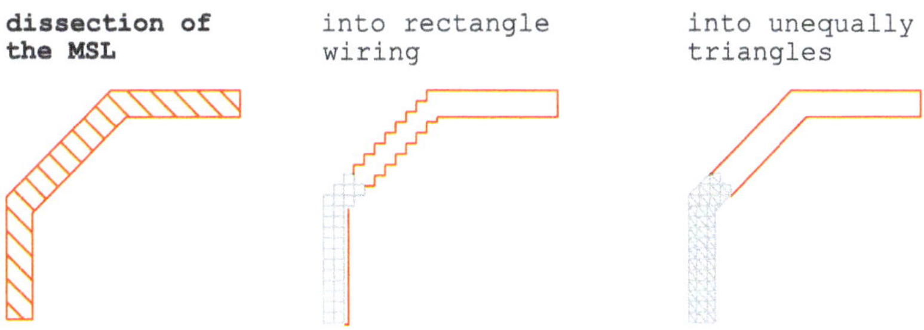

Fig. 2.18 Disassembly of a microstrip line (MSL)

Calculus of variations: Determination of the extreme values of functionals defined on a class of functions. If necessary, it provides a function that must satisfy a differential equation.

A differential equation in implicit form is represented as follows:

$$F(y_{(x)}, y'_{(x)}, x) = F(y, y', x) \; . \tag{2.36}$$

In the variation problem, the functional I is calculated using

$$I = \int_{x_1}^{x_2} F(y, y', x)\, dx \stackrel{!}{=} \text{extreme value} \; . \tag{2.37}$$

The corresponding solution is:

$$\frac{\delta F}{\delta y} - \frac{d}{dx}\frac{\delta F}{\delta y'} = 0 \; . \tag{2.38}$$

In the following, the principle of FE calculation is first presented in a brief overview. The individual steps are then explained in detail.

Principle of the FE Method (Overview)

1. Step: Decomposition of the solution domain into subdomains as shown in Fig. 2.18. For 2D problems often triangulation with triangles is chosen and for 3D objects a tetrahedron formation is chosen.

2. Step: Determination of nodes in the individual sub-areas and definition of approximate solutions using the "node potentials" $U_{i,j}$.

3. Step: Determination of the node potentials using the extreme value properties of the variation integral.

4. Step: Post-processing: Interpolation calculation etc.

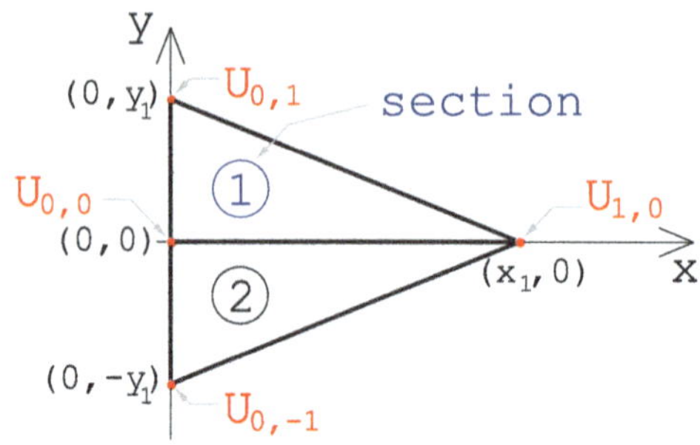

Fig. 2.19 Representation of node potentials U over 2 sub-areas

Principle of the FE Method (Details)

To step 1: For example, the function for node potential $u(x, y)$ should be

$$u(x,y) = a + bx + cy \tag{2.39}$$

with the three auxiliary variables a, b and c can be selected. The node potentials U are calculated at the accounts using this function. The amounts of the vector potentials $|\vec{A}|$ are often used (Fig. 2.19).

To step 2: For sub-area 1, Eq. (2.39) is formulated as follows:

$$u^{(1)}_{(x,y)} = a^{(1)} + b^{(1)}x + c^{(1)}y\,. \tag{2.40}$$

In the following, the auxiliary variables a, b and c are replaced by the various account potentials.

$$U_{0,0} = a^{(1)} \quad \Rightarrow \quad a^{(1)} = U_{0,0} \tag{2.41}$$

$$U_{1,0} = U_{0,0} + b^{(1)}x_1 \quad \Rightarrow \quad b^{(1)} = \frac{1}{x_1}(U_{1,0} - U_{0,0}) \tag{2.42}$$

$$U_{0,1} = U_{0,0} + c^{(1)}y_1 \quad \Rightarrow \quad c^{(1)} = \frac{1}{y_1}(U_{0,1} - U_{0,0}) \tag{2.43}$$

The calculated auxiliary variables a, b and c are now used in the Eq. (2.40).

$$u^{(1)}_{(x,y)} = U_{0,0} + (U_{1,0} - U_{0,0})\frac{1}{x_1}x + (U_{0,1} - U_{0,0})\frac{1}{y_1}y \tag{2.44}$$

$$= U_{0,0}(1 - \frac{1}{x_1}x - \frac{1}{y_1}y) + U_{1,0}\frac{1}{x_1}x + U_{0,1}\frac{1}{y_1}y \tag{2.45}$$

The continuity condition between sub-areas 1 and 2 is as follows:

$$U^{(1)}_{(x,0)} = U^{(2)}_{(x,0)}\,.$$

For sub-area 2, based on the equation

$$u_{(x,y)}^{(2)} = a^{(2)} + b^{(2)}x + c^{(2)}y\,, \tag{2.46}$$

proceed as already shown, whereby the two potentials $U_{0,0}$ and $U_{1,0}$ are already given due to the continuity.

To step 3: Precise numerical calculation of the potentials $U_{(x,y)}^{(m)}$, where the potentials $U^{(m)}$ at the edges are known.

2.2.5 FE Simulation Process

The following table is intended to illustrate the process of a typical finite element simulation:

1st:	Open project	
2nd:	Create 3D drawing	
3rd:	Assign materials	
4th:	Specify boundary conditions and gates	
5th:	Enter simulation parameters	(control port only)
6th:	Have field problem solved	
7th:	Perform post processing	
7.1:	Check mesh	(opt.)
7.2:	Display E-, H-, J-fields	(opt.)
7.3:	De-embedding & renormalize	(opt.)
7.4:	Output scattering parameters	

This process will be illustrated using the example of the calculation of a waveguide. The Ansys program *High Frequency Structure Simulator,* or HFSS for short, was used as the FE simulator to illustrate this.

Example: A rectangular waveguide is to be calculated at 5 GHz, with the H_{10} already propagating at the frequency $f_{cH_{10}}$= 4 GHz. The entire length of the waveguide should correspond to 2 wavelengths and therefore $\beta l = 4\,\pi$. This standard waveguide is illustrated in Fig. 2.20.

The following relationships apply to the design of the waveguide.

$$\frac{b}{a} = \frac{1}{2} \qquad \Rightarrow \qquad \lambda_{cH_{10}} = \frac{a^2}{\sqrt{(1/4)\,a^2}} = 2a \tag{2.47}$$

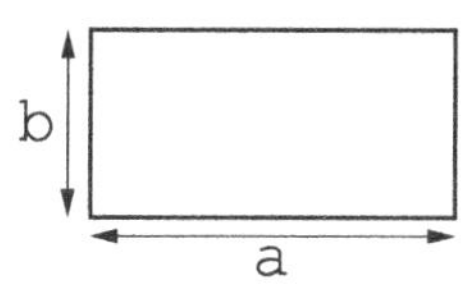

Fig. 2.20 Example: Rectangular waveguide with $a = 2 \cdot b$

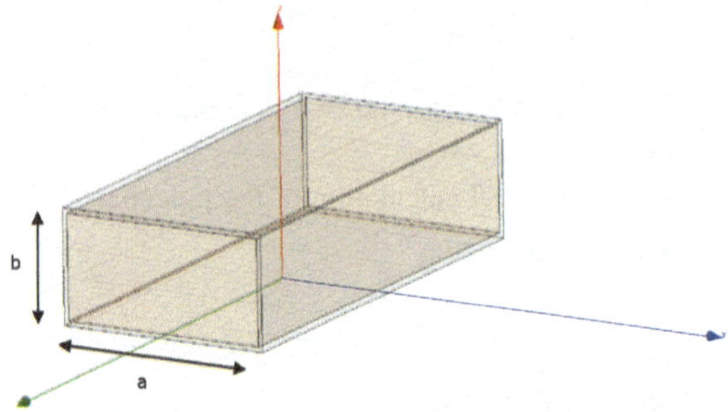

Fig. 2.21 Example: HFSS drawing of a rectangular waveguide with $a = 2 \cdot b$

$$\lambda_{cH_{10}} = \frac{c_0}{f_{cH_{10}}} \qquad \Rightarrow \qquad a = 3.75\,\text{cm} \tag{2.48}$$

$$\gamma_{mn} = \frac{2\pi}{\lambda_0} \cdot \sqrt{(\frac{\lambda_0}{\lambda_{c_{10}}})^2 - 1} \quad \text{with} \quad \lambda_0 = \frac{c_0}{f_{5GHz}} = 6\,\text{cm} \tag{2.49}$$

$$\Rightarrow \beta_{10}|_{5GHz} = 62.6\,\text{m}^{-1} \tag{2.50}$$

$$\beta_{10}\, l_x = 4\pi \quad \Rightarrow \quad l_x = 20.0\,\text{cm} \tag{2.51}$$

$$\text{Controling:}\ \lambda_h = \frac{\lambda_0}{\sqrt{1-(\lambda_0/(2a))^2}} = 10.0\,\text{cm} \tag{2.52}$$

To 2nd: The performance of the 3D drawing interface is an important quality feature for the 3D simulator.

In the case of the waveguide, it is sufficient to draw only the air-filled area, provided the outer metallization is assumed to be perfectly conductive. If this is not the case, draw the waveguide as shown in Fig. 2.21.

Note: The entire volume around the drawn 3D object is a perfectly conductive metal area!

To 3rd: Quality features of a simulator are many material constants (standard: κ, μ, ε) with losses and, if possible, direction- and frequency-dependent.

Air with ε_r=1 is the default setting.

For a clear representation, it is important to assign the appropriate colors to the materials. For example, air should be light blue and transparent. The color red is suitable for metal.

To 4th: The quality characteristics for the boundary conditions and the gates are: As many different boundary conditions as possible (standard: Perfect-E, -H; Radiation, Impedance (should be as complex as possible))[4].

To 4.1: Waveguide Ports

[4] For the 2D solution of the ports (waveguide ports), the R.B. must always be perfect. There can be problems with de-embedding in particular, as the losses are not taken into account.

The gates for mode excitation (waveguide ports) belong to the very important boundary conditions: These ports require additional definitions of the following lines:

a) **Impedance Line**
   ⇒ Specifies points for Z_L definition.
   ⇒ Specifies the integration path for $U = \int_1^2 \vec{E}\, ds$.
b) **Calibration Line**
   ⇒ Specifies the direction of the E-field:
   The example in Fig. 2.22 shows the waveguide with the blue calibration line.
c) **Polarisation Line**
   The polarization line is used less frequently.
   An example with a green polarization line is given in Fig. 2.23.

The (simultaneous) use of calibration lines, polarization and perfect walls is very prone to user error! No transmission takes place if the settings are incorrect. Often only the impedance line is used.

In addition to the impedance line, the area of the port (see blue area in Fig. 2.24) and definition of the port impedance are also specified for the so-called waveguide ports. The following three definitions exist:

a. Z_{PI}: Calculation from Power and Current

$$Z_{PI} = \frac{P}{I \cdot I^*} \quad \text{with} \quad P = \oiint_A E \times H^*\, dA \quad \text{with} \quad I = \oint_S H\, ds \tag{2.53}$$

Z_{PI}: Standard definition of Z_L

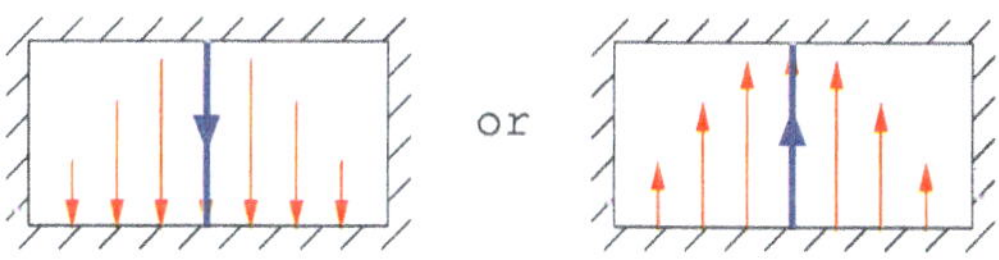

Fig. 2.22 Example: Calibration line in the rectangular waveguide

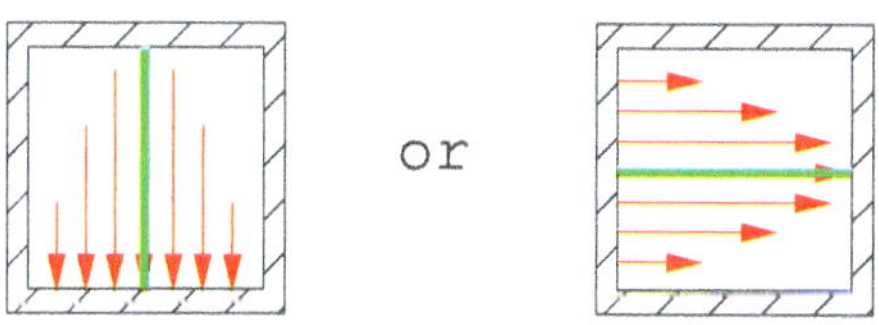

Fig. 2.23 Example: Polarization in a square waveguide

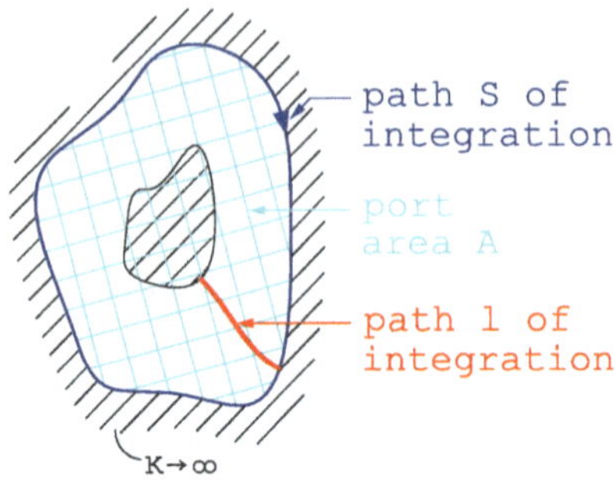

Fig. 2.24 Two-wire system for the introduction of port impedances

b. Z_{PV}: Calculation from Power and Voltage

In European nomenclature, Z_{PV} would be referred to as Z_{PU}[5]. The calculation is done via

$$Z_{PV} = \frac{V \cdot V^*}{P} \tag{2.54}$$

with

$$V = \int_{\ell} E \cdot d\ell \tag{2.55}$$

about the integration path ℓ.

For TEM wave only, the same result is obtained for Z_{PI} and Z_{PV}, Fig. 2.25.

The usual integration path of a microstrip line is shown in Fig. 2.26.

c. Z_{VI}: Calculation from Z_{PI} and Z_{PV}

This is the mean value of the previously calculated characteristic impedances.

$$Z_{VI} = \sqrt{Z_{PI} \cdot Z_{PV}}$$

d. Control

The following must apply to TEM waves: $Z_{PV} \stackrel{!}{=} Z_{PI}$.

To 4.2 Lumped Ports

[5] In this book, the voltage is referred to as the European standardization code U, as this is clearer than the American standardization code. However, the V is used in this section as this is the only way it is used in the simulator.

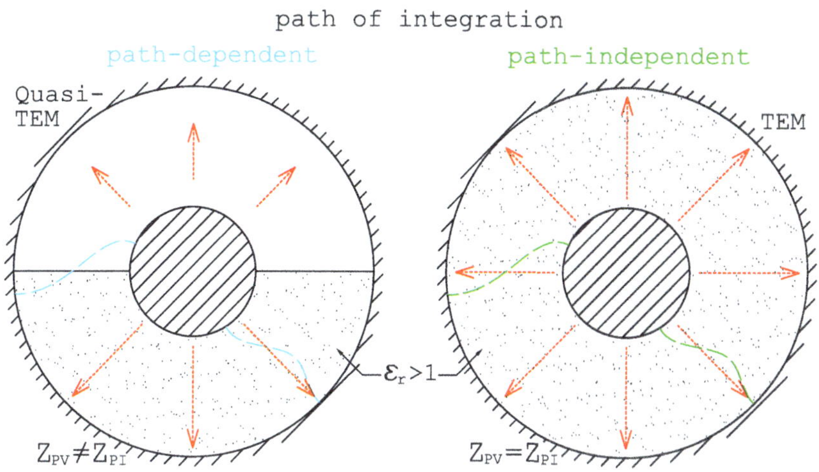

Fig. 2.25 Path-dependent and independent integration path

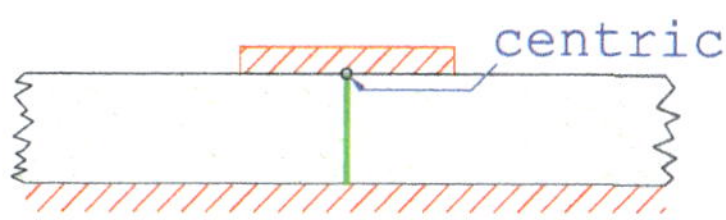

Fig. 2.26 Integration path for a microstrip line

Lumped ports are very simple microwave ports that are only defined by the port impedance value between two selected nodes.

Benefits:

1. These ports are easy to define. There are generally no implementation errors.
2. These lumped ports can also be used inside a 3D structure.

Drawback:

1. These ports generally do not correspond to physics. The excitation looks different in reality. Here you have to estimate whether these ports correspond to the real situation.

In practice, matching values of up to approx. $S_{11}^{dB} \geq -20\,\text{dB}$ can be optimized for microstrip lines (small conductors).

To 5 & 6. Procedure of the HFSS calculation process:

1. Calculate fields for waveguide ports (these are 2D simulations).
2. Calculate first mesh for the 3D structure.
   The so-called adaptive frequency is the basis for the node spacing.
3. Calculate FE solution for adaptive frequency.
   ⇒ Scattering parameter of the 1st solution!
4. Calculate second mesh with (e.g.) 20 % more nodes (calculate 0 % in "cold" areas to 40 % in "hot" areas).

5. Similar to 3. ⇒ 2. solution!
6. Calculation of ΔS error,
   e.g. one-port $\Delta S = S_{11}^{1} - S_{11}^{2}$.
7. Repeat 4.–6. until ΔS criterion is fullfilled.
8. For the refined mesh, calculate the S-parameters at all frequency points as a sweep.

2.2.6 Method of Moments (MoM)

The method of moments (MM) is a numerical method that is only used in high and radio frequency technology.

The numerical method of moments (MoM) is a 2.5D computational technique widely employed in microwave engineering for the analysis of planar electromagnetic structures. MoM discretizes integral equations governing electromagnetic phenomena, enabling the efficient solution of complex scattering and radiation problems. By dividing structures into small elements and representing electromagnetic fields using basis functions, MoM accurately models the interactions between components such as mono- and multi-layer substrates with antennas, filters, couplers, signal dividers, scattering bodies, and transmission lines. This method is particularly valuable for designing and optimizing planar microwave devices and systems, offering engineers insights into electromagnetic behavior and facilitating the development of high-performance technologies.

Basic Idea:

Fig. 2.27 shows the top view of two microstrip lines in which the two partial currents shown flow.

A partial current i_{ku} induces a partial current i_{ev} in a point *ev*. This task can be solved very precisely with the so-called Green's functions.

The calculation times are comparatively short, the simulators are simpler and also cheaper.

The limits in practice are:

- Only 2.5D-capable (only shift models),
- Only perfect-E boundary conditions,
- No E and H fields can be displayed,
- Only gap sources (similar to lumped ports).

Products: –Sonnet
–Momentum(in ADS)

More information on the momentum method can be found in [105]. Many applications of HFSS and the moment method are also shown there in detail.

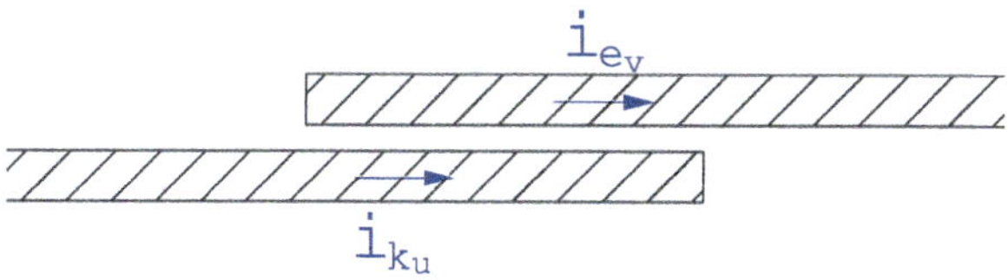

Fig. 2.27 Momentum partial current in microstrip lines

Chapter 3
Fundamentals of Nonlinear RF Technology

This chapter starts with a description of the phenomena that occur in electronics due to non-linear effects. Components with non-linear effects are then described analytically. Furthermore, the simulation technology used in the microwave range for the integration of non-linear effects is described in detail. The mixed-frequency S-parameters are then introduced and their measurement technology and use explained. The derivation of the X-parameters is also included in this chapter. In the application part of this chapter, the harmonic radar is discussed in detail in addition to the IM measurements.

3.1 Classes of Nonlinearities in the Microwave Range

In radio frequency electronics, the simple statement can be made that every circuit is nonlinear. These non-linearities are usually divided into three classes:

Class I: However, many components such as small signal amplifiers and switches are only weakly nonlinear. That means, these quasilinear semiconductor components are first developed as linear circuits and then their small nonlinear effects are investigated in a second step. This was shown, for example, in [38] for a small signal amplifier. With these components, the nonlinear behavior is a parasitic effect. But even purely passive components such as coils with a ferrite core can also exhibit nonlinear effects. It also affects all passive components that exhibit effects due to heating. Such devices and components are referred to as `quasilinear`.

Class II: Other components such as mixers, modulators, detectors and frequency multipliers can only be implemented due to strong nonlinear effects. Although the strong nonlinear behavior of a component is exploited in these components, the transmission behavior is also quasi-linear. Although these linear mixed-frequency components can no longer be calculated approximately based on the physical conditions using a linear circuit simulator, they can be integrated into simulations as simplified models. These components are referred to as `weakly nonlinear`.

Class III: Components that are controlled under large signal conditions form the third class of highly nonlinear circuits. Examples of these large signal components are power amplifiers, oscillators and digital logic gates. These components very often exhibit memory behavior due to charged capacitors. The operating points of

H. Heuermann, *Microwave Technology*,
https://doi.org/10.1007/978-3-658-45686-3_3

internal transistors are dependent on the charge of these capacitors. Such devices and components are referred to as `strongly nonlinear`.

Weakly and strongly nonlinear components can only be analyzed using special nonlinear simulation methods such as the Volterra series or harmonic balance. Quasilinear components are analyzed using the same methods as weakly nonlinear components: S-parameter methods for the transfer function and the aforementioned methods for the small nonlinear variables such as harmonics.

The statement that all RF components are nonlinear can be understood by looking at the individual components. It is generally known that all semiconductor components exhibit nonlinear behavior. However, coils, capacitors and resistors can also exhibit nonlinearities at extreme operating points. For example, a resistor changes its value when the current flow increases. Even RF connectors are known to exhibit nonlinear effects at higher power levels. Normally, any N connector (with 7 mm outer diameter) does not meet the intermodulation requirements of a 50 W base station for modern mobile communication.

The linear transmission behavior of a circuit is investigated by applying two input signals ($x_1(t)$ and $x_2(t)$), which in turn generate two output signals ($y_1(t)$ and $y_2(t)$).

$$x_1(t) \rightarrow y_1(t) \qquad\qquad x_2(t) \rightarrow y_2(t) \tag{3.1}$$

If the two input signals are added together, the result is the sum of the output signals,

$$x_1(t) + x_2(t) \quad \rightarrow \quad y_1(t) + y_2(t) \, , \tag{3.2}$$

and multiplying the input signal by a constant A results in an output signal increased by the factor A,

$$A x(t) \quad \rightarrow \quad A y(t) \, , \tag{3.3}$$

then a circuit is *linear*!

Although there is no standard for nonlinear effects, a description based on the standard work [68], a description of nonlinear description of nonlinear circuit effects based on three nonlinear nonlinear transfer functions has become established:

- Current - Voltage (I/U) ,
- Charge - Voltage (Q/U) ,
- Flow - Current (Φ/I) .

For weakly nonlinear components, truncated polynomial representations can be used with sufficient accuracy.

In this chapter, forms of description for components are first presented, the frequency generation of nonlinear components is mathematically introduced and the nonlinear phenomena are explained based on these results. This is followed by a detailed description of nonlinear components and finally simulation methods for nonlinear circuits are discussed.

In [104] it can be read that the nonlinear description of components is also increasingly being used in the field of electronics. The fundamentals presented here for the description of high-frequency circuits and RF simulation techniques are aimed at methods that are available in modern circuit simulators and are therefore directly applicable.

3.2 Description Forms for Components

Very important nonlinear components are the diodes. In detectors or mixers, diodes are regarded as weakly nonlinear elements and in rectified mode as strongly nonlinear elements. Since the detector, mixer or controllable capacitance operating modes are of great interest for practical high and radio frequency technology, the diodes should be considered as weakly nonlinear components. These diodes have only two terminals or two poles. Fig. 3.1 shows the generalized circuit symbol of a two-terminal nonlinearity, which will be referred to as *terminal element* in the following.

In addition to diodes, which are important in practice, resistors, coils and capacitors can also be described as terminal elements. These nonlinear RLC components are preferably used for parasitic effects in the very complex nonlinear modeling of transistors.

Schottky diodes serve as a simple example of a terminal element. The nonlinear characteristic between current I and voltage U of a Schottky diode is given by

$$I = I_{ss}\left(e^{\left[\frac{q}{nkT}U\right]} - 1\right) . \tag{3.4}$$

In addition to the elementary charge q, it contains the saturation blocking current I_{ss} and the empirically determined ideality factor $n = 1.05 \ldots 1.15$. In practice, the temperature voltage is used

$$U_T = \frac{nkT}{q} = 25.9 \text{ mV} \tag{3.5}$$

for the values $n = 1$ and $T = 300\,\text{K}$.

So-called transfer nonlinearities, or *transfer elements* for short, can be treated analytically somewhat more simply than terminal elements. The most important

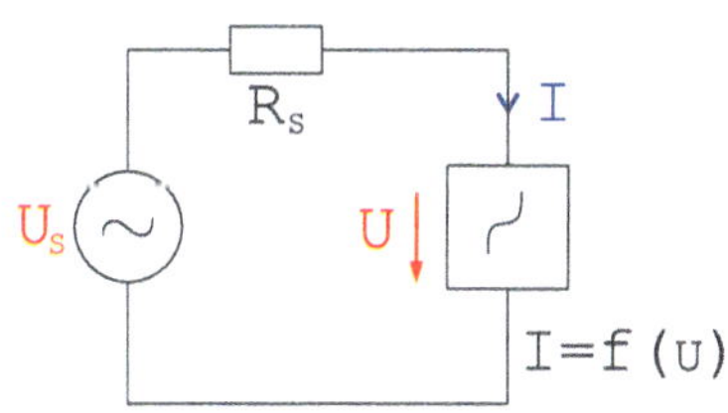

Fig. 3.1 Two-terminal nonlinearity (terminal element for short) in a simple circuit

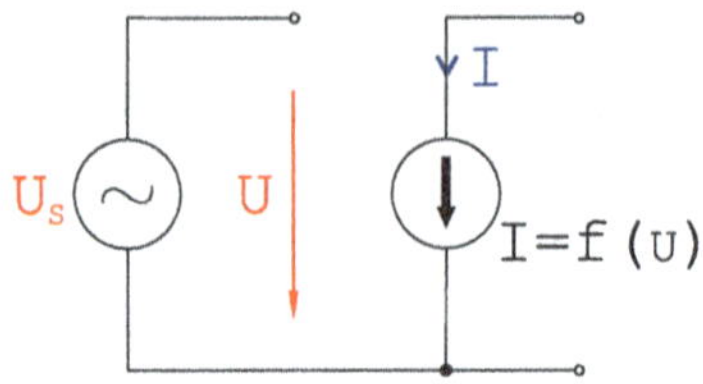

Fig. 3.2 Transfer nonlinearity (transfer element for short) with control by a voltage source

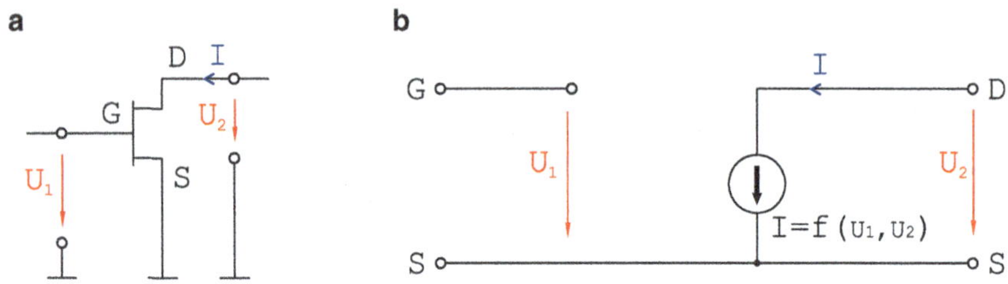

Fig. 3.3 Example of a field-effect transistor as a controllable resistor with dependencies on U_1 and U_2

transfer element is the transistor. Fig. 3.2 shows a transfer element driven by the AC voltage source with the effective value U_S.

The nonlinear properties of the components generate new frequencies, among other things. In the case of terminal elements, the entire circuit is influenced by these nonlinear effects. In contrast, it can be clearly seen that with transfer elements only the subsequent part of the circuit is affected. This non-reactive transmission behavior is the reason why transfer elements are much easier to describe than terminal elements.

However, this simple transfer element is not a sufficient description of a transistor in practice. For example, the output current is not only dependent on the input voltage but also on the output voltage, as shown in Fig. 3.3. In addition, there are other dependencies and a large number of parasitic elements, many of which are also nonlinear in nature.

The practitioner considers additional components for a real transistor in order to better model the parasitic effects of the transistor. The following Fig. 3.4 shows the model of a CMOS transistor for small-signal modeling of the parasitic components up to 170 GHz [52].

All capacitances and resistances are fitted purely from a scattering parameter measurement up to 170 GHz, as shown in detail for a coil in [38], Chap. 9. Only the current is calculated in simplified nonlinear models in modeled as a function of the applied voltages with its nonlinear transfer function. For large signal transistors, the parasitic components (especially the capacitors) are also modeled using nonlinear terminal elements.

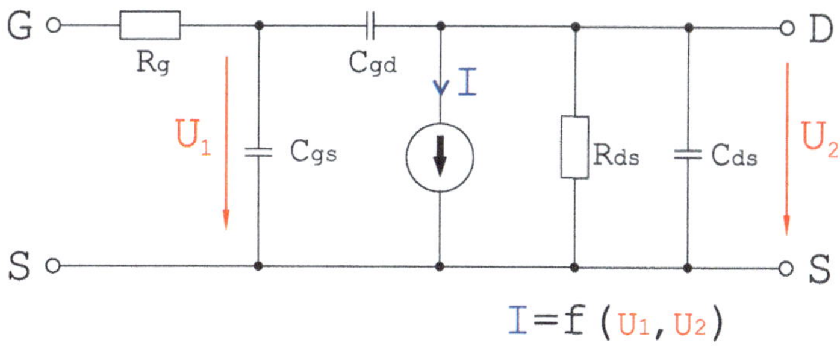

Fig. 3.4 Practical example of a CMOS transistor whose modeling is described in [52]

3.3 Frequency Generation by Nonlinearities

Weakly nonlinear terminal elements and transfer elements can generally be described by a polynomial representation. Such a description shows how new signals arise at other frequencies. In the following, this is shown using the example of the current-voltage dependence of a truncated polynomial representation with the real coefficients *a*, *b* and *c*:

$$i_{(t)} = a u_{(t)} + b u_{(t)}^2 + c u_{(t)}^3 \,. \tag{3.6}$$

The voltage $u_{(t)}$ is excited by a source with two frequency components and different amplitudes:

$$u_{(t)} = \hat{u}_1 \cos(\omega_1 t + \alpha) + \hat{u}_2 \cos(\omega_2 t + \beta) \,. \tag{3.7}$$

Both signal components of the voltage $u_{(t)}$ should have the fundamental phase of 0° and $\hat{u}_i$ should be replaced by the corresponding amplitude values in the frequency range (here also peak values) U_i (with the fundamental phase of 0°):

$$u_{(t)} = U_1 \cos(\omega_1 t) + U_2 \cos(\omega_2 t) \,. \tag{3.8}$$

According to Eq. (3.6), there are three terms for the current: one linear, one quadratic and one cubic. In general, the three partial currents should be introduced:

$$i_{(t)} = i_{a(t)} + i_{b(t)} + i_{c(t)} \,. \tag{3.9}$$

After inserting Eq. (3.7) into Eq. (3.6) and multiplying out, the following terms result.

1. **Linear term:**

$$i_{a(t)} = a u_{(t)} \,, \tag{3.10}$$

$$i_{a(t)} = a U_1 \cos(\omega_1 t) + a U_2 \cos(\omega_2 t) \,. \tag{3.11}$$

2. **Quadratic term:**

$$i_{b(t)} = b\,u_{(t)}^2, \tag{3.12}$$

$$i_{b(t)} = \frac{1}{2} b\{U_1^2 + U_2^2 + U_1^2 \cos(2\,\omega_1 t) + U_2^2 \cos(2\,\omega_2 t) \\ + 2U_1U_2 [\cos((\omega_1 + \omega_2)t) + \cos((\omega_1 - \omega_2)t)]\}\,. \tag{3.13}$$

3. **Cubic term:**

$$i_{c(t)} = c\,u_{(t)}^3\,, \tag{3.14}$$

$$i_{c(t)} = \frac{1}{4} c\{U_1^3 \cos(3\,\omega_1 t) + U_2^3 \cos(3\,\omega_2 t) \\ + 3U_1^2U_2 [\cos((2\,\omega_1 + \omega_2)t) + \cos((2\,\omega_1 - \omega_2)t)] \\ + 3U_1U_2^2 [\cos((2\,\omega_2 + \omega_1)t) + \cos((2\,\omega_2 - \omega_1)t)] \\ + 3(U_1^3 + 2U_1U_2^2) \cos(\omega_1 t) \\ + 3(U_2^3 + 2U_1^2U_2) \cos(\omega_2 t)\}\,. \tag{3.15}$$

In the example above, the polynomial representation was terminated at the third element. If you were to work without truncation, this would result in signals at the discrete frequencies

$$\omega_{m,n} = |m\,\omega_1 + n\,\omega_2| \quad \text{with} \quad m, n = \ldots, -3, -2, -1, 0, 1, 2, 3, \ldots \quad . \tag{3.16}$$

These newly resulting signals at the different frequencies are generally referred to as *mixed products*. This spectrum is also known as a mixed spectrum.

Circuitry can be used to ensure that only odd-numbered mixed products in the form of

$$i_{(t)} = a\,u_{(t)} + c\,u_{(t)}^3 + e\,u_{(t)}^5 \tag{3.17}$$

or only even-numbered mixed products, as in

$$i_{(t)} = a\,u_{(t)} + b\,u_{(t)}^2 + d\,u_{(t)}^4 \tag{3.18}$$

must be taken into account. To achieve this, balanced structures are used, which will be discussed below.

The cubic term of Eq. (3.16) consists only of third-order mixed products. The contained signal component at the fundamental frequency ω_1 therefore results from

$$\omega_1 = (\omega_1 + \omega_1) - \omega_1 \quad \text{and} \quad \omega_1 = (\omega_1 + \omega_2) - \omega_2 \quad . \tag{3.19}$$

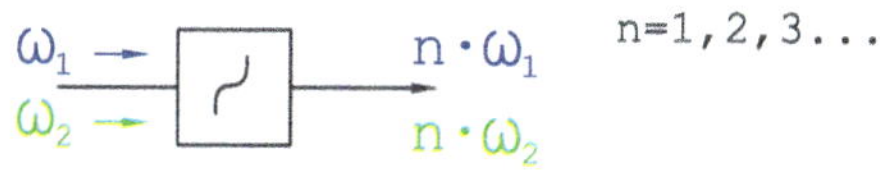

Fig. 3.5 Schematic illustration of harmonic generation by a nonlinear component

3.4 Nonlinear Phenomena

In this section, the nonlinear phenomena are explained in more detail and more specific properties are assigned to the mixed products.

Harmonic Generation

Every nonlinear component generates harmonics through excitation with a fundamental wave. Fig. 3.5 illustrates this relationship schematically.

If only a mono-frequency signal ω_1 is present at the nonlinear element *and* all harmonics are terminated without reflection, then no further mixing products result. However, if the harmonics are not terminated without reflection, mixed products result from the reflected components at the frequencies $n\,\omega_1$.
Following situation is possible about the relationship

$$\omega_1 = 3\,\omega_1 - 2\,\omega_1 \ . \tag{3.20}$$

Signal energy is converted from the higher radio frequency range in the lower RF range.

Mixed Frequencies

An upward mixer is controlled with a large signal at the frequency ω_{LO} and should linearly convert a low-frequency small signal (intermediate frequency: IF) with the frequency ω_{IF} to a high-frequency small signal (radio frequency: RF) with the frequency ω_{RF}. Eq. (3.16) generally applies to the mixer, so that the mixer also converts the low-frequency signal by ω_{IF} for $m > 1$ at the discrete mixing frequencies according to

$$\omega_{RF} = m\,\omega_{LO} \pm \omega_{IF} \tag{3.21}$$

linear mixes up into the high frequency range. The two mixing frequencies in Eq. (3.13) are the two most frequently used high-frequency signals. If the signal components around the harmonics of the local oscillator signal are used, this is referred to as a harmonic mixer. Mixing frequencies can only be found in even-numbered terms (e.g. Eq. (3.13)).

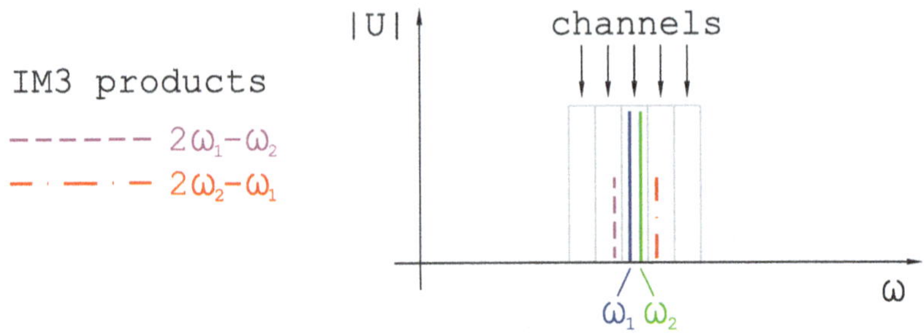

Fig. 3.6 Crosstalk in adjacent channels

In Eq. (3.16) you can clearly see that the prefactors before the $\cos(\omega_x t)$ signal components of the odd cubic term are nonlinear.

Intermodulation

All mixed products in the Eq. (3.16), which are caused by two or several *tones* and have a nonlinear amplitude dependence, are

Intermodulation products (IM products).

Thus, the portion $3U_1^3 \cos(\omega_1 t)$ is also a third-order IM product of the tones at ω_1 and $2\omega_1$. There are only IM products of third and higher orders. The IM products contained in the cubic term are all third-order products or IM_3 products for short (or IM3 products).

The biggest problem with nonlinear phenomena in practice is crosstalk in adjacent channels, as illustrated in Fig. 3.6.

Due to the cubic or higher amplitude dependence of the IM products, their output power does not increase linearly with increasing input power, but much more strongly. In Eq. (3.16), these IM products can be found in the 2nd and 3rd row.

Saturation and Compression

If the modulation of a component is so strong that the operating voltage values are reached, then this component goes into saturation. Saturation or compression can be recognized by the fact that the output power no longer increases proportionally to the increasing input power. This is illustrated in Fig. 3.7.

Such effects occur not only in amplifiers, but also in electronic switches and mixers. In data sheets, the saturation behavior is illustrated by specifying the so-called IP_3 or IM_3 and P_{-1dB} points. The P_{-1dB} point indicates the input power at which the forward power transfer function (e. g. gain) has decreased by 1 dB compared to a linear power transfer function (gain). In Fig. 3.7, the deviation of the

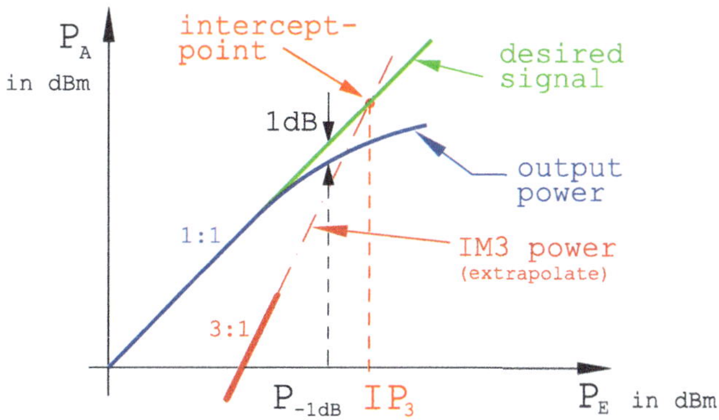

Fig. 3.7 Saturation phenomenon and associated IP_3- or IM_3- and P_{-1dB} points of the output power P_A plotted against the input power P_E

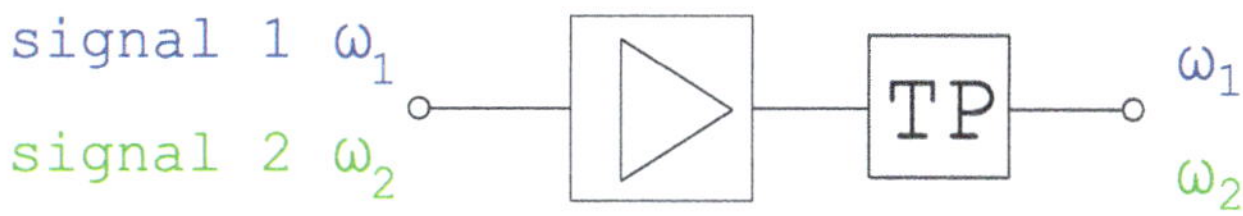

Fig. 3.8 Amplifier and low-pass filter

useful power from the ideal linear curve caused by saturation is characterized by the 1 dB compression point. At this point, the deviation is just 1 dB.

The IP_3 or IM_3 point indicates the calculated input power at which the nonlinear mixing product of the second and third harmonics corresponds to the input power of the fundamental wave. At the third-order intercept point (IP_3), the extrapolated useful power and the extrapolated interference power are therefore just equal. Mathematically, the saturation effect can be described by the linear amplification of the fundamental wave and a cubic term with opposite sign that increases with increasing modulation (here $U_2 = 0\,\mathrm{V}$):

$$i_{1(t)} = \left(aU_1 + \frac{3}{4}cU_1^3\right)\cos(\omega_1 t) \quad . \tag{3.22}$$

It should be noted that the sensitivity *generally* decreases for components that show signs of saturation.

An example of this problem is shown in Fig. 3.8 and 3.9. An amplifier with two signals with different levels is to be controlled.

The amplifier should have a small signal gain of 20 dB and a maximum output power of 30 dBm. Signal 1 at the frequency ω_1 is a small signal that does not drive the amplifier into compression. Signal 2 should be varied via the input power as shown in Fig. 3.9.

You can clearly see that the small signal is also significantly less amplified due to decreasing selectivity when the amplifier is saturated by a completely different signal.

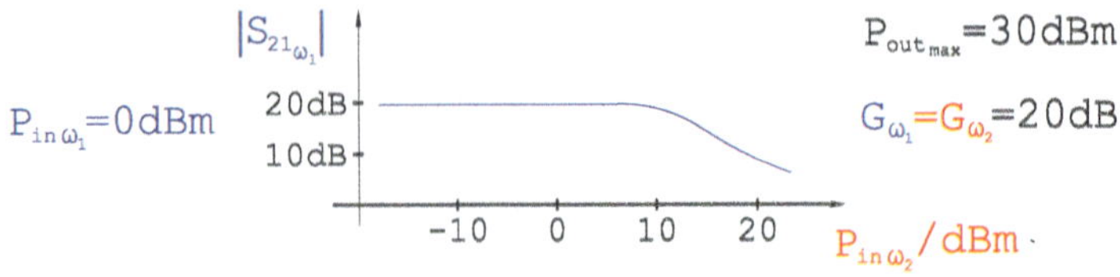

Fig. 3.9 Transmission gain for the small signal power at ω_1 as a function of of the signal power at ω_2

Cross Modulation

Cross modulation is the conversion of modulated information from one frequency band to another. A simple example is an amplitude-modulated signal for the voltage $U_2 = (1+m_{(t)})\,U_2'$. If this is inserted into Eq. (3.16), the partial current is obtained:

$$i_{c(t)}^{part} = \frac{3}{2}\,c\,U_1\,U_2'^2\,[1+2m_{(t)}+m_{(t)}^2]\cos(\omega_1 t)\ . \tag{3.23}$$

This means that the modulated information also appears around the carrier ω_1.

AM/PM Conversion

AM/PM conversion is the phenomenon that amplitude information (or amplitude noise) is converted into a modulation of the phase. Especially since modern radio systems are phase modulated, AM/PM conversion is a source of interference that should not be neglected.

From the Eq. (3.16), the partial current

$$i_{(t)}^{part} = \left(a\,U_1 + \frac{3}{4}\,c\,U_1^3\right)\cos(\omega_1 t) \tag{3.24}$$

must be considered. In the high and radio frequency range in particular, it is not unusual for the nonlinear effects to exhibit a so-called memory effect[1]. In such a case, there is a phase shift in the frequency range ($i_{(t)}^{part} \circ\!\!-\!\!\bullet\ I^{part}$),

$$I^{Teil} = A\,U_1 + \frac{3}{4}\,C\,U_1^3\,e^{j\phi}\ . \tag{3.25}$$

If the amplitude of U_1 now fluctuates, the phase of I^{part} (i. h. the ϕ component, which is normally zero) also changes.

[1] This memory effect will be discussed below.

Spurious Responses

The response to spurious frequencies is a given problem for mixers, which will be explained using the following example.

A receive mixer is controlled with a large signal at the frequency ω_{LO} and should convert a high-frequency small signal with the frequency ω_{RF} to a low-frequency signal with the frequency ω_{IF}.

However, Eq. (3.16) applies to the mixer, so that the mixer also converts all signals for $m > 1$ at the discrete secondary frequencies according to

$$\omega_{IF,m} = m\,\omega_{LO} \pm m\,\omega_{RF} \tag{3.26}$$

mixes down into the low-frequency receive band around ω_{IF}.

3.5 Nonlinear Components

The following shows how to describe nonlinear components analytically. As transistor models are beyond the scope of this chapter, the description is limited to diodes.

3.5.1 Nonlinear Conductance and Resistance

A nonlinear component is most frequently described by the dependence of the current I on the voltage U:

$$I = f(U)\,. \tag{3.27}$$

This current-voltage function usually contains various physical constants.

One example is the Schottky diode. The nonlinear characteristic curve between current I and voltage U of a Schottky diode is described by

$$I = I_{ss}\left(e^{\left[\frac{U}{U_T}\right]} - 1\right)\,. \tag{3.28}$$

The temperature voltage $U_T = 25.9\,\mathrm{mV}$ and the saturation blocking current I_{ss} are included.

The nonlinear current-voltage characteristic of a Schottky diode is shown in Fig. 3.10.

The diffusion voltage U_D is also referred to as the diffusion potential Φ and depends on the semiconductor and metal materials selected for the junction. For the standard Si diode, U_D is around 0.7 V and for the Si Schottky diode around 0.4 V.

The diode is often activated at the set small-signal AC conductance value

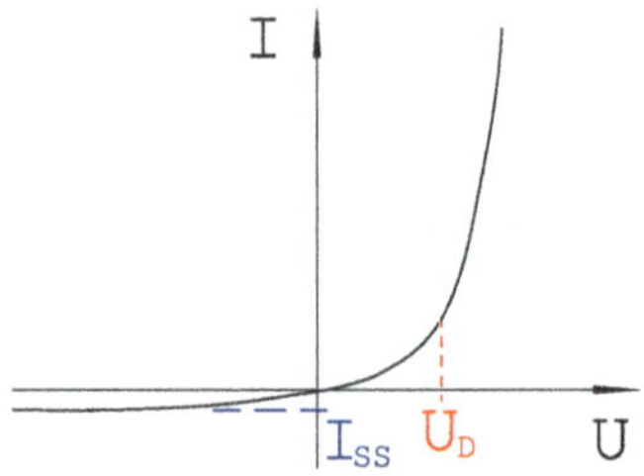

Fig. 3.10 Current-voltage characteristic of a Schottky diode with the saturation blocking current I_{ss} and the diffusion voltage U_D

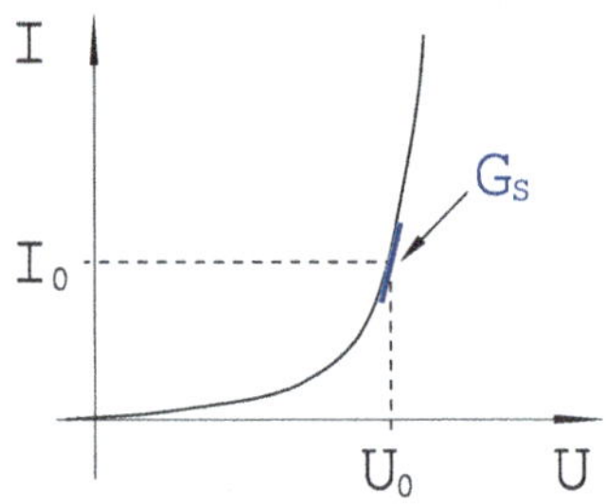

Fig. 3.11 Small signal conductance G_S of a Schottky diode for the operating point with the DC voltage U_0 and the DC current I_0

$$G_S(I_0) = \frac{I_0 + I_{ss}}{U_T} \tag{3.29}$$

for the operating point with the direct current I_0 as shown in Fig. 3.11.

A general derivation is presented below, which also results in Eq. (3.29). To describe a nonlinear component, the $f(U)$ is to be considered in the area around the working point U_0. A Taylor series is used as an aid for this:

$$f(U_0+u) = f(U_0) + \underbrace{\left.\frac{df(U)}{dU}\right|_{U_0} u + \frac{1}{2}\left.\frac{d^2f(U)}{dU^2}\right|_{U_0} u^2 + \frac{1}{6}\left.\frac{d^3f(U)}{dU^3}\right|_{U_0} u^3 + \ldots}_{=i \text{ (AC part)}} \tag{3.30}$$

The current I is divided into the direct component I_0 for the operating point setting and the alternating component i:

$$I = I_0 + i. \tag{3.31}$$

Consequently, the same applies to the alternating current (AC) part:

$$i = f(U_0+u) - f(U_0)\,. \tag{3.32}$$

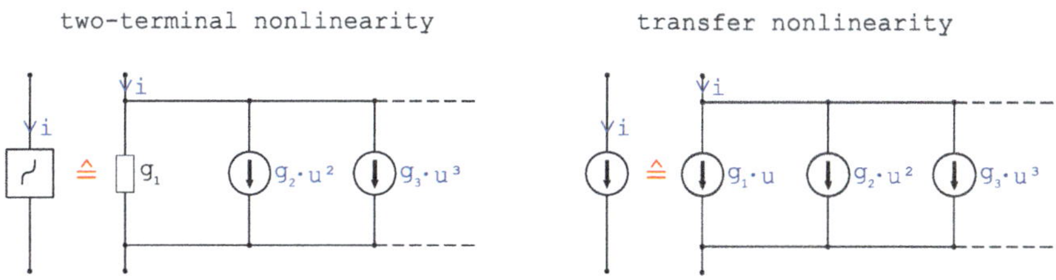

Fig. 3.12 Small signal equivalent circuit diagrams for weakly nonlinear terminal and transfer elements

Using Eq. (3.30), the following nonlinear dependence on the voltage u generally results for the current i:

$$i = \left.\frac{df(U)}{dU}\right|_{U=U_0} u + \frac{1}{2}\left.\frac{d^2f(U)}{dU^2}\right|_{U=U_0} u^2 + \frac{1}{6}\left.\frac{d^3f(U)}{dU^3}\right|_{U=U_0} u^3 + \dots \quad . \tag{3.33}$$

A comparison with a link via the conductance in the form of

$$i = g_1 u + g_2 u^2 + g_3 u^3 + \dots \tag{3.34}$$

results in their calculation rule. The corresponding results for terminal and transfer elements are shown in Fig. 3.12.

If the differentiation of Eq. (3.28) is now carried out for $g_1 = G_S$, the result is

$$G_S = \frac{I_{ss}}{U_T} e^{\frac{U_0}{U_T}} . \tag{3.35}$$

The use of

$$I_0 = I_{ss} \left(e^{\frac{U_0}{U_T}} - 1 \right) \tag{3.36}$$

for the operating point current then provides the simple Eq. (3.29) for the description of a Schottky diode, as used for the design of detectors, for example. It can be seen that the quadratic component g_2 is already neglected here.

By selecting the conductance $g_1 = G_S$, the diode can either be set to matching (often 50 Ω) or (when selecting small conductance values) to maximum sensitivity.

By optimizing g_2 in the direction of matching at the corresponding harmonic, the optimum efficiency for the frequency conversion at the diode can be set.

3.5.2 Nonlinear Capacitances

Semiconductors in general and transistors in particular have parasitic memory elements, which are usually nonlinear. These nonlinear memory effects pose a notice-

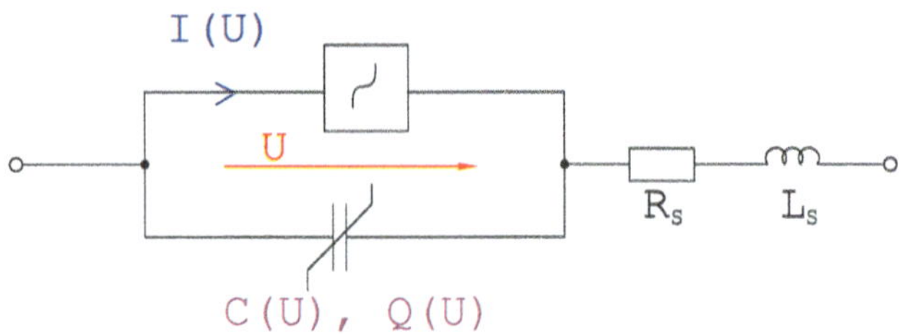

Fig. 3.13 Nonlinear ECD of the Schottky diode

able problem in the application of linear power amplifiers, for example. In general, this functionality can be described via

$$Q = f_Q(U) \quad . \tag{3.37}$$

With the renewed split into an operating point component Q_0 and a small signal component q, for the charge applies $Q = Q_0 + q$ and again applies to the small signal component

$$q = f_Q(U_0 + u) - f_Q(U_0) \tag{3.38}$$

and in analogy to Eq. (3.33)

$$q = \left.\frac{d f_Q(U)}{dU}\right|_{U=U_0} u + \frac{1}{2}\left.\frac{d^2 f_Q(U)}{dU^2}\right|_{U=U_0} u^2 + \frac{1}{6}\left.\frac{d^3 f_Q(U)}{dU^3}\right|_{U=U_0} u^3 + \ldots \quad . \tag{3.39}$$

In simulation technology, however, one would also like to represent the current as a function of the voltage. The following generally applies in the first step:

$$i = \frac{dq}{dt} \quad . \tag{3.40}$$

In the second step, the current can therefore be calculated from the voltage using Eq. (3.39):

$$i = \left.\frac{d f_Q(U)}{dU}\right|_{U=U_0} \frac{du}{dt} + \left.\frac{d^2 f_Q(U)}{dU^2}\right|_{U=U_0} u \frac{du}{dt} + \frac{1}{2}\left.\frac{d^3 f_Q(U)}{dU^3}\right|_{U=U_0} u^2 \frac{du}{dt} + \ldots \, . \tag{3.41}$$

A comparison with the link via the capacities in the form of

$$i = \left(C_1(U_0) + C_2(U_0)\, u + C_3(U_0)\, u^2 + \ldots\right) \frac{du}{dt} \tag{3.42}$$

results in the calculation rule.

Once again, the results are to be applied to the Schottky diode. Fig. 3.13 shows the resulting nonlinear Equivalent Circuit Diagram (ECD) of a Schottky diode.

With the diffusion potential Φ (or the diffusion voltage U_D) and the blocking capacitance C_{j0} at a voltage of 0 V, the voltage-dependent charge for a Schottky diode with a uniformly distributed doping profile according to [68] can be specified as follows:

$$Q(U) = -2C_{j0}\Phi\sqrt{(1-U/\Phi)}\,. \tag{3.43}$$

The generally used voltage-dependent capacitance of a Schottky diode corresponds to $C_1(U)$ and can be calculated by differentiation according to Eq. (3.41).

$$C(U=U_0) = C_1(U_0) = \left.\frac{df_Q(U)}{dU}\right|_{U=U_0} = \frac{C_{j0}}{\sqrt{(1-U_0/\Phi)}} \tag{3.44}$$

Finally, the Schottky diode at the operating point can be treated as a linear component to a first approximation. Fig. 3.14 shows the equivalent circuit diagram.

In the example of the simplest description of a Schottky diode, the series was broken off after the first element of the Taylor series. In practice, the series is not broken off so early. This always results in new frequency components.

The mass of the models for semiconductor components is based on physical assumptions, as with the Schottky diode. Fixed nonlinear equations are available for each model to describe the nonlinear properties.

The actual "modeling" of nonlinear components is carried out using numerous DC and RF measurements at various operating points via a so-called operating points via a so-called "fitting" (more: [38], Chap. 9). The user of nonlinear simulators must be aware of the finiteness of their models. Models do not always cover all relevant nonlinearities.

Inadequate models or inadequate fitting are the biggest sources of error in nonlinear simulation.

3.6 Overview: Simulation Methods for Nonlinear Components

There are a large number of simulation methods for nonlinear components with all their advantages and disadvantages available to high-frequency engineers. The following brief overview should help the user to choose the right simulation method.

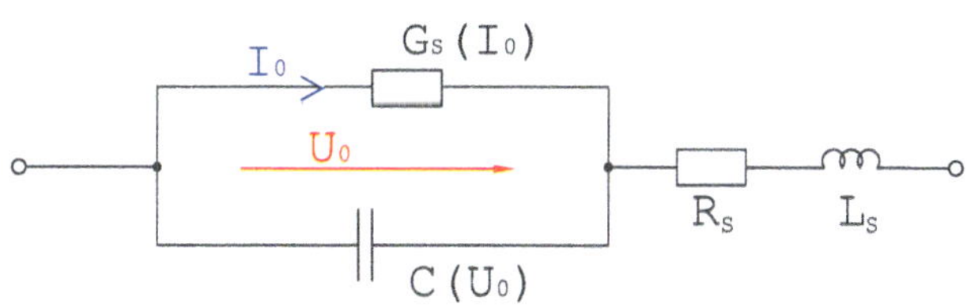

Fig. 3.14 Linear ECD of the Schottky diode in operation point (U_0, I_0)

Time Domain Methods

SPICE is the most frequently used time domain method in the electronics sector. Many RF simulators have now implemented time domain methods as an alternative method.

In this method, the difference equations of the associated network differential equations are solved numerically step by step over short time intervals.

This method calculates the transient behavior of currents and voltages. Very long calculation times and convergence problems can occur with strong nonlinear elements. Another disadvantage is that complex distributed components cannot be easily taken into account. For example, with most products it is not possible to include a lossy or dispersive line in the simulation.

Since a very large number of time intervals must be calculated in the radio frequency case until the steady state is calculated and each linear reactive component must also be described by a differential equation, the calculation time is indisputably long for many applications.

Basics of Harmonic Balance (HB)

Harmonic balance methods analyze a circuit in the frequency range. To do this, the circuit is divided into a linear and a nonlinear part. The linear part is calculated using the well-known linear network analysis. The entire mathematical treatment of both parts is demonstrated below using one method.

The harmonic balance methods presented below do not have the disadvantages listed for the time domain methods. For this reason, these methods have become established in high-frequency simulators for nonlinear circuits. Some HB methods are subject to the restriction that at least one source has such a high signal power that noticeably different frequency components are generated in the circuit. In practice, there must be at least one harmonic with a separation from the fundamental signal of at least 60–80 dBc.

The differences between the various harmonic balance methods are highlighted below. The main difference to the time domain methods is that the harmonic balance methods must be differentiated according to the requirements with regard to the necessary sources.

Harmonic Balance (HB) for Large Signal-Single Tone

With this variant of the HB method, only one signal source with a monofrequency large signal can be introduced in the simulation. This source is usually located at the circuit input. This HB method is used to calculate the signals present in the frequency and time domain for the fundamental wave and a specified number of harmonics.

Applications include power amplifiers and frequency multipliers.

This method was often the only one used in older simulation programs. This HB method is even included in the freely available student version (SV) Serenade from Ansoft.

Harmonic Balance (HB) for Large and Small Signal Tone

In addition to the purely sinusoidal large signal, a mono-frequency small signal source is also permitted in this HB variant. For the treatment of the small signal, the small signal conductance values of the nonlinear elements resulting at the operating point are calculated in advance and taken into account as the only substitute variable. This means that only the linear circuit simulator is used for the small signal. These methods may also allow more than one small signal.

Components such as mixers, modulators as well as parametric amplifiers and converters can be analyzed using this method.

Generalized Harmonic Balance Analysis

Using this simulation variant, several large signal sources can be taken into account in the circuit. This HB method is useful for the IM analysis of each component.

Volterra Series Method

The optimum nonlinear analysis method for applications with one or more small signals is the Volterra series analysis. This method provides the best results even for very weakly nonlinear circuits.

3.7 Derivation of the Harmonic Balance Method for Large-Signal Single Tones

Amplifiers are certainly the most frequently developed components with more or less pronounced nonlinear behavior.

Small signal amplifiers are initially optimized using a purely linear circuit simulation. This is based on the (measured) S-parameter values of the transistor and an otherwise purely passive circuit. The wiring of the DC voltage paths are only taken into account with the input impedance for the radio frequency signal. In a second step, the nonlinear behavior using a nonlinear simulator. The entire procedure is explained in detail in [38], chapter 8 has been explained in detail.

Power amplifiers are developed exclusively using a nonlinear circuit simulator.

The standard method for analyzing the nonlinear properties of amplifiers is the HB method for large-signal single tones. The general block diagram for analyzing amplifiers or frequency doublers with a DC supply source is shown in Fig. 3.15.

The coil and the capacitor should only be used to carry the direct current. In practical circuits, the DC voltage is often supplied to several points by splitting it through a resistor network. The impedances Z_S and Z_L represent the possibly complex and frequency-dependent characteristic impedances of the input port (source port) and the output port (load port).

For investigations in the reverse direction, the AC voltage source can also be located at the output.

A practical circuit clearly shows that most of the components are passive in nature.

For a general analysis of this problem, the linear networks are summarized as shown in Fig. 3.16.

Both matching networks as well as the impedance connections of both gates and the components for coupling the DC voltage are combined in a linear network.

The linear circuit is described with S or Y multiport parameters and analyzed according to the laws of network theory. The (often smaller) nonlinear part of the circuit contains elements that are modeled their I/U or Q/U characteristics. This part is calculated in the time domain, as previously described using the example of the diode. The results of the time domain analysis flow into the frequency domain analysis via Fourier analysis. Time domain and frequency domain signals are available for the steady state at all ports shown in Fig. 3.16.

The number of harmonics is limited by input to the Kth harmonic. Consequently, current and voltage have signal components at the frequencies:

$$\omega = 0, \omega_p, 2\,\omega_p, 3\,\omega_p, \ldots\, K\,\omega_p \quad .$$

The current associated with the DC component should be designated as I_{i0} and the current associated with the frequency ω_p should be designated as I_{i1}.

A vector is introduced for each gate that contains all current components from the direct current to the frequency $K\,\omega_p$ of the total current I_i:

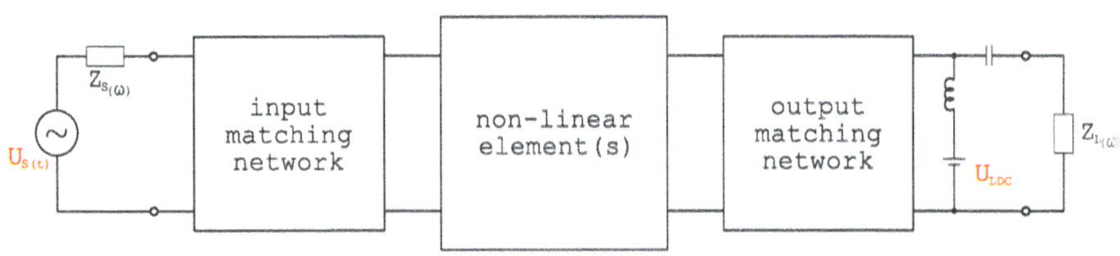

Fig. 3.15 General ECD for networks and circuits of the HB method for large signal single tones

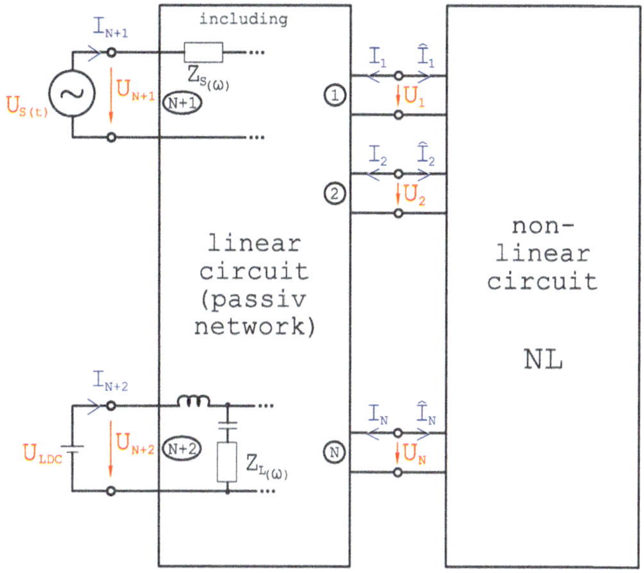

Fig. 3.16 Initial model for networks and circuits of the HB method for large signal single tones

$$\vec{\mathbf{I}}_i = \begin{pmatrix} I_{i0} \\ I_{i1} \\ \vdots \\ I_{iK} \end{pmatrix} \quad \text{with} \quad i = 1, 2, \ldots, N+2 \quad . \tag{3.45}$$

Kirchhoff's current law applies to the terminals between the linear and nonlinear network:

$$\begin{pmatrix} \vec{\mathbf{I}}_1 \\ \vec{\mathbf{I}}_2 \\ \vec{\mathbf{I}}_3 \\ \vdots \\ \vec{\mathbf{I}}_N \end{pmatrix} + \begin{pmatrix} \vec{\hat{\mathbf{I}}}_1 \\ \vec{\hat{\mathbf{I}}}_2 \\ \vec{\hat{\mathbf{I}}}_3 \\ \vdots \\ \vec{\hat{\mathbf{I}}}_N \end{pmatrix} = \begin{pmatrix} 0 \\ 0 \\ 0 \\ \vdots \\ 0 \end{pmatrix} . \tag{3.46}$$

These currents are referred to as interface currents.

The following applies to the linear network:

$$\begin{pmatrix} \vec{\mathbf{I}}_1 \\ \vec{\mathbf{I}}_2 \\ \vec{\mathbf{I}}_3 \\ \vdots \\ \vec{\mathbf{I}}_N \\ \vec{\mathbf{I}}_{N+1} \\ \vec{\mathbf{I}}_{N+2} \end{pmatrix} = \begin{bmatrix} \begin{bmatrix} \boldsymbol{Y}_{1,1} \\ \boldsymbol{Y}_{2,1} \end{bmatrix} & \begin{bmatrix} \boldsymbol{Y}_{1,2} \\ \boldsymbol{Y}_{2,2} \end{bmatrix} & \cdots & [\boldsymbol{Y}_{1,N+2}] \\ \vdots & & \ddots & \vdots \\ [\boldsymbol{Y}_{N+2,1}] & & \cdots & [\boldsymbol{Y}_{N+2,N+2}] \end{bmatrix} \begin{pmatrix} \vec{\mathbf{U}}_1 \\ \vec{\mathbf{U}}_2 \\ \vec{\mathbf{U}}_3 \\ \vdots \\ \vec{\mathbf{U}}_N \\ \vec{\mathbf{U}}_{N+1} \\ \vec{\mathbf{U}}_{N+2} \end{pmatrix} . \tag{3.47}$$

The admittance matrices included are pure diagonal matrices. These admittance values of the linear network should all be known.

$$[\boldsymbol{Y}_{m,n}] = \begin{bmatrix} Y_{m,n(0)} & 0 & \dots & 0 \\ 0 & Y_{m,n(\omega_p)} & & \vdots \\ \vdots & & \ddots & 0 \\ 0 & \dots & 0 & Y_{m,n(K\omega_p)} \end{bmatrix} \tag{3.48}$$

The interface currents can be calculated from Eq. (3.47) directly from the voltage values.

$$\underbrace{\begin{pmatrix} \vec{\mathbf{I}}_1 \\ \vec{\mathbf{I}}_2 \\ \vec{\mathbf{I}}_3 \\ \vdots \\ \vec{\mathbf{I}}_N \end{pmatrix}}_{=\vec{\mathbf{I}}} = \underbrace{\begin{bmatrix} [\boldsymbol{Y}_{1,N+1}] & [\boldsymbol{Y}_{1,N+2}] \\ [\boldsymbol{Y}_{2,N+1}] & [\boldsymbol{Y}_{2,N+2}] \\ \vdots & \vdots \\ [\boldsymbol{Y}_{N,N+1}] & [\boldsymbol{Y}_{N,N+2}] \end{bmatrix} \begin{pmatrix} \vec{\mathbf{U}}_{N+1} \\ \vec{\mathbf{U}}_{N+2} \end{pmatrix}}_{=\vec{\mathbf{I}}_s} + \underbrace{\begin{bmatrix} [\boldsymbol{Y}_{1,1}] & \cdots & [\boldsymbol{Y}_{1,N}] \\ [\boldsymbol{Y}_{2,1}] & & \vdots \\ & \ddots & \\ \vdots & & \\ [\boldsymbol{Y}_{N,1}] & \cdots & [\boldsymbol{Y}_{N,N}] \end{bmatrix}}_{=[Y_{N*N}]} \underbrace{\begin{pmatrix} \vec{\mathbf{U}}_1 \\ \vec{\mathbf{U}}_2 \\ \vec{\mathbf{U}}_3 \\ \vdots \\ \vec{\mathbf{U}}_N \end{pmatrix}}_{=\vec{\mathbf{U}}} \tag{3.49}$$

The result of the interface current calculation can also be summarized as follows:

$$\vec{\mathbf{I}} = \underbrace{\vec{\mathbf{I}}_s}_{\text{known}} + \underbrace{[\boldsymbol{Y}_{N*N}]}_{\text{known}} \vec{\mathbf{U}} \stackrel{!}{=} -\hat{\vec{\mathbf{I}}} = -\vec{\mathbf{I}}_C - \vec{\mathbf{I}}_G . \tag{3.50}$$

The current vector $\vec{\mathbf{I}}_s$ and the matrix $[\boldsymbol{Y}_{N*N}]$ are known from linear circuit simulation. Eq. (3.50) shows once again that Kirchhoff's current law must be observed for the interface currents. The interface current of the nonlinear block is divided into a resistive component $\vec{\mathbf{I}}_G$ and a capacitive component $\vec{\mathbf{I}}_C$.

The nonlinear equations (of the physical model for the transistor or other nonlinear component) are known from the nonlinear block. Nonlinear conductance and capacitance are the core elements of this model.

The resistive and capacitive current components are to be calculated below given the voltage vector $\vec{\mathbf{U}}$ will be calculated.

Calculation of the Resistive Part $\vec{\mathbf{I}}_G$

First, the voltage values from the frequency domain must be transformed into the time domain for each port using the inverse Fourier transformation:

$$u_{i(t)} \quad \circ\!\!-\!\!\bullet \quad U_i \,. \tag{3.51}$$

The resistive current component for the port i can be calculated directly from the model for the nonlinear component.

$$i_{gi(t)} = f_{gi}(u_{1(t)}, u_{2(t)}, \ldots, u_{N(t)}) \tag{3.52}$$

From the time-dependent current $i_{gi(t)}$, the corresponding current in the frequency domain is obtained using the Fourier transformation for the port i:

$$\vec{\mathbf{I}}_{Gi} \quad \bullet\!\!-\!\!\circ \quad i_{gi(t)} \tag{3.53}$$

with the frequency components:

$$\vec{\mathbf{I}}_{Gi} = \begin{pmatrix} I_{Gi0} \\ I_{Gi1} \\ \vdots \\ I_{GiK} \end{pmatrix} . \tag{3.54}$$

The vector $\vec{\mathbf{I}}_G$ should be introduced to summarize all resistive interface currents.

$$\vec{\mathbf{I}}_G = \begin{pmatrix} \vec{\mathbf{I}}_{G1} \\ \vec{\mathbf{I}}_{G2} \\ \vdots \\ \vec{\mathbf{I}}_{GN} \end{pmatrix} \tag{3.55}$$

Calculation of the Capacitive Part $\vec{\mathbf{I}}_C$

First, the voltage values from the frequency domain must also be transformed into the time domain using the inverse Fourier transform for each port according to Eq. (3.51).

From the model for the nonlinear component, the charge components for the port i can first be calculated for the capacitive current component.

$$q_{i(t)} = f_{ci}(u_{1(t)}, u_{2(t)}, \ldots, u_{N(t)}) \tag{3.56}$$

From the time-dependent charge $q_{i(t)}$, using the Fourier transform for the port i, the charge is

$$\vec{\mathbf{Q}}_{\mathbf{i}} \quad \bullet\!\!-\!\!\circ \quad q_{i(t)} \tag{3.57}$$

with the frequency components

$$\vec{\mathbf{Q}}_i = \begin{pmatrix} Q_{i0} \\ Q_{i1} \\ \vdots \\ Q_{iK} \end{pmatrix} \quad . \tag{3.58}$$

The vector $\vec{\mathbf{Q}}$ should be introduced to summarize all charges at all interface ports.

$$\vec{\mathbf{Q}} = \begin{pmatrix} \vec{\mathbf{Q}}_1 \\ \vec{\mathbf{Q}}_2 \\ \vec{\mathbf{Q}}_3 \\ \vdots \\ \vec{\mathbf{Q}}_N \end{pmatrix} \tag{3.59}$$

The consideration of the current for the port i provides:

$$i_{ci(t)} = \frac{dq_{i(t)}}{dt} \quad \circ\!\!-\!\!\bullet \quad \sum_{k=0}^{K} jk\,\omega_p\,Q_{ik} = I_{Ci} \quad . \tag{3.60}$$

Thus, the capacitive current component in the frequency range can generally be expressed by

$$\vec{\mathbf{I}}_C = j\,[\Omega]\,\vec{\mathbf{Q}} \tag{3.61}$$

using the diagonal matrices

$$[\Omega] = \begin{bmatrix} [\Omega_1] & 0 & \dots & 0 \\ 0 & [\Omega_2] & & \vdots \\ \vdots & & \ddots & 0 \\ 0 & \dots & 0 & [\Omega_N] \end{bmatrix} \tag{3.62}$$

$$\text{with} \quad [\Omega_m] = \begin{bmatrix} 0 & 0 & \dots & 0 \\ 0 & \omega_p & & \vdots \\ \vdots & & \ddots & 0 \\ 0 & \dots & 0 & K\omega_p \end{bmatrix} \quad \text{whereby: } m = 1,2,\dots,N. \tag{3.63}$$

It was thus shown how to use the nonlinear equations in the time domain to calculate the two frequency domain variables $\vec{\mathbf{I}}_G$ and $\vec{\mathbf{I}}_C$ were calculated.

Numerical Solution of the HB Problem

All variables of Eq. (3.50) except for the voltage vector $\vec{\mathbf{U}}$ are known. For a numerical solution, Eq. (3.50) is advantageously reformulated as follows:

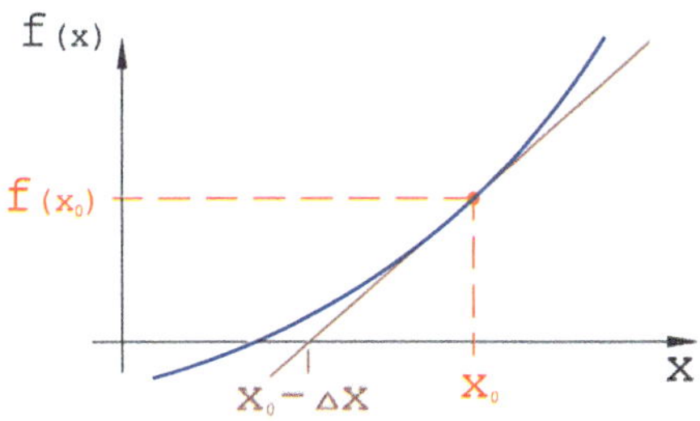

Fig. 3.17 Function diagram to illustrate the Newton method for finding zeros

$$\vec{\mathbf{F}}(\vec{\mathbf{U}}) = \vec{\mathbf{I}}_s + [Y_{N*N}]\vec{\mathbf{U}} + j\,[\Omega]\,\vec{\mathbf{Q}} + \vec{\mathbf{I}}_G \stackrel{!}{=} \vec{\mathbf{0}} \quad . \tag{3.64}$$

$\vec{\mathbf{F}}(\vec{\mathbf{U}})$ is the error function or the HB error that is displayed for every nonlinear simulation. simulation is displayed. A threshold value for this error function, which is ideally zero, can be changed by the user of an HB method if required. The HB calculation is terminated when the value falls below this threshold.

The Newton method is typically used to search for the unknown stress vector.

Fig. 3.17 is used for general consideration.

The zero of an arbitrary function $f(x)$ is to be found. To do this, select an arbitrary value x_0, calculate the function at this point with the result $f(x_0)$ and the corresponding derivative. The derived function has the zero at the point $x_0 + \Delta x$. This zero point is the new starting value for the improved calculation of the zero crossing of the function $f(x)$. The improved starting point is calculated from the context:

$$f(x_0) - \left.\frac{df}{dx}\right|_{x=x_0} \Delta x = 0 \qquad \text{resp.} \qquad \Delta x = \frac{f(x_0)}{df/dx} \quad . \tag{3.65}$$

When applied to the HB problem, the correction vector can be found via

$$\vec{\mathbf{F}}(\vec{\mathbf{U}}^p) - \left.\frac{\delta\,\vec{\mathbf{F}}(\vec{\mathbf{U}})}{\delta\vec{\mathbf{U}}}\right|_{\vec{\mathbf{U}}=\vec{\mathbf{U}}^p} \Delta\vec{\mathbf{U}} = 0 \quad \text{resp.} \quad \Delta\vec{\mathbf{U}} = \frac{\vec{\mathbf{F}}(\vec{\mathbf{U}}^p)}{\left.\frac{\delta\,\vec{\mathbf{F}}(\vec{\mathbf{U}})}{\delta\vec{\mathbf{U}}}\right|_{\vec{\mathbf{U}}=\vec{\mathbf{U}}^p}} \quad . \tag{3.66}$$

The improved voltage vector for the next iteration step is then simply calculated using

$$\vec{\mathbf{U}}^{p+1} = \vec{\mathbf{U}}^p - \Delta\,\vec{\mathbf{U}} \quad . \tag{3.67}$$

3.8 Introduction to Scattering Parameters for Mixed-Frequency Measurements

In the past, nonlinear effects were represented with individual special measurements. The simplest measurement, which has already been described several times, is the power measurement. As soon as the power transfer function changes with changing input power, there is a nonlinear effect (usually compression).

Using a spectrum analyzer, it is very easy to identify weak nonlinear effects. Modern analyzers allow a dynamic range between fundamental frequency and harmonics of significantly more than 100 dB. In addition to the harmonics, it is also common to use the spectrum analyzer to measure the signal strengths of intermodulation products and mixed products and to relate these to the input power.

Other special measurements that have been used for a long time are AM/AM and AM/PM conversion. What all these measurements have in common is that only scalar measurement values are available.

With modern modulation methods for mobile radio technology, the Error Vector Magnitude measurement (EVM measurement) has become established. This rather complex and very specific measurement allows a statement to be made about the magnitude and phase errors of a component for the specific modulation.

However, what all these measurements have in common is that they do not support optimal modeling of a slightly or strongly nonlinear component.

For more than 20 years, intensive research has been carried out on network analyzers that measure generally mixed-frequency scattering parameters[2] (short MF scattering parameters). In the beginning, these were also be referred to as "frequency conversation scattering parameters".

Individual solutions have been developed under the most diverse names such as NonLinear Network Analyzer (NLNA), Nonlinear Vectorial Network Analyzer (NVNA), Vectorial NonLinear Network Analyzer (VNLNA), Nonlinear Network Analyzer System (NNAS), Nonlinear Network Measurement System (NNMS), Large-Signal Network Analyzer (LSNA) and NonLin-S. Many of the solutions listed here solutions listed here only measure the transfer function from the fundamental to the harmonics. These are therefore only better (as calibrated) sampling oscillographs, as the latter can also be used to calculate the transfer function in the frequency domain via an FFT.

In the following, a network analyzer measuring station that can measure the vectorial S-parameters of objects that convert the signals in frequency is simply referred to as a network analyzer (NWA or VNA). These special MF-S parameters can also be referred to as `frequency conversion parameters`.

To introduce the S-parameters for MF measurements, an example of an intermodulation measurement should first be used, Fig. 3.18.

On the side at input port 1, there are the excitations a_1^I and a_1^{III} at two different frequencies (f^I and f^{III}). The incident waves at port 2 are only waves reflected at the measurement port, which can be determined by calibration and additional error correction of the network analyzer.

Reflected and transmitted waves occur at the input frequencies f^I and f^{III}. Furthermore, there are also reflected and transmitted waves at other frequencies, which are based on the difference of $n \cdot \left|f^I - f^{III}\right|$, generated reflected and trans-

[2] This name is derived from the mixed-mode scattering parameters ([38])

mitted waves. Fig. 3.18 shows one of the many possible transfer functions of the S-parameters for MF measurements.

Fig. 3.18 shows the complex transfer function $S_{21}^{II,I}$ for two different frequencies. Regarding the magnitude, this transfer function is known from the scalar measurements. It is more difficult to imagine how to define a phase for the transmission of a signal from a frequency f^I to a frequency f^{II}.

Definition of the Complex Mixed-Frequency Conversion Parameter

To understand the definition of this transfer function, please refer to Fig. 3.19. It shows the XCO reference signal (10 MHz signal with high phase stability) and sinusoidal signals at n times the frequencies of the XCO signal.

The zero crossings shown at the times 0 µs, 0.1 µs and 0.2 µs areas represent the phase origin. In Fig. 3.19 all harmonics are exactly in phase with the fundamental frequency. This means that with a mixed frequency S-parameter measurement, a measured value can only be recorded every 0.1 µs. This measurement process is triggered by the reference signal. Furthermore, only signals with a frequency spacing of 10 MHz can be measured. Some commercially available measuring stations even only offer 600 MHz frequency spacing. By dividing the reference signal, smaller step widths can be achieved, whereby the measuring time increases and the sensitivity is often lower.

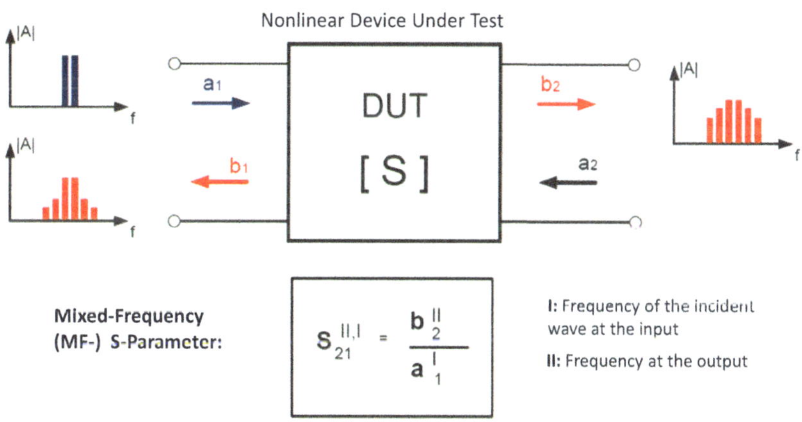

Fig. 3.18 MF-S parameters for nonlinear measurements and frequency components in an intermodulation measurement

Absolute Error Coefficients for Mixed-Frequency Measurements

The general structure of a network analyzer for a two-standard measurement is shown in Fig. 3.20. The radio frequency (RF for short) test signal is generated by a frequency-tunable oscillator. Part of the outgoing wave is detected by a coupler and shifted to the intermediate frequency range (IF for short, here ZF) by means of a mixer stage (consisting of the local oscillator [LO] and the mixer). There, the signal is digitized using A/D converters and processed further in the computer. The same process is also carried out with all other wave quantities occurring in the Fig. 3.20.

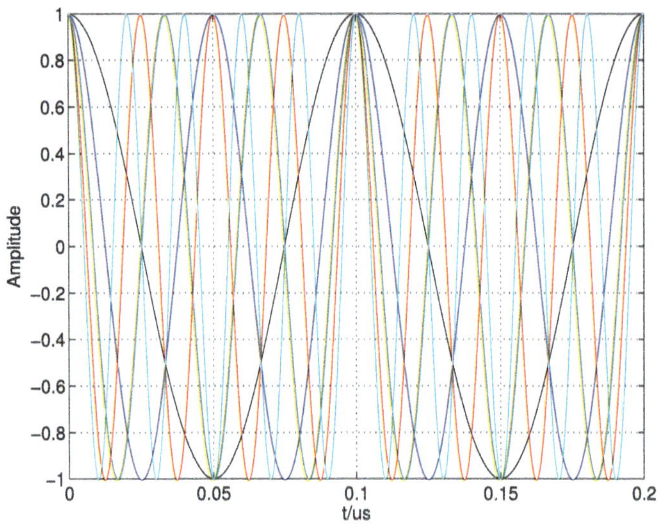

Fig. 3.19 10 MHz quartz signal and its harmonics

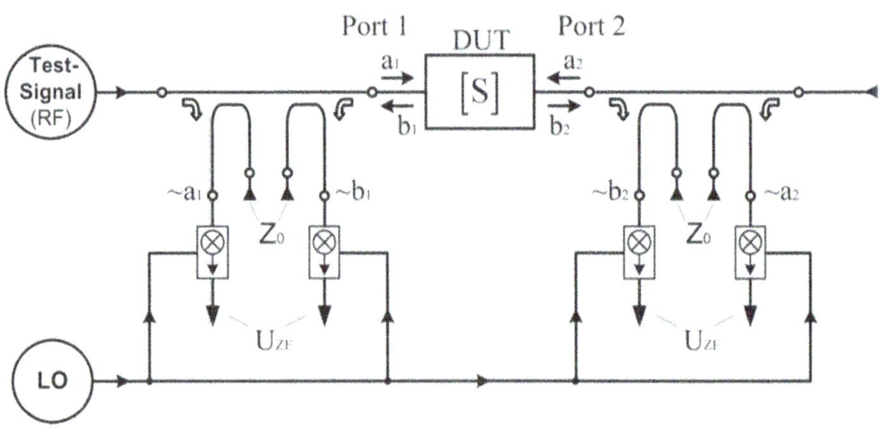

Fig. 3.20 Setup of a two-port network analyzer

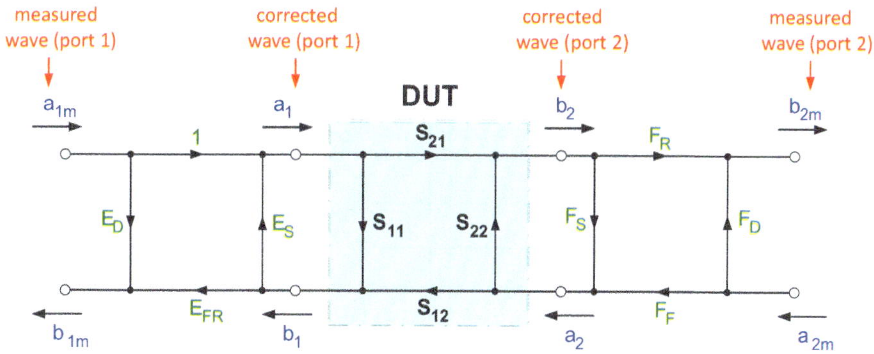

Fig. 3.21 7 term error model of a two-port network analyzer for linear S-parameter measurements

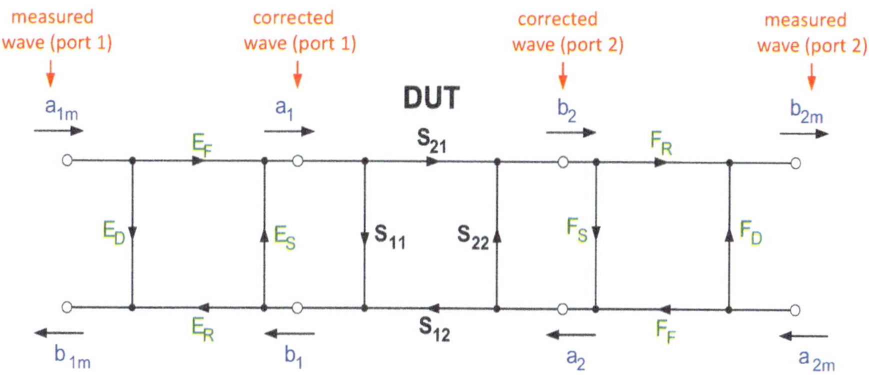

Fig. 3.22 8 term error model of a two-port network analyzer for frequency conversion S-parameter measurements

Before measuring classical (purely linear) scattering parameters, such a vector network analyzer (VNA for short) must be calibrated. That means, there is an associated error model (Fig. 3.21) whose error coefficients are determined with a certain number of calibration measurements on partially or completely known standards.

For these linear S-parameter measurements, is sufficient if all error coefficients are only known up to one factor. This means that for the 8 error coefficients in Fig. 3.21 one may be set to 1. This calibration method is therefore also referred to as the 7 term method [15, 18–20, 29, 37]. This is permitted, as an equal error in all wave sizes is always reduced in the scattering parameter calculation $S_{ij} = \frac{b_i}{a_j}$.

However, if you want to measure scattering parameters for the frequency conversion, the b and a waves occur at two different frequencies. The corresponding S-parameters can only be calculated if the wave sizes are absolutely known [40]. This can only be achieved by also determining the absolute error coefficients over the frequency (Fig. 3.22).

All new calibration methods for MF measurements therefore have two new calibration standards:

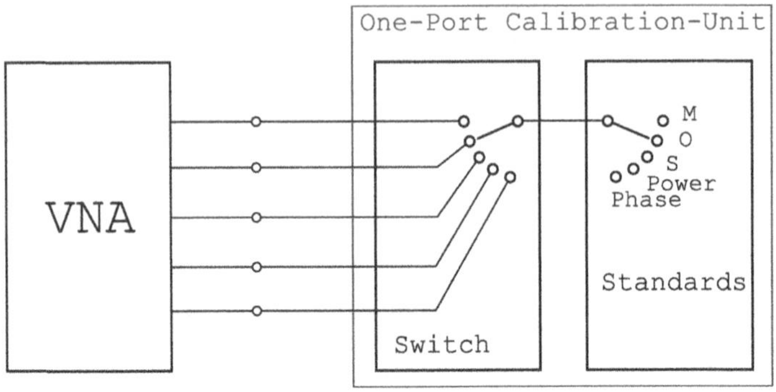

Fig. 3.23 VNA (Vector Network Analyzer) with the automated calibration via five single-port standards and the `Without Thru`, [42]

Power sensor and Phase reference.

The power sensor is used to measure the magnitude of an error coefficient transfer function. With phase reference, signals are made available at the measurement frequencies whose phase is known in to the reference oscillator is known. This can be, for example, a comb generator that generates harmonics with a known phase relationship to the fundamental. The phase to the fundamental and whose phases have been measured in relation to the reference signal. Recently, even synthesizers can be used for this purpose, [32].

The now simplest calibration method for mixed frequency measurements is called `Without Thru` [42] . The corresponding calibration standards are, in addition to the two new standards, the wave termination (match or load, M or L), the short circuit (short, S) and the open circuit (open, O). If these five standards are measured on a measurement gate, its absolute error coefficients can be calculated. This method can be used on n gates and is therefore directly a multi-port method. A through-connection is never necessary and can therefore also be used for verification [40].

These calibration methods can also be used to calibrate multi-port VNAs, as the calibration measurements can be carried out easily. Either connect the calibration box with the 5 standards from Fig. 3.23 behind it to each gate in turn or connect the standard via a switching matrix, as shown in Fig. 3.23. The third and most convenient solution is to use an additional switching matrix for the n measurement gates so that the calibration box does not have to be reconnected, which is interesting for automated measurements.

If a Vector Network Analyzer (VNA) has been calibrated in this way, it can be used to measure the error-corrected absolute wave quantities and therefore also current and voltage. The associated time functions of periodic signals can be calculated very precisely from the transfer function using an FFT. The VNA also performs the function of a sampling oscilloscope, Fig. 3.24.

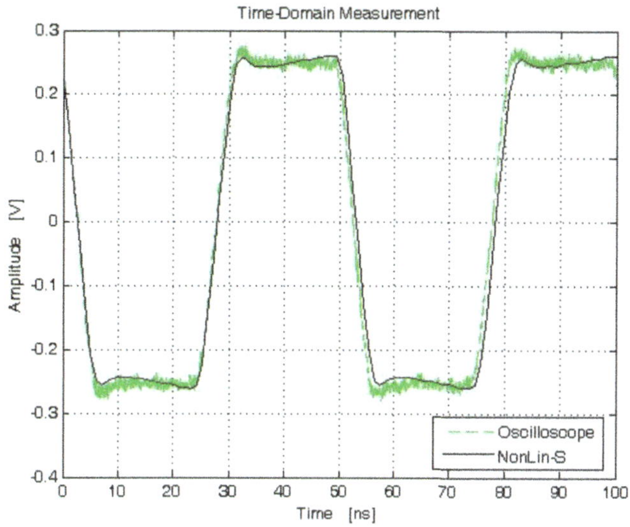

Fig. 3.24 Time signal from the oscilloscope and from a mixed-frequency VNA measurement

In the following, the S-parameters for MF measurements will be introduced in detail, the mixed-frequency S-parameters will be explained in more detail using the example of vector intermodulation measurement and the most general form of the X-parameters for power amplifiers will be presented.

3.8.1 S-Parameters for Mixed-Frequency Networks

First, the complex frequency conversion or so-called mixed frequency S parameter value for a one-port device will be considered for a simplified overview.

One-Port with an Excitation Signal

If, for example, a single large signal a_1 is applied to a diode at the frequency f^1, this wave is advantageously referred to as a_1^1. In this case, harmonics are generated. In addition to the 1st harmonic with the wave a_1^1 and the reflected wave b_1^1, two further harmonics at the frequencies f^2 and f^3 are to be considered. The two corresponding reflected waves are referred to as b_1^2 and b_1^3. As these reflected waves can be reflected again at the input, the two further waves a_1^2 and a_1^3 also occur.

The complete frequency conversion S-parameter matrix for the fundamental wave (at f_1) and the two harmonic waves (at f_2 and f_3) is:

$$b_1^1 = S_{11}^{11}\, a_1^1 + S_{11}^{12}\, a_1^2 + S_{11}^{13}\, a_1^3\,, \tag{3.68}$$

$$b_1^2 = S_{11}^{21}\, a_1^1 + S_{11}^{22}\, a_1^2 + S_{11}^{23}\, a_1^3 \,, \tag{3.69}$$
$$b_1^3 = S_{11}^{31}\, a_1^1 + S_{11}^{32}\, a_1^2 + S_{11}^{33}\, a_1^3 \,. \tag{3.70}$$

Energy may also be converted as a DC component. An S-parameter can also be introduced for this at the frequency 0 Hz. Thus S_{11}^{01} describes the complex transfer function from the excitation frequency f_1 to the DC component f_0. In this case, it is often advantageous to switch back to a current/voltage approach (as with harmonic balance). This case is generally not supported by the network analyzers.

The meaning of the scattering parameters S_{ij}^{mn} can be illustrated using the following examples:

$S_{11}^{11} = \frac{b_1^1}{a_1^1}|_{a_1^2=a_1^3=0}$ Input reflection coefficient S_{11} with matched harmonics, corresponds for f^1

$S_{11}^{21} = \frac{b_1^2}{a_1^1}|_{a_1^2=a_1^3=0}$ Mixed frequency conversion input reflection coefficient with matched harmonics

$S_{11}^{12} = \frac{b_1^1}{a_1^2}|_{a_1^1=\text{fixed},a_1^3=0}$ Mixed frequency conversion input reflection coefficient with existing fundamental a_1^1 and matched harmonic a_1^3

$S_{11}^{22} = \frac{b_1^2}{a_1^2}|_{a_1^1=\text{fixed},a_1^3=0}$ Input reflection coefficient S_{11} with existing fundamental a_1^1 and matched harmonic a_1^3 corresponds to f^2

It is easy to see that the S-parameters of the track correspond to the linear S-parameters at the associated frequencies (e.g.: $S_{11}^{33} = S_{11}$ at f^3). However, this only applies to the associated `operating point.`

The definition of S_{11}^{12} reveals a (metrological) problem. How should S_{11}^{12} be measured if a_1^1 cannot be applied? The solution lies in the two-step procedure. First determine S_{11}^{11}. Then determine S_{11}^{12} from $a_1^3 = 0$:

$$S_{11}^{12} = b_1^1 \,/\, a_1^2 - S_{11}^{11}\, a_1^1 \,/\, a_1^2 \,. \tag{3.71}$$

This excitation takes place with a fundamental signal and a harmonic. Two excitation signals are also required for mixer and intermodulation measurements.

Frequency conversion scattering parameters are not only frequency dependent, but also a function of the associated large signal or signals. In general, the following applies to Eq. (3.68):

$$b_1^1 = S_{11}^{11}\left(a_1^1,\omega\right)\, a_1^1 + S_{11}^{12}\left(a_1^1,\omega\right)\, a_1^2 + S_{11}^{13}\left(a_1^1,\omega\right)\, a_1^3 \,, \tag{3.72}$$

where a_1^1 must even be considered as a complex number in special cases. In many cases, however, it is sufficient to consider only the amount $\left|a_1^1\right|$. The MF-S parameters are also defined here, which is an approximation that can be used for many practical cases and changes Eq. (3.72) as follows:

$$b_1^1 = S_{11}^{11}\left(\left|a_1^1\right|,\omega\right)\, a_1^1 + S_{11}^{12}\left(\left|a_1^1\right|,\omega\right)\, a_1^2 + S_{11}^{13}\left(\left|a_1^1\right|,\omega\right)\, a_1^3 \,. \tag{3.73}$$

Scalar or Vector Reference of the Excitation Signals

For the exact X-parameters[3], the literature often only specifies the dependence on the vectorial reference of the excitation signal a_1 ($= a_1^1$) and assumes the frequency dependence. In the following, these dependencies are only occasionally explicitly emphasized[4].

At the end of the chapter, also the exact relationship between the X-parameters and the mixed-frequency S-parameters presented here is also derived in section 3.8.4.

The prerequisite for the mixed-frequency S-parameters presented here is the simple condition that for the excitation signal

$$\angle a_1^1 = 0^\circ \tag{3.74}$$

must apply! Details and the general case are explained in section 3.8.4.

The phase position of the excitation waves is important for some components. There are two classes of nonlinear effects here: The amplifiers in saturation and the rest.

a) For amplifiers in saturation, the phase position of the excitation signals slightly changes the operating point and thus the frequency conversion S-parameters.
b) For the rest (amplifiers, mixers, intermodulation, etc.), the phase angle of the excitation signals has no influence on the MF-S-parameters. Here, for example, the following applies to a single gate with the two higher-frequency waves a_1^2 and a_1^3 as excitation waves:

$$b_1^1 = S_{11}^{11}\left(\left|a_1^2\right|,\left|a_1^3\right|,\omega\right) a_1^1 \;,+ S_{11}^{12}\left(\left|a_1^2\right|,\left|a_1^3\right|,\omega\right) a_1^2 + S_{11}^{13}\left(\left|a_1^2\right|,\left|a_1^3\right|,\omega\right) a_1^3 \,. \tag{3.75}$$

However, these results are only valid for the selected operating point. Only measurements at different operating points allow a nonlinear model of the DUT to be created.

Details on One-Ports with Two Excitation Signals

For the following theory, it is the same whether, the two excitation signals are of the same or different size. Eq. (3.68) to (3.70) remain unchanged. However, the two higher-frequency waves a_1^2 and a_1^3 should now be the excitation waves. In this case, it is only possible to set the wave a_1^1 to zero at the frequency f^1 using a $50\,\Omega$ termination (Fig. 3.25).

If you now want to determine S_{11}^{12} again, you have to record the measured values for Eq. (3.68) twice (measurements *a* and *b*) with `different phase ratios`

[3] The X-parameters are presented at the end of the chapter

[4] The associated large-signal measurements are carried out on a so-called Hot-S-parameter measuring station

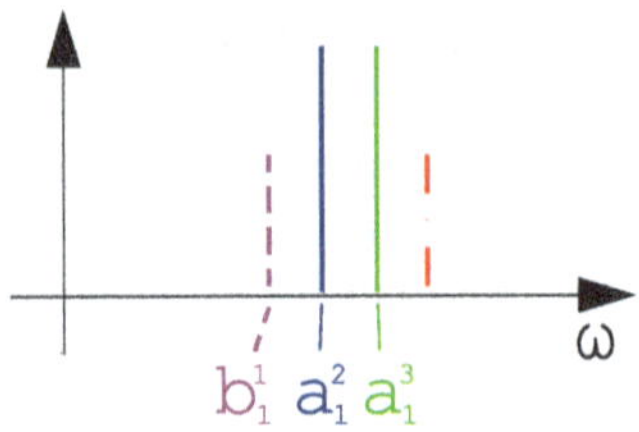

Fig. 3.25 Representation of the two input signals (a-waves) and the one output signal (b-wave)

of the excitation waves a_1^2 and a_1^3:

$$b_{1a}^1 = S_{11}^{12}\, a_{1a}^2 + S_{11}^{13}\, a_{1a}^3\,, \tag{3.76}$$

$$b_{1b}^1 = S_{11}^{12}\, a_{1b}^2 + S_{11}^{13}\, a_{1b}^3\,. \tag{3.77}$$

The simple matrix conversion provides the quested MF-S-parameter S_{11}^{12} and at the same time S_{11}^{13}:

$$\begin{pmatrix} S_{11}^{12} \\ S_{11}^{13} \end{pmatrix} = \begin{pmatrix} a_{1a}^2 & a_{1a}^3 \\ a_{1b}^2 & a_{1b}^3 \end{pmatrix}^{-1} \begin{pmatrix} b_{1a}^1 \\ b_{1b}^1 \end{pmatrix}. \tag{3.78}$$

Multi-Ports

As the number of increases, the matrices for MF-S-parameters become significantly more complex. The Eq. (3.79–3.82) describe a two-port in which the fundamental and a harmonic are to be taken into account.

$$b_1^1 = S_{11}^{11}\, a_1^1 + S_{11}^{12}\, a_1^2 + S_{12}^{11}\, a_2^1 + S_{12}^{12}\, a_2^2 \tag{3.79}$$

$$b_1^2 = S_{11}^{21}\, a_1^1 + S_{11}^{22}\, a_1^2 + S_{12}^{21}\, a_2^1 + S_{12}^{22}\, a_2^2 \tag{3.80}$$

$$b_2^1 = S_{21}^{11}\, a_1^1 + S_{21}^{12}\, a_1^2 + S_{22}^{11}\, a_2^1 + S_{22}^{12}\, a_2^2 \tag{3.81}$$

$$b_2^2 = S_{21}^{21}\, a_1^1 + S_{21}^{22}\, a_1^2 + S_{22}^{21}\, a_2^1 + S_{22}^{22}\, a_2^2 \tag{3.82}$$

The extension to three harmonics looks as follows in matrix notation:

$$\begin{pmatrix} b_1^1 \\ b_1^2 \\ b_1^3 \\ b_2^1 \\ b_2^2 \\ b_2^3 \end{pmatrix} = \begin{pmatrix} S_{11}^{11} & S_{11}^{12} & S_{11}^{13} & S_{12}^{11} & S_{12}^{12} & S_{12}^{13} \\ S_{11}^{21} & S_{11}^{22} & S_{11}^{23} & S_{12}^{21} & S_{12}^{22} & S_{12}^{23} \\ S_{11}^{31} & S_{11}^{32} & S_{11}^{33} & S_{12}^{31} & S_{12}^{32} & S_{12}^{33} \\ S_{21}^{11} & S_{21}^{12} & S_{21}^{13} & S_{22}^{11} & S_{22}^{12} & S_{22}^{13} \\ S_{21}^{21} & S_{21}^{22} & S_{21}^{23} & S_{22}^{21} & S_{22}^{22} & S_{22}^{23} \\ S_{21}^{31} & S_{21}^{32} & S_{21}^{33} & S_{22}^{31} & S_{22}^{32} & S_{22}^{33} \end{pmatrix} \begin{pmatrix} a_1^1 \\ a_1^2 \\ a_1^3 \\ a_2^1 \\ a_2^2 \\ a_2^3 \end{pmatrix}. \tag{3.83}$$

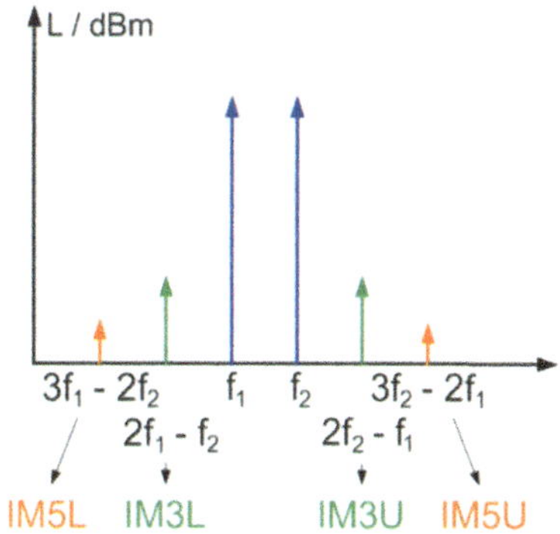

Fig. 3.26 IM3 and IM5 products for the designation of a mixed frequency S-parameter measurement

3.8.2 Vector Intermodulation Measurements

Intermodulation measurements are an important point in the development as well as in the production of mobile radio modules to ensure compliance with the system specification.

For the sake of clarity, the following will be limited to reflection measurements (vector MF-S_{11} measurements). The necessary effort, the results and the new advantages of these vectorial measurements will be highlighted.

Fig. 3.26 shows the possible frequencies for IM3 and IM5 products.

For these two IM products, there are already 8 MF-S-parameters:

$S_{11}^{IM3L,f1}$, $S_{11}^{IM3U,f1}$, $S_{11}^{IM3L,f2}$, $S_{11}^{IM3U,f2}$ and
$S_{11}^{IM5L,f1}$, $S_{11}^{IM5U,f1}$, $S_{11}^{IM5L,f2}$, $S_{11}^{IM5U,f2}$.

These MF-S parameters can be defined as in the previous chapter or based on the scalar IM measurement.

In a scalar IM measurement, two excitation signals of the same size are applied to the measurement object. The intermodulation distance is calculated from the difference between the signal power of an excitation signal (e.g. at f_1) and the IM power (e.g. at $IM3L$). This procedure ignores the energy input at the second frequency (e.g. f_2). Mathematically, the Eq. (3.68) is reduced to (3.70):

$$b_1^{IM3L} = S_{11}^{IM3L,IM3L} \; a_1^{IM3L} + S_{11}^{IM3L,f1} \; a_1^{f1}, \tag{3.84}$$

$$b_1^{f1} = S_{11}^{f1,IM3L} \; a_1^{IM3L} + S_{11}^{f1,f1} \; a_1^{f1}. \tag{3.85}$$

This procedure is permissible as there is no phase dependency in intermodulation measurements. It is also advantageous in practice as it saves a second measurement (see Eq. (3.76) and (3.77)). In addition, this procedure provides the same results for the magnitudes as a scalar IM measurement. Due to these advantages, these slightly different frequency conversion S-parameters are used in practice for intermodulation measurements.

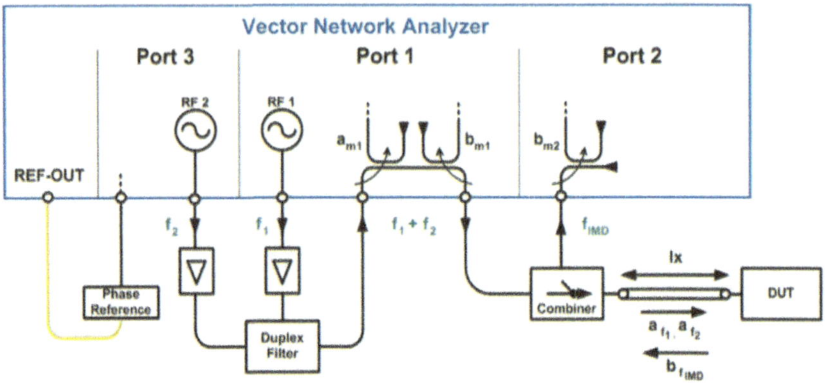

Fig. 3.27 Network analyzer with associated test set for vectorial IM reflection measurement

The corresponding circuit of a network analyzer is illustrated in Fig. 3.27.

Here, the comb generator with known phase values between the fundamental and harmonics is connected from a receiver as a phase reference. This was often in former times necessary, because very often the internal RF and LO oscillators could not be reproduced in phase lock. The absolute phase value of the LO signal was determined at this measuring point.

In modern NWAs such as the ZVA from Rohde&Schwarz, the two RF sources required for IM measurements are integrated. Their signals are amplified to increase the measurement dynamic range. In practice, IM measurements of passive components (antennas, cables) for mobile radio technology are carried out with 100 % measurements with a dynamic range of 160 dBc. The best IM measuring station in Europe for the future traceability of measurements at mobile stations uses the MF-S measuring technology presented here is presented in ??.

To ensure that this high dynamic range can be achieved, a duplex filter is used to suppress the harmonics of the amplifiers. This is followed by external or internal couplers. The internal couplers of the NWA may not be able to cope with the power. The DUT is reached via a combiner. For test sets with large measurement dynamics, this is again a duplex filter. A receiving measuring point for vectorial IM detection is connected to the third port of the combiner.

With a scalar measuring system, the test set is identical, only the measuring points are replaced by scalar detectors.

A corresponding result of this new measurement technique is shown in Fig. 3.28.

A frequency sweep is shown in this example. What is new is that two x-axes are shown for the a and b wave. The right part shows the phase progression of the reflection over the frequency. In the upper part you can see all setting values.

This additional information about the phase helps with modeling the component and can also be used directly to calculate the distance between the measuring plane and the interference point and thus specify the location of the interference point.

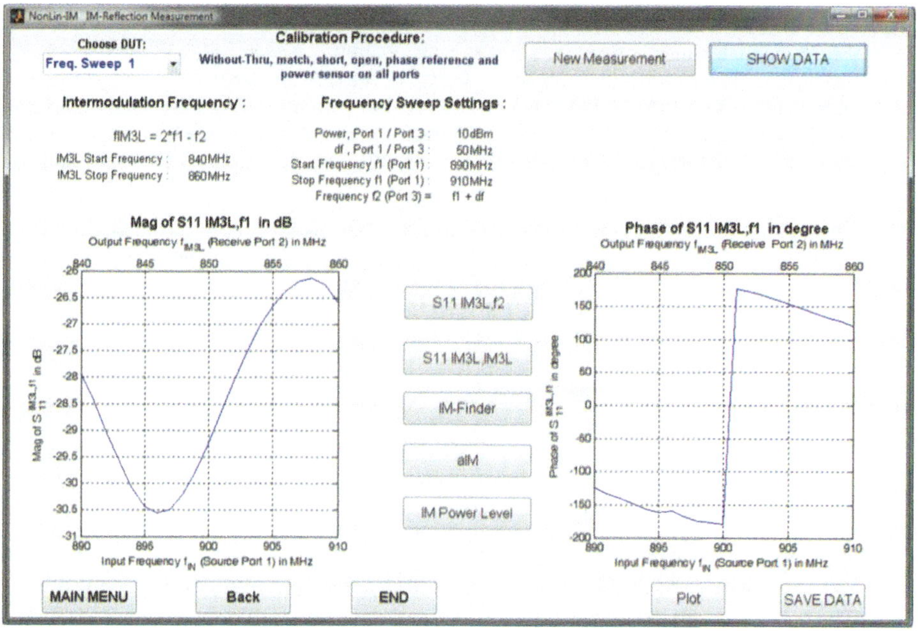

Fig. 3.28 Measurement result of a vectorial IM reflection measurement

This data can also be further processed in a circuit simulator. The file format required for this is presented below.

Optimal Data Format for Mixed Frequency S-Parameters

The consistent further development of the s1p, s2p or snp formats allows a relatively clear form of storing the mixed frequency S-parameters.

In addition to the number of measurement ports, the number of harmonics must now also be specified. It is also necessary to include the incoming large signals in the file.

For a one-port, the control string must be extended by the new commands

- Hn: Number of harmonics (n is a number)
- An: Consideration of the large signal at port n

must be taken into account. So that not only harmonics can be displayed, it makes sense to list the harmonics individually as frequencies.

The following applies to a one-port with the two signals a_1^2 and a_1^3 as excitation waves and three harmonics:

```
# frequency_unit  H3    A2     A3     S      DB     Ang     R     impedance
freq1   freq2   freq3   dbmA2   angA2    dbmA3   angA3              !1st row
dbS11(11) angS11(11)  dbS11(12) angS11(12)  dbS11(13) angS11(13) !2nd row
dbS11(21) angS11(21)  dbS11(22) angS11(22)  dbS11(23) angS11(23) !3nd row
dbS11(31) angS11(31)  dbS11(32) angS11(32)  dbS11(33) angS11(33) !4nd row
```

The format presented here can be seen as a further development of the already used P2D format, which, however, is limited to the magnitude values of the excitation waves and pure (possibly mixed-frequency) two-port transmission.

Data Format for IM-Measurements

In HF simulation programs such as ADS, the IMT format is available for the integration of vectorial intermodulation measurements. The name stems from the English term "intermodulation table".
This format exists in three variations:

* `O-type` IMT-format, * `A-type` IMT-format, * `B-type` IMT-format.

The `O-type` only provides scalar values and therefore will not be further discussed here. The `A-type` handles 2 input signals and is well-suited for standard IM measurements. The `B-type` presents a variant especially interesting for complex mixer measurements.

Often, the IM format is only illustrated using mixer measurements as an example. Here, a combined representation will be chosen for the `A-type`. Furthermore, in the following:

$$f_1 \text{ will be replaced by } f_{LO} \qquad \text{and} \qquad f_2 \text{ by } f_{RF} \quad .$$

Just as it was depicted in Fig. 3.26, only the IM products around the two frequencies will be considered.

In principle, the nomenclatures for the file formats bear great similarities to the well-known s2p format. In the case of the IMT format, the definition line typically reads:

```
# IMT ( GHz S DBM R 50.0 )
```

The frequencies are specified in the familiar manner in GHz, and the reference impedance is 50 Ω. However, the S-parameters are presented in a special format. They are given in dBm with a phase in degrees. This is possible because the powers of the input signals at the frequencies f_{LO} and f_{RF} are also known.

The table is provided for a frequency scenario. That is, f_{LO} and f_{RF} are fixed. The format is chosen to represent the mixing products around the harmonics for mixers as well. In the following example, only the two third-order IM products, IM3L and IM3U, will be shown.

These values are represented in matrix form, where the rows M denote multiples of the LO frequency and the columns N (not labeled in the file) denote multiples of the RF frequency. The following example will illustrate this file format.

```
   ! A-type IMT file
BEGIN IMTDATA
! Option line for reference power at:
! RF = +10 dBm, LO = +7 dBm
# IMT ( GHz S DBM R 50.0 )
! Format line for RF frequency - Eingangssignale bei N=1
und M=1  % FRF
1.1  ! Format line for LO frequency
% FLO
1.0
! Format line for reference RF power
% PRF
10
! Format line for reference LO power
% PLO
7
! Format line for LO-side harmonics - Übertragungsfkt. zu
2*LO-1*RF = 2*M-1*N
% M -1 2
-1 0 0 -55 139
2 -45 -33 0 0
END IMTDATA
```

Therefore, for IM3L, it holds: $-45\,\text{dB}$ and -33° phase shift. Unmeasured values must be specified as zero.

Furthermore, the definition presented under frequency conversion or mixed-frequency S-parameters would be highly applicable. This allows for easy processing of frequency sweep and power sweep measurements.

3.8.3 X-Parameters

X-Parameters[5] were introduced by Agilent and represent a special class of mixed-frequency scattering parameters specifically designed for characterizing power amplifiers. The model of X-parameters is based on Polyharmonic Distortion Modeling (PDM) and was published in [110], forming the basis of the following discussion. However, for clarity, the presentation here is only loosely based on the nomenclature of Agilent X-parameters. The chosen more compact representation aligns with that of the previously introduced mixed-frequency scattering parameters.

[5] X-parameters is a registered trademark of Agilent Technologies.

In the following, the following indices are used:

i : Index of the output port k : Index of the output frequency
j : Index of the input port l : Index of the input frequency

For k and l, DC components at frequency zero are also allowed. Generally, the mixed-frequency parameters can be represented with the multivariable complex description function F_i^k for a multiport for any number of harmonics as follows:

$$b_i^k = F_i^k(a_1^1, a_1^2, \ldots, a_2^1, a_2^2, \ldots)\,. \tag{3.86}$$

The dependence on ω is no longer represented in (3.86). Otherwise, the frequency conversion S-parameters according to Eq. (3.83) can be specified as a solution to this equation.

The X-parameters bear a close resemblance to the Harmonic Balance method for a single excitation signal and are therefore based on a large signal a_1^1. Furthermore, to derive the X-parameters, Eq. (3.86) considers the phase Θ of the complex excitation signal a_1 at frequency f_1. It holds that

$$a_1^1 = \left|a_1^1\right| e^{-j\Theta}\,. \tag{3.87}$$

This phase rotation Θ is to be interpreted as time delay. Therefore, this phase Θ appears with the opposite sign in the fundamental frequency and in the harmonics with the corresponding factor.

$$b_i^k e^{jk\Theta} = F_i^k(a_1^1 e^{j\Theta}, a_1^2 e^{j2\Theta}, \ldots, a_2^1 e^{j\Theta}, a_2^2 e^{j2\Theta}, \ldots)\,. \tag{3.88}$$

In the next step, this phase Θ should correspond to the inverse of the phase value of a_1^1 ($\angle a_1^1$ for short) and the following relationship should apply for P:

$$P = e^{j\angle a_1^1} = e^{-j\Theta} \qquad \text{resp.} \qquad P^{-1} = e^{j\Theta}\,. \tag{3.89}$$

Inserting into Eq. (3.88) and multiplying by P^{+k} gives the following result:

$$b_i^k = F_i^k\left(\left|a_1^1\right|, a_1^2 P^{-2}, a_1^3 P^{-3}, \ldots, a_2^1 P^{-1}, a_2^2 P^{-2}, \ldots\right) P^k\,. \tag{3.90}$$

The next step is to decompose the description function into the two vectorial components of the excitation:

$$\begin{aligned} b_i^k =& XF_i^k\left(\left|a_1^1\right|\right) P^k \\ &+ \sum_{jl}\left[G_{ij}^{kl}\left(\left|a_1^1\right|\right) P^k \operatorname{Re}\left\{a_j^l P^{-l}\right\} + H_{ij}^{kl}(\left|a_1^1\right|) P^k \operatorname{Im}\left\{a_j^l P^{-l}\right\}\right] \end{aligned} \tag{3.91}$$

with

$$XF_i^k\left(\left|a_1^1\right|\right) = F_i^k\left(\left|a_1^1\right|,0,\ldots,0\right), \tag{3.92}$$

$$G_{ij}^{kl}\left(\left|a_1^1\right|\right) = \frac{\delta F_i^k}{\delta\,\mathrm{Re}\left\{a_j^l P^{-l}\right\}}\Big|_{\left|a_1^1\right|,0,\ldots,0}, \tag{3.93}$$

$$H_{ij}^{kl}\left(\left|a_1^1\right|\right) = \frac{\delta F_i^k}{\delta\,\mathrm{Im}\left\{a_j^l P^{-l}\right\}}\Big|_{\left|a_1^1\right|,0,\ldots,0}. \tag{3.94}$$

It can be seen that the transfer function parameters XF, G and H are derived for the operating point with the excitation signal a_1^1. The transmission parameters in XF correspond to the large-signal S-parameters at the associated frequency. In contrast, G and H describe the nonlinear properties around the operating point.

The next step is to use

$$\mathrm{Re}\left\{a_j^l P^{-l}\right\} = \frac{a_j^l P^{-l} + \left(a_j^l P^{-l}\right)^*}{2} \quad \text{and} \tag{3.95}$$

$$\mathrm{Im}\left\{a_j^l P^{-l}\right\} = \frac{a_j^l P^{-l} - \left(a_j^l P^{-l}\right)^*}{2} \tag{3.96}$$

to derive the further X-parameters for $(j,l) \neq (1,1)$:

$$XS_{ij}^{kl}\left(\left|a_1^1\right|\right) = \frac{1}{2}\left(G_{ij}^{kl}\left(\left|a_1^1\right|\right) - jH_{ij}^{kl}\left(\left|a_1^1\right|\right)\right), \tag{3.97}$$

$$XT_{ij}^{kl}\left(\left|a_1^1\right|\right) = \frac{1}{2}\left(G_{ij}^{kl}\left(\left|a_1^1\right|\right) + jH_{ij}^{kl}\left(\left|a_1^1\right|\right)\right). \tag{3.98}$$

With these X-parameters, Eq. (3.91) can be converted into

$$b_i^k = XF_i^k\left(\left|a_1^1\right|\right)P^k + \sum_{(j,l)\neq(1,1)}\left[XS_{ij}^{kl}\left(\left|a_1^1\right|\right)P^{k-l}a_j^l + XT_{ij}^{kl}\left(\left|a_1^1\right|\right)P^{k+l}a_j^{l*}\right]. \tag{3.99}$$

In the nomenclature chosen by Agilent, the same equation is

$$b_{ik} = X_{ik}^F\left(\left|a_{11}\right|\right)P^k + \sum_{(j,l)\neq(1,1)}\left[X_{ik,jl}^S\left(\left|a_{11}\right|\right)P^{k-l}a_{jl} + X_{ik,jl}^T\left(\left|a_{11}\right|\right)P^{k+l}a_{jl}^*\right] \tag{3.100}$$

with the same meaning of the indices.

The term X^T only appears when an amplifier is operated strongly in saturation. The other terms X^F and X^S show a relationship with the S-parameters and transition into the scattering parameters under linear excitation. The X^F term describes the large-signal S-parameters, entirely without frequency conversion.

More details regarding the X-parameters can be found in [83].

3.8.4 Relationship Between the X-Parameters and the Mixed-Frequency-S-Parameters

The derivation presented below is taken from the scientific publication [17], which the first equation shows the X-parameters of Eq. (3.100). Here, the more compact nomenclature according to Eq. (3.99) is continued.

In the first step, the XF-parameters can be integrated into the XS-parameters by introducing

$$XF_i^k\left(\left|a_1^1\right|\right)P^k = XS_{i1}^{k1}(\left|a_1^1\right|)P^{k-1}\,. \tag{3.101}$$

Thus

$$b_i^k = \sum_{(j,l)}\left[XS_{ij}^{kl}\left(\left|a_1^1\right|\right)P^{k-l}a_j^l + XT_{ij}^{kl}\left(\left|a_1^1\right|\right)P^{k+l}a_j^{l*}\right]\,. \tag{3.102}$$

In the second step, in [17] *XS* is renamed to *S* and *XT* is renamed to *T*. The latter is unfavorable as it can lead to confusion with the known T-parameters.

However, the *XT* term only occurs if an amplifier is driven strongly into saturation. If this is not the case and there is the special case that the excitation wave is in phase with the reference signal, then $\angle a_1^1 = 0^\circ$ and consequently $P^{k-l} = 1$ and therefore only the previously presented mixed frequency S-parameters remain in the general form

$$b_i^k = \sum_{(j,l)}\left[S_{ij}^{kl}\left(\left|a_1^1\right|\right)a_j^l\right]\,. \tag{3.103}$$

The fact that the other known network parameters (e.g. Z, Y) can also be calculated for the mixed-frequency cases is also derived in [17].

3.9 The Harmonic Transfer Function and the Application of the MF-Measurement Technique in Localization

A vector network analyzer (VNA) can also be used as a multifunctional device for measuring materials or distances. The precision of the VNA stems from its ability to be calibrated using the GSOLT method[6] or similar techniques. This full calibration allows for highly accurate transmission and reflection measurements regardless of the number of measurement ports.

For a simple transmission measurement, a reference measurement is sufficient, which is referred to in the VNA menu as *response calibration*. Mathematically, the measurement of a device under test (DUT) with the measured S-parameters $S_{21\,meas}$

[6] This method, now widely used, was developed by the author and published in 2003 [39].

is referenced to the through measurement with the S-parameters $S_{21\,thru}$:

$$S_{21\,DUT} = S_{21\,meas} / S_{21\,thru} \,. \tag{3.104}$$

These transmission measurements (with an open test set) allow good vectorial S_{21} measurements if the measurement object is well adapted, as is the case with a good coaxial cable, for example. Comparatable distance reflection measurements presented below can be carried out with a similar quality!

Hardware Basis of Mixed-Frequency Measurement Technology in Localization

The network analyzer solutions previously developed at the IMP up to the purchasable PNA-X from Keysight are far too complex to use for localization applications, too large, too heavy, too slow and far too expensive.

A IMP spin-off company developed transmit (TX) and receive (RX) modules specifically for mixed-frequency measurement technology, which, among other things, show reproducible phase states in the oscillators after switching on. These modules shown in Fig. 3.29 can be used as VNA modules for standard S-parameter measurements and have an extremely phase-stable coherence due to the special sampling system for MF-S parameter measurements, which results in the extremely high sensitivity of $-180\,\text{dBm}$, among other things.

The fact that the procedure with these two modules (see next page) for MF measurements is much easier, also shows the block switching Fig. 3.30 vividly. The hardware effort with these modules is no greater than the effort for a standard S_{21} measurement using the reference method.

The hardware architecture is also referred to as a Stepped-Frequency Continuous-Wave Harmonic Radar system (SFCW-HR system for short).

3.9.1 Cooperative Localization Using Nonlinear Backscatter Tag and Stepped Frequency Harmonic Radar

A cooperative target is a target that changes an incident signal is modified differently than it is with natural targets. A cooperative target is referred to in many areas as a "transponder" (a portmanteau of *transmitter* and *responder*) and in practice as a "tag".

In GHz localization technology, a very simple solution for this is a dual-band antenna that is adapted at f_1 and at $f_2 = 2 \cdot f_1$, and a diode to generate a signal with twice the frequency of the input signal. An example of a scalar measuring Harmonic Radar is the Recco system, whose tags are already sewn into ski jackets and can be purchased in normal shops in order to find the wearer even under avalanches [98]. Scalar systems only measure power and therefore do not allow any real distance

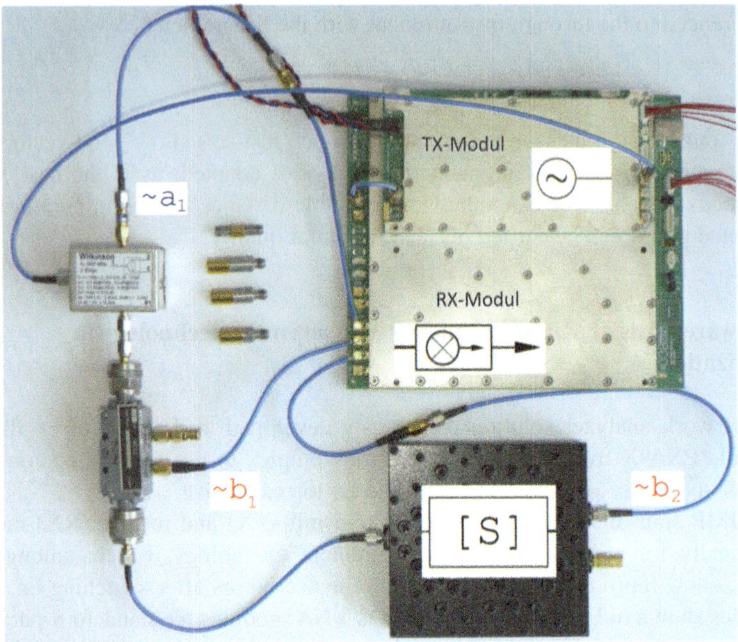

Fig. 3.29 Unidirectional vector network analyzer based on the switched receiver concept [31] with only one switched receiver and TX and RX module from RX module from [34]

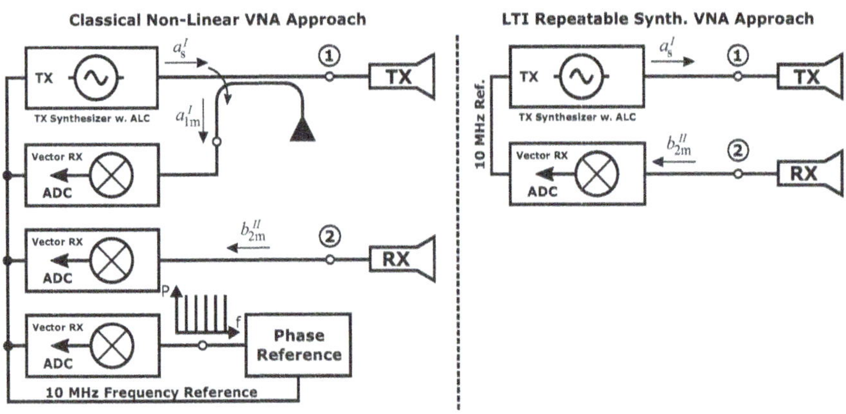

Fig. 3.30 Comparison of the known fully calibratable network analyzer solutions with the hardware solution for vectorial mixed-frequency measurements for the reference measurement used for localization.

determination. With Recco system, a directional antenna is only used to detect the tag and determine its direction. The measured power values only provide a very rough estimate of the distance.

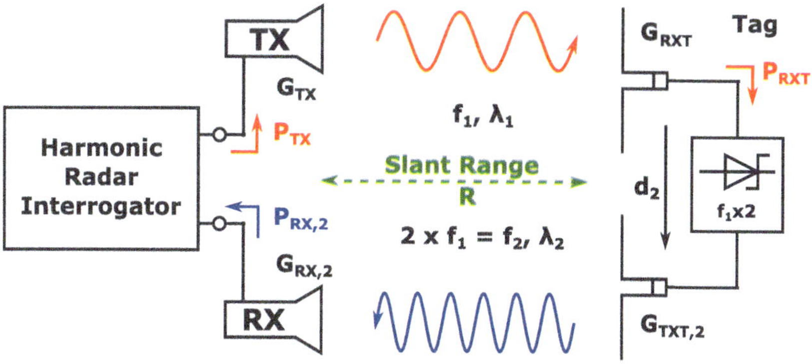

Fig. 3.31 Simplified block diagram for a harmonic radar with one tag containing two antennas and a Schottky diode as frequency doubler

Fig. 3.31 shows the cooperative localization system using a nonlinear backscatter tag.

Based on the scientific work from [24] and the further scientific work up to the implementation of the MF-S measurement system for a cooperative localization system using nonlinear backscatter tag and stepped frequency harmonic radar, which are presented in great detail in [31], result in for the performance during the day:

$$P_{RXT} = \underbrace{\frac{P_{TX} \cdot G_{TX}}{4\pi \cdot R^2}}_{\text{Energy Density}} \cdot \underbrace{\frac{G_{RXT} \cdot \lambda_1^2}{4\pi}}_{\text{Antenna Aperture}} = \frac{P_{TX}\, G_{TX}\, G_{RXT}\, \lambda_1^2}{(4\pi R)^2}\,. \tag{3.105}$$

The first term in Eq. (3.105) describes the energy density at the distance R, taking into account the transmission power P_{TX} and the antenna gain G_{TX}. The second term in Eq. (3.105) indicates the effective antenna effective area of the tag antenna.

The power at the receiver of the radar set can also be specified from the work cited above:

$$P_{RX,2} = \frac{G_{RX,2}\, \lambda_2^2 \cdot (P_{TX} \cdot G_{TX})^2 \cdot \sigma_2}{(4\pi)^4 \cdot R^6}\,. \tag{3.106}$$

This contains the reflection cross section or radar cross section σ_2 of the tag, which in harmonic radar is referred to as "nonlinear pseudo harmonic radar cross section (RCS)" and is calculated as follows:

$$\sigma_2 = d_2 \cdot G_{TXT,2} \cdot \left[\frac{G_{RXT}\, \lambda_1^2}{4\pi}\right]^2. \tag{3.107}$$

A new parameter is d_2, which takes into account the conversion losses at the diode.

The difference between this harmonic radar and a standard radar with transmit power P_{TX} and receive power P_{RX} is very remarkable.

$$P_{RX,2} \sim \frac{P_{TX}^2}{R^6}, \text{ compared to } \quad P_{RX} \sim \frac{P_{TX}}{R^4} \tag{3.108}$$

In practice, a harmonic radar (HR) with a passive tag cannot achieve the ranges of a normal radar. However, this is also possible if an active tag is used, which will be explained later.

3.9.2 The HR Equation and HR Transfer Function

To understand the basic operation of an SFCW HR-system that measures MF-parameters in the frequency domain like a network analyzer, it is beneficial to examine this HR in terms of the transfer function. Here, the two components of a single discrete tone of the measured complex harmonic received signal, the amplitude and the phase argument containing the distance information, are analyzed together. Fig. 3.31 shows the typical scenario of a HR for a reference measurement without material under test (MUT) between the antennas. The power $P_{RX,2}$ of the returning harmonic can be calculated at the output of the receiving antenna of the query system using the HR equation by evaluating Eq. (3.106).

The phase component $\angle(P_{RX,2})$ of the measured harmonic return-signal phasor can be determined by tracing the course of the signal phase through all system components, which is shown in detail in [23, 33]. In this context, it is only important to know that the phase of the received harmonic signal can be calculated with

$$\angle(P_{RX,2}) = \underbrace{e^{2j(\omega^I t+\varphi_0)}}_{\substack{\text{Freq. Doubled} \\ \text{Illum. Signal}}} \cdot \underbrace{e^{-j4\omega^I v_p^{-1} \cdot R}}_{\substack{\text{Slant Range} \\ \text{Information}}}, \tag{3.109}$$

where ω^I denotes the frequency of the illumination signal, φ_0 describes an arbitrary but repeatable phase shift, the phase propagation velocity $v_p = c_0\left(\sqrt{\varepsilon_r \mu_r}\right)^{-1}$ through the dispersion-free medium and R describes the distance traveled during propagation (slant range) when a memoryless power series model is assumed for the frequency doubler.

For this application, it is important to note that the frequency doubling process leads to an increased sensitivity of the propagation phase to the propagation medium or material compared to the material compared to the standard SFCW radar with fundamental frequency. The phase response is optimized in a dispersion-free environment compared to an otherwise simple penetration is increased by a factor of 4.

A similar observation can be made for the loss contributions of the material under test due to the quadratic dependence of the return magnitude on the power of the received illumination signal through the day. Both observations together lead to a greater sensitivity of such a system to small variations in dielectricity and loss variations in the material or object under test. These statements are further substantiated by the derivation of the HR transfer function.

In reality, however, nonlinear frequency doublers are anything but ideal components and must be designed and optimized for a range of conditions related to the particular measurement setup. For most users, it is sufficient to note that the use of HR leads to additional boundary conditions for the system resulting from the transmission through a lossy medium between the HR interrogation system and the tag. These conditions must be met for the Eq. (3.106) and (3.109) to be valid. Ideally, the HR tag in its well-defined range of the quadratic transfer function without saturation by the lighting signal and still delivers sufficient harmonic output power to achieve a good SNR at the receiver. In order to describe the actual mixed-frequency measurement of the HR setup, it is useful to describe the transfer functions with the help of the MF S-parameters.

In accordance with the definition of linear S-parameters, the measured MF reflection parameter $S_{11,M}^{II,I}$ based on the Fig. 3.31 from the received harmonic complex power wave b_1^{II} at point D can be compared with the transmitted fundamental frequency power parameter transmitted fundamental frequency power wave a_1^I at point a measured by a VNA, formally described as

$$S_{11,M}^{II,I} = \frac{b_1^{II}}{a_1^I} . \tag{3.110}$$

If this measurement is performed without a material or object to be tested, as shown in Fig. 3.31, the result obtained is the so-called reference measurement result is the so-called reference measurement $S_{II,I}^{11,Ref}$ and must be performed and saved for each HR tag element. This mixed-frequency reflexion parameter can be further subdivided into various system contributions. In addition to the MF conversion parameter, the mismatch reflections may also occur. The waves between individual MF parameter blocks can often be ignored, as only the superposition of several reflection effects is taken into account in the forward transmission. This first-order approximation is applicable and valid here due to the large free-space wave propagation losses.

Based on the interrogation system, the first relevant MF parameter block is the linear forward propagation of the fundamental frequency stimulus signal through the medium to the input of the frequency doubler, which is denoted by $S_{21,\text{Ref}}^I$. The nonlinear MF parameter of the doubler, which depends on the input power, can be described by the parameter $S_{43}^{II,I}|_{P_{Ref}}$, while the linear return path through the medium at the harmonic frequency is described by harmonic frequency is captured by $S_{12\text{Ref}}^{II}$. A graphical representation of the HR signal flow can be seen in Fig. 3.32. The complete reference measurement $S_{11,\text{Ref}}^{II,I}$ can therefore be described by

$$S_{11,\text{Ref}}^{II,I} = S_{21,\text{Ref}}^I \cdot S_{43}^{II,I}|_{P_{Ref}} \cdot S_{12,\text{Ref}}^{II} \quad . \tag{3.111}$$

Eq. (3.111) describes the HR transfer function for the reference measurement.

This reference result is subsequently used to normalize the measurement results. As a second measurement, a material or object to be tested is introduced into the path. When the second measurement of the material or object is referenced to the

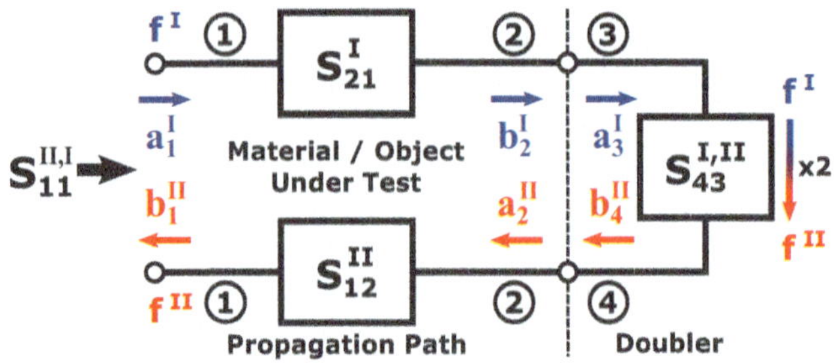

Fig. 3.32 Representation of the HR transmission function with the two-port S-parameters of the material under test (MUT, ports 1 and 2) and the nonlinear part of the doubler (ports 3 and 4).

first reference measurement, it corresponds to the mixed-frequency equivalent of the reflection normalization method used in linear VNAs (see [14]). This measurement implicitly contains the propagation properties of the channel at the two frequencies as well as the specific conversion behavior of the HR tag. Additionally, these data are also used to correct systematic and repeatable phase deviations in the signal generation and receiver parts of the system (details see [33]).

A Device Under Test (DUT) or Material Under Test (MUT) has scattering parameters $S^{I}_{21,\mathrm{MUT}}$ and $S^{II}_{21,\mathrm{MUT}}$ at the two frequencies I and II, which describe the superposition of all changes in transmission wave propagation caused by the material. When the material or object to be tested is inserted into the propagation path, replacing a section of the free-space reference measurement, and a measurement $S^{II,I}_{11,\mathrm{MUTz}}$ is performed, diffraction and refraction effects occur, as well as an observed change in phase velocity v_p and losses caused by the dielectric and magnetic properties of the material, influencing wave transmission through the material. Since the material replaces an equivalently shaped free-space section contained in the reference measurement, the corresponding free-space section needs to be decoupled from the material properties, resulting in scattering parameters $S^{I}_{21,\mathrm{MUTz}}$ and $S^{II}_{21,\mathrm{MUTz}}$ that only capture the relative change compared to an equivalent section of free-space propagation. The corresponding HR transmission function for the measurement involving the MUT is therefore

$$S^{II,I}_{11,\mathrm{MUTz}} = S^{I}_{21,\mathrm{Ref}} \cdot S^{I}_{21,\mathrm{MUTz}} \cdot S^{II,I}_{43}|_{P_{MUTz}} \cdot S^{II}_{12,\mathrm{Ref}} \cdot S^{II}_{12,\mathrm{MUTz}} \quad . \tag{3.112}$$

The mechanical distance between the HR and the tag is unchanged compared to the reference measurement. The added material/object causes the observed additional attenuation and phase rotation $S^{I}_{21,\mathrm{MUTz}}$ and $S^{II}_{12,\mathrm{MUTz}}$.

The nonlinear change in the transmission behavior of a frequency doubler $\Delta S^{II,I}_{43}$ only influences the magnitude according to the following formulas:

$$\Delta S^{II,I}_{43} = \frac{S^{II,I}_{43}|_{P_{MUTz}}}{S^{II,I}_{43}|_{P_{Ref}}} \quad , \quad \Delta S^{II,I}_{43} = S^{I}_{21,\mathrm{MUTz}} \quad , \tag{3.113}$$

when the doubler is operated in its quadratic domain where the memoryless power series approximation is applicable [24, 33, 70].

The full normalized measurement with a material or object under test can therefore be described by

$$S^{II,I}_{11,\mathrm{Norm.}} = \frac{S^{II,I}_{11,\mathrm{MUTz}}}{S^{II,I}_{11,\mathrm{Ref}}} \text{, which can be separated into three two-port blocks}$$

$$S^{II,I}_{11,\mathrm{Norm.}} = \Delta S^{I}_{21} \cdot \Delta S^{II,I}_{43} \cdot \Delta S^{II}_{12} \,, \tag{3.114}$$

and finally using Eqs. (3.111)–(3.113), we achieve the HR transfer function of normalized measurements:

$$\boxed{S^{II,I}_{11,\mathrm{Norm.}} = (S^{I}_{21,\mathrm{MUTz}})^2 \cdot S^{II}_{12,\mathrm{MUTz}}} \tag{3.115}$$

While the increased amplitude sensitivity of such a system is immediately apparent from Eq. (3.115), the increased phase sensitivity of the HR principle is not directly apparent and is hidden in the formal definition of the phase between different frequencies in the MF-parameters. It becomes more obvious when considering the range information conveyed in the harmonic phase progression shown in Eq. (3.109):

While a normal transmission measuring system would only detect an increase in electrical length of Δx ($\Delta S^{I}_{21} = S^{I}_{21,\mathrm{MUTz}}$), a reflective measuring radar system (for materials with good matching and low transmission loss) would detect an increase of $2\Delta x$ ($\Delta S^{I}_{21} = \Delta S^{I}_{12}$) due to the double transmission through the material. Meanwhile, in an HR system, the measured path length increases by $4\Delta x$ because the instantaneous phase of ΔS^{I}_{21} at the input of the frequency doubler doubles and the return path at doubled fundamental frequency is via ΔS^{II}_{12} (see [33] for the derivation). In addition, the reflection energy has just as little influence here as with the normal transmission measurement.

Eq. (3.115) can be used to calculate the transfer function of different materials and objects over frequency using the results of a mathematical model or 3D finite element simulation. A database of such expected transfer function can then be used in addition to the actual measurements to train an AI model for object or material recognition and to classify the results of an actual unknown measurement. At the end of the related publication [43], a traceable measurement in a closed, non-radiative, coaxial system is shown to prove the validity of Eq. (3.115) and thus also of Eq. (3.113) in magnitude and phase.

3.9.3 Range Determination with a Stepped Frequency Harmonic Radar with Nonlinear Backscatter Tag

The major advantage of a harmonic radar over the standard radar system (pulse or FMCW radar) is the so-called clutter freedom. Reflections from (large) objects located between the radar device and the tag have no influence on the measurement[7].

In principle, this harmonic radar can be implemented using various technologies (microwave technology, optics, ultrasound, etc.). In the radio frequency and microwave range, the propagation coefficient for propagation in the cable or in free space is

$$\beta(f_i) = \beta_i = (2\,\pi\,f_i)\,/\,c\,, \tag{3.116}$$

where $c = c_0$ is the speed of light for free space and $c = c_0/\sqrt{\varepsilon_r}$ is to be used [38].

The distance determination l_x for a radio link or a perfect matched line can be performed directly via a VNA measurement and the S_{21} results on a larger number of measurement points over the frequency via

$$S_{21} = e^{-j\,\beta\,\ell_x}\,. \tag{3.117}$$

For the transmission measurement via two cables or two antennas, the double phase shift or slope of the phases of S_{21} occurs, as Fig. 3.33 in the middle section clearly illustrates. If, in addition, as in the bottom line of Fig. 3.33, a frequency doubler circuit (also known as a transponder or tag) is also used, the phase response of the output signal from the doubler is twice as large as that of the feed line and, in addition, the phase response of the return line is twice as large as that of the feed line, as the frequency is twice as high. It is known from the fundamentals of nonlinear frequency conversion that both the quadratic term, which must be used for frequency conversion with doublers, and the cubic term, which must be used for 3rd order IM products, change the phase of the incoming signal by a fixed factor of 2 as a result of the frequency conversion. The following therefore applies to the entire phase rotation of the frequency-converting reflection measurement over the distance ℓ_x for the frequency point f_1 at the input and the frequency point f_2 at the receiver

$$\angle t_g^{21} = \angle S_{21\,g}^{21} = \beta_1\,\ell_x + \underbrace{\beta_1\,\ell_x + t_{ref}^{21}}_{\text{Complete Tag}} + \beta_2\,\ell_x\,. \tag{3.118}$$

Here, t_x^{21} corresponds to the design-dependent S_{ij}^{kl} transmission factor of the tag at the frequency f_1 to the frequency f_2. The following applies for twice the frequency: $f_2 = 2\,f_1$ or $\beta_2 = 2\,\beta_1$. In the following, identical tags are to be used for the reference measurement and as a DUT and the reference tag is to be measured at the location $\ell_{ref} = 0\,\text{m}$.

---

[7] The exception is, of course, overdriving by the large signal, but this does not occur in practice.

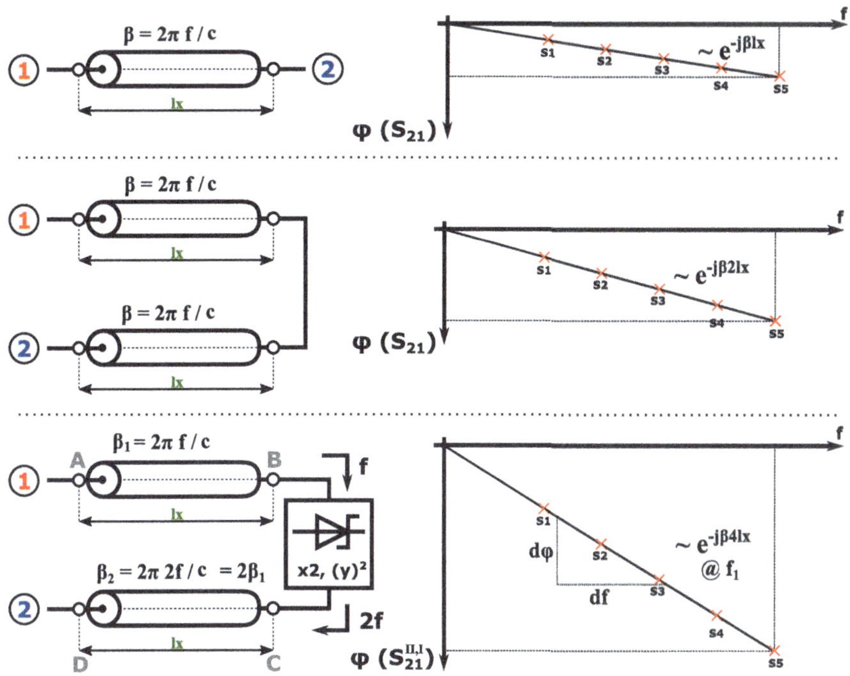

Fig. 3.33 Different phase responses of standard VNA measurements and a MF S-parameter measurement (with tag)

The phase of the frequency-converting component is only measured in the uniqueness range of 360°. Consequently, the total phase value $\angle t_g^{21}$ depends on the measured value $\angle t_{gm}^{21}$:

$$\angle t_g^{21} = \angle t_{gm}^{21} + n \cdot 2\pi \,, \tag{3.119}$$

where n is an integer that will be determined below.

For the reference tag, which is measured directly at the measuring ports, the simple relationship applies:

$$\angle t_{ref}^{21} = \angle t_{refm}^{21} \,. \tag{3.120}$$

The measured values of the S_{ij}^{kl} transmission factor are available from a lower frequency f_{1u} to an upper frequency f_{1o} as $\angle t_{gm}^{21u}$ and $\angle t_{gm}^{21o}$. There should be no phase jumps between the different measurement points, which can also be easily eliminated in practical mathematics. The slope of the phase over the frequency can be determined from these two measured values and the known associated frequency points:

$$S = \frac{\Delta\varphi}{\Delta f} = \frac{\angle t_{gm}^{21o} - \angle t_{gm}^{21u}}{f_{1o} - f_{1u}} . \tag{3.121}$$

At any frequency f_i, there is a fundamental phase rotation of $\varphi_i = f_i S$ due to the slope. This equation can also be derived from Eq. (3.121) with $f_o = f_i$ and $f_u = 0\,\text{Hz}$ and the corresponding phase values. With the following condition

$$180^\circ \leq \left| \angle t_{gm}^{21i} + n \cdot 2\,\pi - f_i S \right| \tag{3.122}$$

the correct value for n is found. To determine the length of the distance ℓ_x, t_{gm}^{21} is measured via a frequency sweep from f_{1u} to f_{1u} and the slope S is determined from Eq. (3.121). Eq. (3.122) then allows the determination of n and (3.119) the determination of t_g^{21}. Based on Eq. (3.118), the distance for each frequency point can be calculated using

$$\ell_x = \frac{\angle t_g^{21} - \angle t_{ref}^{21}}{4\beta_1} . \tag{3.123}$$

In practice, Eq. (3.124) is evaluated for each measured frequency point so that stochastic errors are minimized by subsequent averaging.

The accuracy of this measurement increases with the bandwidth $(f_{1o} - f_{1u})$ and the number of measurements, i.e. with the measurement time. For the following example of sea rescue, we will work out a compromise between accuracy and measurement time, which can also depend on distance, among other things.

3.9.4 Cooperative Localization Using Harmonic Radar for the Example of a Sea Rescue System

In contrast to mountain rescue, there is no modern solution for sea rescue that can effectively locate individual people or the smallest boats.

The overall objective of the research project, which was funded by the BMBF under the abbreviation SEERAD[8]. The aim of the project was to develop and test a new type of sea rescue system (SNRS) that uses the ship radar (S-band at 2.9–3.1 GHz) as a transmission source and is therefore only an extension of this radar. The SNRS is based on a new type of radar and reflectors (tags) similar to those used in avalanche rescue (the Recco system).

The necessary frequency conversion is carried out by very inexpensive mixed-frequency reflectors ("tags" or "transponders"), which are implemented in jackets, on life jackets and life rings as well as in lifeboats. These reflectors or tags are either passive or active. The passive reflectors have a range of just under 1 km. Active reflectors, on the other hand, can be detected up to the visual horizon.

[8] The project partners were: Raytheon-Anschütz in Kiel, Fraunhofer Gesellschaft FHR in Wachtberg and FH Aachen, IMP in Aachen.

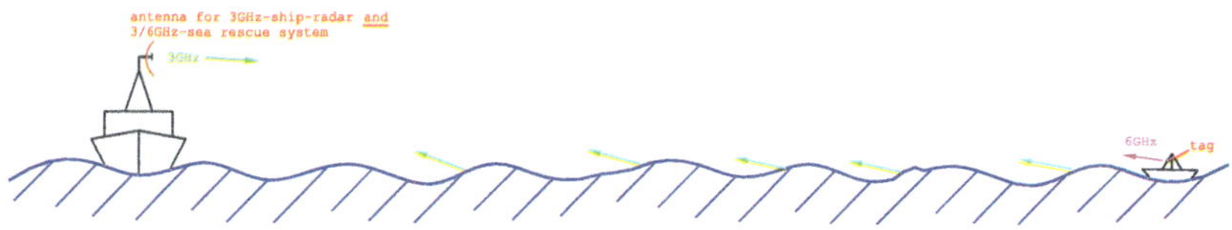

Fig. 3.34 Illustration of the rescue scenario with the new sea rescue system, which only detects detects signals at 6 GHz

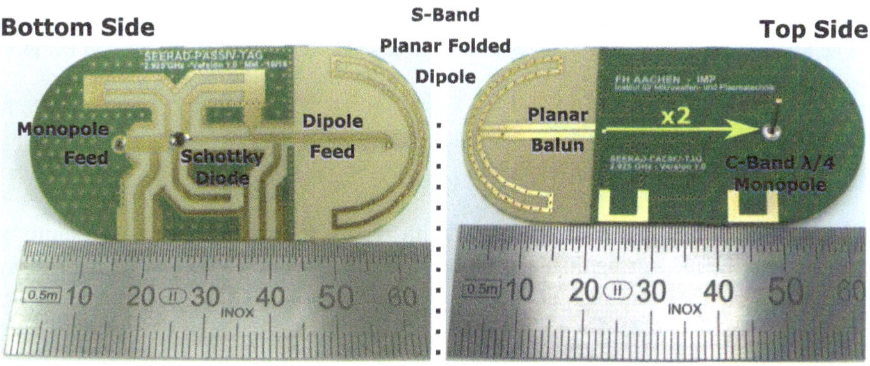

Fig. 3.35 Illustration of the two sides of the passive SEERAD tag, which only has two antennas, matching lines and a diode

The outstanding feature of this innovative radar system architecture is that the receiver only detects signals at the second harmonic of the transmitter. This makes the system extremely resistant to interference. Experts speak of immunity to clutter, which is otherwise the dominant reflection due to waves at sea and possibly precipitation. This is the only reason why small objects such as people and lifeboats can be detected. In the following Fig. 3.34 this scenario for a 3 GHz ship radar with the special 6 GHz receiver and the 3-to-6 GHz converter (reflector in the lifeboat) included in this application is shown visually.

Compared to other solutions that have already been developed, this approach also stands out because the reflectors, which have to be used in large numbers, are very inexpensive. The passive reflectors (tags) consist of a dual-band foil antenna and a diode, Fig. 3.35. These can be manufactured in large quantities for well under 10 € and can be brought to market at a correspondingly low price. The active tags contain small signal amplifiers and a battery that is activated by water. These active tags are therefore much less than 20 € more expensive to produce than passive tags, but this is not a major obstacle for mass use.

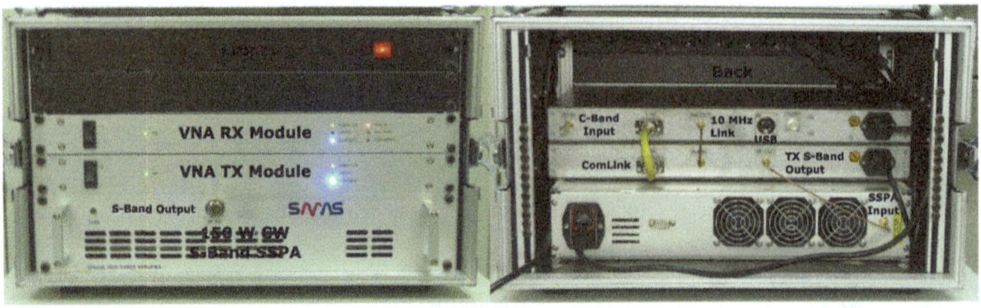

Fig. 3.36 Front and rear view of the SEERAD rack, which contains the entire TX and RX electronics for the sea rescue system

System Dynamics and Level Plan

In this block, only the system design with regard to. The necessary transmitting power and the measuring distance is presented completely and in a way that is easy to understand for every high and radio frequency technician. The simplified conclusion is that with a transmission power of around 200 W, a range of the desired 17 km can be achieved.

For the received power P_{E,f_k} at the transceiver for the signal converted by the tag with the frequency f_k, the following applies

$$P_{E,f_k} = P_{S,f_i} G_{s,f_i} G_{e,f_i} \left(\frac{\lambda_{f_i}}{4\pi R} \right)^2 L_{f_k f_i} G_{e,f_k} G_{s,f_k} \left(\frac{\lambda_{f_k}}{4\pi R} \right)^2, \tag{3.124}$$

with

- P_{S,f_i} the transmission power of the transceiver at f_i,
- G_{s,f_i} the antenna gain of the transmitting antenna for f_i,
- G_{e,f_i} the antenna gain of the receiving antenna during the tag for f_i,
- λ_{f_i} the wavelength at f_i,
- R the simple distance between the transceiver and the tag,
- $L_{f_k f_i}$ the conversion factor for the frequency shift from f_i to f_k,
- G_{s,f_k} the antenna gain of the transmitting antenna during the tag for f_k,
- G_{e,f_k} the gain of the receiving antenna at the transceiver for f_k,
- and λ_{f_k} the wavelength at f_k.

Results: With a targeted transmission power of 200 W (53 dBm) at 3.0 GHz (f_i) and ship antenna gain of 25 dBi for (f_i) and (f_k) and a tag antenna gain of 0 dBi, a passive tag with conversion loss of 20 dB a range of around 1 km and for an active tag (30 dB gain) a range of over 17 km (line-of-sight horizon for ship antennas of 30 m) with a receiver sensitivity of −140 dBm (with very low averaging).

In practice, these values were verified with measurements at a transmission power of 100 W and the transmission/reception technology with the previously presented VNA modules, which is explained in the following text.

Fig. 3.37 Installation of the transmitting and receiving antenna for carrying out measurements on the linear path

The immense necessary suppression of the crosstalk between the transmitter and receiver units, which are located very close to each other units is ensured on the one hand by the frequency offset and on the other by the consistently differential design of the and transmitting unit of the HHF electronics. Related work has been published in [38, 40], among others.

Measurement Results of the Sea Rescue System

The first series of measurements with the SNRS was used for the basic robustness analysis of this system and also provides important insights for other applications such as indoor localization and navigation.

These measurements were carried out in the laboratory at a transmission power of only 100 mW using only the hardware as shown in Fig. 3.36. The transmitting and receiving antenna as well as the tag, which was moved on a linear path, are shown in Fig. 3.37.

In the detailed presentation of these investigations, which is printed in [31], it was among other things, it was also shown that it offers advantages that the transmitting

Fig. 3.38 Setup of the transmitting and receiving antenna for carrying out measurements on the linear track from the perspective of the transmitting and receiving antenna

antenna is polarized differently than the receiving antenna. Also in the case of the Fig. 3.37 with external screw-on antennas, the two different polarizations are clearly visible.

The most demanding measurement scenario in this series of scientific investigations is shown in Fig. 3.38.

A series of measurements was carried out over this 5 m long linear path, with three triple mirrors already generating large reflections at a short distance and the wall directly behind the linear path also reflecting strongly. The large triple mirror has a backscatter cross-section of $\sigma_{3\,\text{GHz}} = 13.67\,\text{m}^2$ at a distance of 2.5 m and the two small ones each have $\sigma_{3\,\text{GHz}} = 0.16\,\text{m}^2$ at distances of 1.6 m and 1.8 m. In addition, there are many other sources of reflection interference that also strongly interfere with normal wave propagation.

The measurement results, which were determined using various evaluation methods [31], are shown in Fig. 3.39.

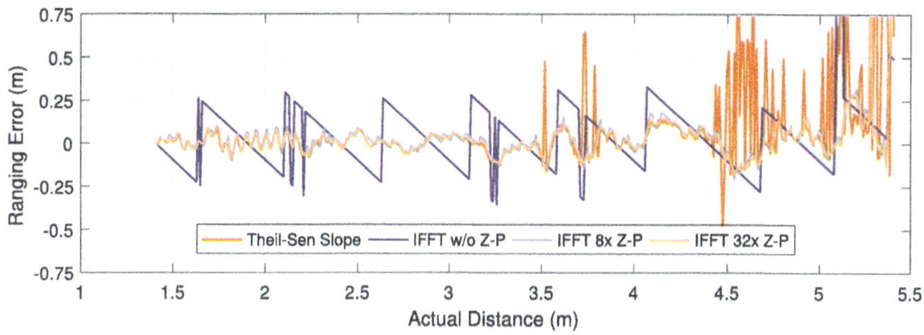

Fig. 3.39 Representation of the measurement error to the actual distance over the length of the linear track with measured values in 1 cm steps

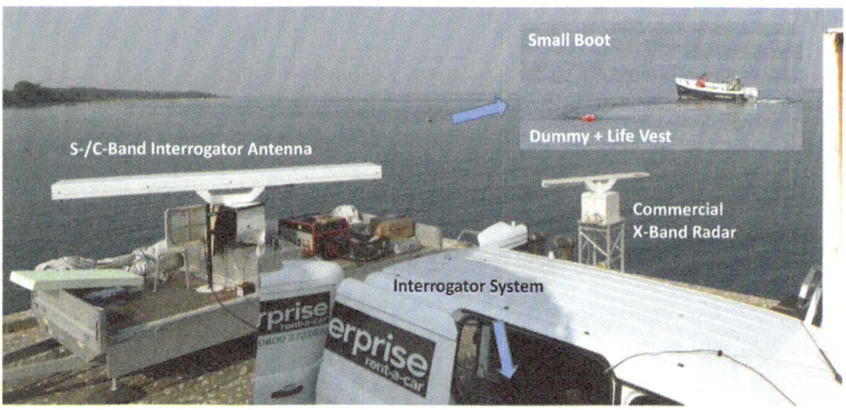

Fig. 3.40 Illustration of the complete rescue system installed on a pier in the Baltic Sea rescue system, including the dual-band large antenna installed on a rotary joint (HF rotary coupler) (at a height of 7 m height), the lifeboat and the dummy in a lifejacket equipped with the passive SEERAD tag

Outdoor Results of the Sea Rescue Project

The dual-band antenna developed as part of the BMBF project SEERAD[9] as well as the overall demonstrator system (created by Raytheon-Anschütz and the IMP) contains many innovations, details of which can be found in scientific publications from mid-2020.

A large number of measurements were carried out on land with the overall system (Fig. 3.40), which corresponded fairly well with the preliminary calculation.

The maximum measurement range of 800 m achieved by the passive tag at a transmission power of only 100 W exceeds the state of the art many times over the state of the art, which is 750 m range at a transmission power of 25 kW [70].

[9] This antenna was developed by Dr. Thomas Bertuch's group at FHR in Wachtberg

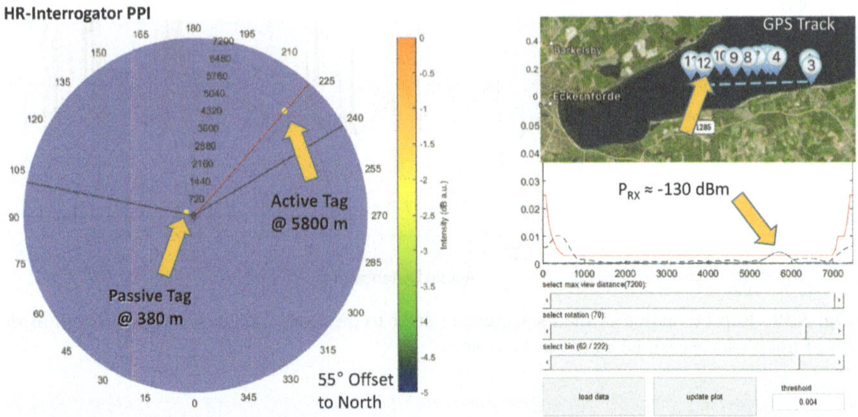

Fig. 3.41 Measurement results of the world record measurement with the active tag

Fig. 3.42 Visualization of the later use of this SEERAD sea rescue system

The active tag was measured up to the visual horizon of 5800 m. The antenna was only 7 m above the water surface. Fig. 3.41 shows the absolutely clutter-free measurement results of the active tag, as well as those of a passive tag that was installed in the shore area of the approach route of this mole.

The subsequent use in practice is best illustrated in Fig. 3.42. However, such a system will only be implemented in practice when it is required by law, as was once the case with compulsory seat belts in cars and compulsory helmets for motorcyclists. Furthermore, the system must be linked to the ship's radar, which also requires further innovations.

Applications Coherent Mixed-Frequency-Systems	Status	Turn-around times	Per so-nal years	Comments	IMP-project partners
Vektorial Mixed-Frequency-Measurement Technique	Basics were done by the spin-off www.HHFT.de	2004 – 2010	3	Calibration procedure Without-Thru as one result	VNA-producer are welcome
Vector-PIM-measurement technique	Several measurement systems from HHFT including one for the PTB, Braunschweig	2007 -	6	For 5G/ 6G very important	Company Rosenberger Fridolfing
Sea rescue system	Rescue system with inexpensive tags (2-15€) and perfect performance	2016 – 2019	5	Demonstrator from BMBF-project SEERAD available	Radar manufactor are welcome
Mountain rescue system	Power-off smartphones can be found by using a drone	2020 -	0	A research proposal is done	Rescue company is welcome
Localisation	Robust and accurate indoor- and outdoor-localisation without the normal clutter problems	2022 -	6	Auch für das autonome Fahren bestens einsetzbar	Project partner is existing
Rear reflector	Cat's-eye 2.0 will save many lives in public traffic	future	0	IMP needs a larger partner	Large project partner
Mobile communication	The hardware design for XG on the basis of the coherent MF-technique allows 10 dB more signal sensitivity	open	0	This technology has a large potential	Very big project partner

Fig. 3.43 Research projects at the Institute for Microwave and Plasma technology (IMP) at FH Aachen

These measurements show that the IMP has a great deal of scientific work in the field of calibrating mixed-frequency S-parameters and the associated vectorial network analyzers, including the innovative without-thru method [40], has opened the door to a new device architecture.

The new hardware produced by Heuermann HF-Technik, which requires only one measuring point for a reflectometer, gave this scientific research project an excellent basis for scientific work in maritime rescue.

This radar project surpasses all previously realized harmonic radar systems in range many times over (including the RECCO mountain rescue system) and is unique due to its fully vectorial measurement capability.

3.10 Coherent Mixed-Frequency Measurement Technology as the Basic Technology

The two application examples of PIM measurement technology and maritime rescue presented in the previous subsections show that coherent mixed-frequency measurement technology is a basic technology that also enables many other applications, which is why this technology can also be referred to as a basic technology. Similar basic technologies that define the hardware architectures are the superheterodyne receiver and the IQ receiver in GHz electronics.

The following overview presents the applications carried out at the author's institute and those that are still open/planned (Fig. 3.43).

Chapter 4
Power Amplifier

Catchwords such as Software Defined Radio, (SDR) suggest that the development of hardware components appears to be on the decline. However, if you analyze implementations of this SDR, you can see that the complexity of the necessary analog circuit components has even increased.

In particular, the need for amplification of the emitted radio frequency power is constantly increasing. The new digital modulation methods such as WLAN, Edge, UMTS and 5G also require a highly linear transmission behavior of these power amplifiers.

The increasing use of radio modules in mobile devices places the greatest demands on the efficiency of power amplifiers in order to maximize transmission times on the one hand and to use as little material as possible for cooling purposes on the other.

Since transistor amplifiers are increasingly being used even in the kW range range, this chapter also contains a restriction to transistor amplifiers. The principles and architectures presented here can of course also be applied to the numerous amplifier designs that are used in many classic books such as [116] can be applied.

High-frequency power amplifiers differ from small-signal amplifiers in that the transistors are nearly or completely driven through the entire output characteristic field. These Power Amplifiers (PA) can only be analyzed using nonlinear simulation techniques. However, all the basic considerations necessary for implementing a linear small-signal amplifier are also necessary for a PA. Therefore, this chapter is based on the contents of Chapter 8 from [38]. All design aspects such as operating voltage constraints, maximum supply currents, required power gain, MAG, k-factor, linearity requirements, noise requirements must necessarily be initially fulfilled according to the criteria of [38]. The method for synthesizing passive matching networks is also explained there.

This chapter first covers the fundamentals of developing a PA, including output impedance matching. Furthermore, the classical operating modes A, B, and C, which are solely determined by the bias point setting, are introduced in detail. Subsequently, a variety of concepts for improving efficiency are presented. Following that, combined amplifiers are introduced. Finally, a detailed discussion on linearization techniques is provided.

H. Heuermann, *Microwave Technology*,
https://doi.org/10.1007/978-3-658-45686-3_4

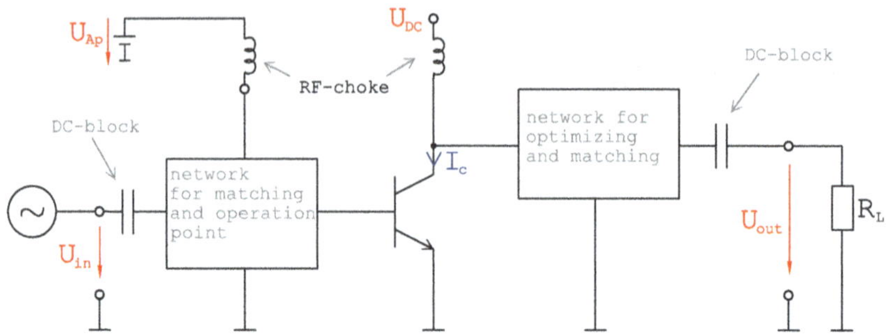

Fig. 4.1 Block diagram of a single-stage power amplifier

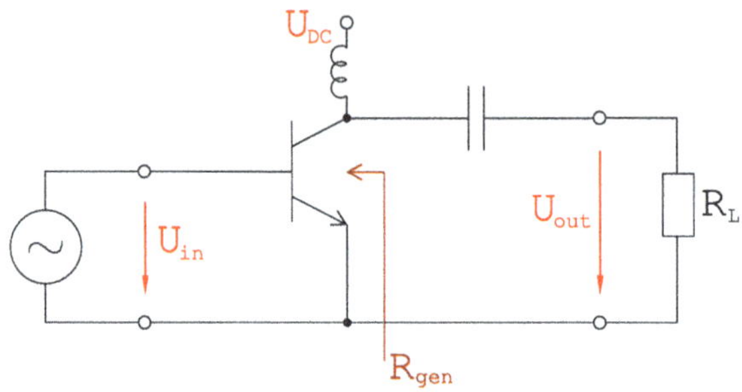

Fig. 4.2 Simplified block diagram of a single-stage power amplifier

Simplified Representation of Power Amplifiers

The block diagram of a single-stage Power Amplifier (PA) is shown in Fig. 4.1. The two included networks contain an impedance matching or an impedance transformation to achieve the greatest possible power amplification. These two matching networks as well as the network for setting the operating point are no longer shown below. For reasons of improved clarity, a PA according to Fig. 4.1 is shown in simplified form according to Fig. 4.2.

The supply coil for the DC voltage and the block capacitor, which are still included, should be ideal components. In real setups, these are usually components of the matching circuits, as presented in [38].

Although only one bipolar transistor will be shown in the following, the theories and concepts presented here also apply to field-effect transistors.

The detailed design of the matching and optimization networks will be discussed later.

4.1 Basics for the Development of Power Amplifiers

The development of transistors is driving high-frequency electronics into the lower THz range. The latest CMOS transistors in 45 nm technology have cut-off frequencies of 250 GHz or 0.25 THz.

The development of power transistors in the single and double-digit range progressed at a similar pace. RF developers are constantly faced with the decision of which transistor technology is best for their application. The following section is intended to help with this.

4.1.1 Transistor Technology

In 1947, AT&T Bell Telephone Laboratories invented the point-contact transistor, consisting of a small piece of germanium, a metallic base, and two metal points. Since then, many other construction technologies in combination with various materials have been developed.

The choice of the best transistor technology for the desired application is crucial for the behavior of an amplifier circuit. Each semiconductor technology has its advantages and disadvantages. These must be precisely understood to choose the best technology for the intended application. For low-noise amplifiers, P/HEMTs (Pseudomorphic/High Electron Mobility Transistors) offer the best characteristics, followed by MESFETs (Metal-oxide Semiconductor Field Effect Transistors). In the power amplifier domain, MESFETs are also suitable. Especially for mobile applications, HBTs (Heterojunction-Bipolar-Transistors) are the best choice as they achieve high efficiency and good linearity.

The technological development of heterostructures in the 1980s resolved the conflicting requirements for high frequency and high gain in bipolar transistors. Hetero-bipolar transistors have now replaced conventional transistors and are found in silicon (Si), gallium arsenide (GaAs), and indium phosphide (InP) technologies. Si-HBTs are preferred for low and medium powers. GaAs-HBTs, along with the classical MESFET, HEMT, and LDMOSFET, are widely used for power transistors.

Heterojunction Bipolar Transistors (HBTs)

The heterojunction bipolar transistor (HBT) is a bipolar transistor in which two different materials are chosen for the emitter and base (hence also "heterojunction"). It corresponds to the bipolar implementation of a High-Electron-Mobility Transistor (HEMT).

HBTs meet the increased requirements for power amplifiers in modern radio systems such as WLAN or UMTS in many respects. In compact structures with only one supply voltage, such as mobile phones, they offer high gain, efficiency, linearity, and can be produced cost-effectively.

Compared to SiGe components, GaAs-HBTs have higher electron mobility and benefit from the availability of semi-insulating substrates. These properties result in low U_{ce} residual voltages at high current densities. Accordingly, GaAs-based HBTs are widely used in the commercial mobile phone market, [53].

HBTs operate on the same principle as "normal" bipolar transistors. The base-emitter diode is forward-biased, and the collector-base junction is reverse-biased. Electrons are injected from the emitter into the base region and accelerated into the collector by the strong electric field in the base-collector junction.

While bipolar transistors are made from a single material, HBTs use a material with a larger bandgap than the base semiconductor in the emitter during production. The base semiconductor itself is heavily doped and made as thin as possible, resulting in short transit times and low base resistance. Additionally, the emitter is lightly doped to minimize the base-emitter capacitance. All of this leads to a transit frequency that is significantly higher than that of ordinary bipolar transistors.

A disadvantage of all GaAs transistors is the significantly higher substrate cost and less efficient production lines compared to Si products, making Si products much more affordable.

GaAs-HBTs replaced MESFETs in mobile phones because the complex negative gate voltage and "normally-on" behavior were eliminated. GaAs in HBTs also currently enjoys considerable popularity as power amplifiers.

With HBTs, cutoff frequencies of 600 GHz and more can now be achieved.

Especially based on silicon germanium (SiGe), HBTs have very low noise figures.

Metal Oxide Semiconductor Field Effect Transistors (MESFETs)

GaAs MESFETs were the most innovative components in the high-frequency range in the 1970s. For power amplifiers, however, the channel area must be very highly doped, which leads to a reduction in electron speed. This was remedied by the introduction of a heterostructure, which culminated in the HEMT.

The disadvantages were the negative gate voltage, the normally-on behavior and the associated additional electronics. MESFETs are now rarely used.

High Electron Mobility Transistors (HEMTs)

The HEMT (high electronmobility transistor[1]) is a special type of field-effect transistor optimized for the GHz range.

As a small GaAs transistor with an extremely low noise figure, the HEMT was the breakthrough for satellite television in the 1980s. This special design is now also frequently used in power transistors.

[1] transistor with high electron mobility

The most common other name for the HEMT is heterojunction field-effect transistor (HFET).

The HEMT consists of layers of different semiconductor materials with different band gaps (hence heterostructure). Initially, gallium arsenide (GaAs) and aluminum gallium arsenide (AlGaAs) were used, but now also gallium nitride (GaN) and aluminum gallium nitride (AlGaN). As the band gap of AlGaAs is larger than that of GaAs, a two-dimensional electron gas (2DEG) forms at the interface of these two materials on the GaAs side, which serves as a conductive channel. The electron mobility is very high in this channel.

Due to the pure 2D movement, this type of transistor has a very low inherent noise. The HEMT is described in great detail in [4].

LDMOS-Feld Effekt Transistors

LDMOS is the abbreviation for: Laterally Diffused Metal Oxide Semiconductor.

Due to its relatively low price, this power transistor manufactured in silicon is most frequently used in the lower GHz range for power values in the 2 and 3-digit watt range.

The input has a very large capacitive component and therefore this transistor is not suitable for broadband power amplifiers. Relatively good efficiencies can now be achieved for narrowband applications (up to 60 % at 2.45 GHz and 20 W).

The application note AN1226 from STMicroelectronics explains the difference between the structure of the LDMOS transistor and a normal DMOS structure.

GaN Transistors

The first gallium nitride RF transistors (GaN for short) have been commercially available since 2006. GaN is a III-V semiconductor with a wide electronic bandgap, which is used in optoelectronics (blue and green light-emitting diodes) and for high-power, high-temperature and high-power, high-temperature and RF field-effect transistors.

GaN-HEMT power transistors are now used most frequently. The advantages of the transistors lie in the highly customizable input and output resistances, the very high efficiencies (>80 % at 2.45 GHz and 20 W) and the very high power levels that can be achieved.

The disadvantages are the price, the negative gate voltage, the normally-on behavior and the associated additional electronics. The high price is currently being reduced by manufacturing GaN transistors on inexpensive silicon carbide wafers (SiC).

For 1 GHz, transistors are already available in this technology that can deliver 500 W RF power with an efficiency of 60 %. Further information on transistor technologies can be found in [53].

The almost 50-pages application note AN10896 from Ampleon reflects just how elaborate and complex the professional installation of RF PA transistors is.

InP Transistors

InP (Indium Phosphide) transistors have a rich history dating back to the mid-20th century when semiconductor researchers began exploring alternative materials for electronic devices. Developed as a successor to earlier semiconductor materials like silicon and gallium arsenide, InP offered superior performance characteristics, particularly in high-frequency and high-speed applications due to its excellent electron mobility.

Over the years, InP transistors have found widespread use in telecommunications, satellite communications, radar systems, and high-frequency amplifiers, where their ability to operate at millimeter-wave frequencies and handle high-power densities is invaluable. The unique properties of InP, such as its high electron mobility and low noise characteristics, make it well-suited for applications requiring low-noise amplification, such as in microwave receivers and sensitive communication systems.

Additionally, InP-based transistors have played a crucial role in the development of next-generation wireless communication technologies, including 5G networks, where their high-speed capabilities and efficiency are essential for achieving faster data rates and lower latency. As semiconductor technology continues to advance, InP transistors are expected to remain at the forefront of high-frequency and high-speed applications, driving innovation in various fields of electronics and telecommunications.

4.1.2 The Signal Flow Method and its Application

Signal flow diagrams have already been introduced in [38]. The signal flow diagram describes a linear network using the S-parameters and the wave sizes a_i and b_i. Fig. 4.3 illustrates the diagram of an amplifier.

From a mathematical point of view, the signal flow method is a universal method for solving linear equations. The Manson rule can be used to describe a transfer function between a dependent and an independent variable. The general form is

$$T = \frac{P_1\,[1-\sum L(1)^{(1)}+\sum L(2)^{(1)}-...]+P_2\,[1-\sum L(1)^{(2)}+...]+...}{1-\sum L(1)+\sum L(2)-\sum L(3)+...}\ . \tag{4.1}$$

Fig. 4.3 is used to explain the application of Manson's rule:

The signal flow diagram of the amplifier has only one independent variable b_s. As an example, the transfer function $T = \frac{b_1}{b_s}$ is created here.

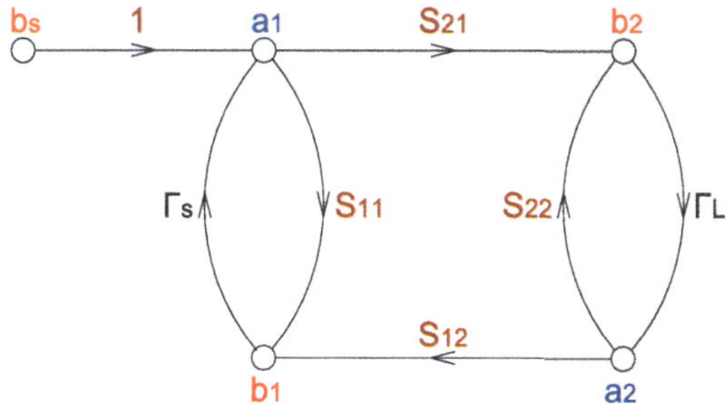

Fig. 4.3 Signal flow diagram of an amplifier with load impedance Γ_L and source b_s

The terms P_i describe the different paths that connect the dependent and the independent variable whose transfer function is to be determined. Nodes may only be traversed once by a path and only in the direction of the arrow from the independent to the dependent variable. For the required transfer function there are two paths $P_1 = S_{11}$ and $P_2 = S_{21}\Gamma_L S_{12}$.

$\sum L(1)$ is the sum of all 1st order loops. A 1st order loop is the product of all branches that are traversed in the direction of the arrow to get from one node back to the same node. In the Fig. 4.3 are $S_{11}\Gamma_S$, $S_{21}\Gamma_L S_{12}\Gamma_S$ and $S_{22}\Gamma_L$ loops of 1st order.

$\sum L(2)$ is the sum of all 2nd order loops. A 2nd order loop is the product of two non-touching 1st order loops. In Fig. 4.3 is $S_{11}\Gamma_S S_{22}\Gamma_L$ a loop 2nd order.

$\sum L(3)$ is the sum of all 3rd order loops. A 3rd order loop is the product of three 2nd order loops that do not touch each other. In the Fig. 4.3 there are no 3rd order loops.

In general: $\sum L(i)$ is the sum of all loops i. order. A loop i. order is the product of i loops $(i-1)$. order that do not touch each other.

The term $\sum L(1)^{(P)}$ is the sum of all loops of 1st order that do not touch the path P. In the Fig. 4.3 there are two paths. For P_1 the following applies $\sum L(1)^{(1)} = \Gamma_L S_{22}$ and for P_2 the following applies $\sum L(1)^{(2)} = 0$.

The term $\sum L(2)^{(P)}$ is the sum of all 2. order loops that do not touch the path P. For $P_{1,2}$ holds $\sum L(2)^{(1,2)} = 0$.

In general: The term $\sum L(1)^{(P)}$ is the sum of all loops i. order that do not touch the path P. Using the Manson rule, this results in

$$T = \frac{b_1}{b_s} = \frac{S_{11}(1-\Gamma_L S_{22}) + S_{21}\Gamma_L S_{12}}{1-(S_{11}\Gamma_S + S_{22}\Gamma_L + S_{21}\Gamma_L S_{12}\Gamma_S) + S_{11}\Gamma_S S_{22}\Gamma_L} . \tag{4.2}$$

Using the signal flow diagram, Manson's rule and the knowledge that the incoming wave squared minus the reflected wave squared equals the power at a load ($P_L = \frac{1}{2}|a_2|^2 - \frac{1}{2}|b_2|^2 = \frac{1}{2}|b_2|^2(1-|\Gamma_L|^2)$), all parameters for the linear consideration of an amplifier can be calculated.

Applying the Manson rule to to determine the input reflection coefficient of an amplifier, as shown in Fig. 4.4, results in the following equation

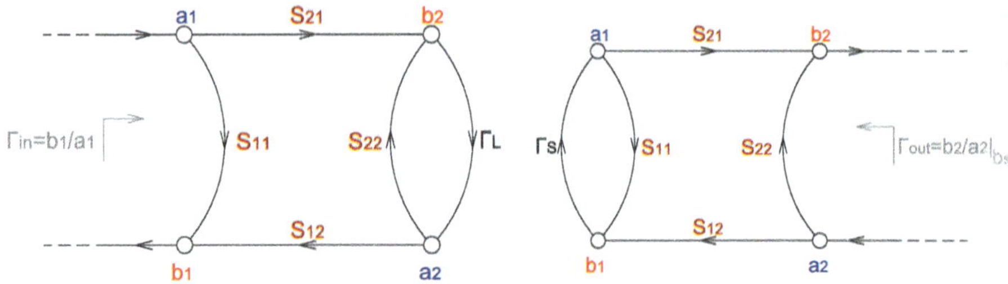

Fig. 4.4 Signal flow diagrams of the input reflection coefficient of an amplifier with load impedance and the output reflection coefficient of an amplifier

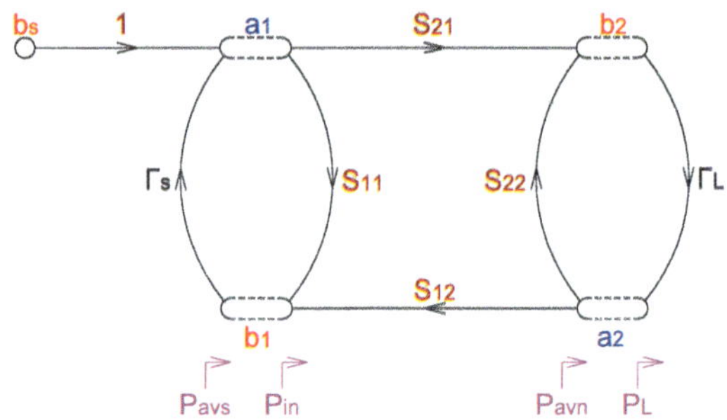

Fig. 4.5 Various power terms in the signal flow diagram of an amplifier

$$\Gamma_{in} = \frac{S_{11}(1-\Gamma_L S_{22}) + S_{21}\Gamma_L S_{12}}{1-S_{22}\Gamma_L} = S_{11} + \frac{S_{12}S_{21}\Gamma_L}{1-S_{22}\Gamma_L} \tag{4.3}$$

and for the output reflection coefficient (Fig. 4.4) the result is

$$\Gamma_{out} = S_{22} + \frac{S_{12}S_{21}\Gamma_S}{1-S_{11}\Gamma_S}\,. \tag{4.4}$$

4.1.3 Performance Terms

There are various power terms in amplifier development, which must be explained using Fig. 4.5 and Fig. 4.6.

These terms only apply without restriction to small signals. Since non-linearities and physical limits must be taken into account in large-signal operation, validity must be checked on a case-by-case basis.

P_{avs} is the power that the source would deliver with conjugate complex matching to a load.

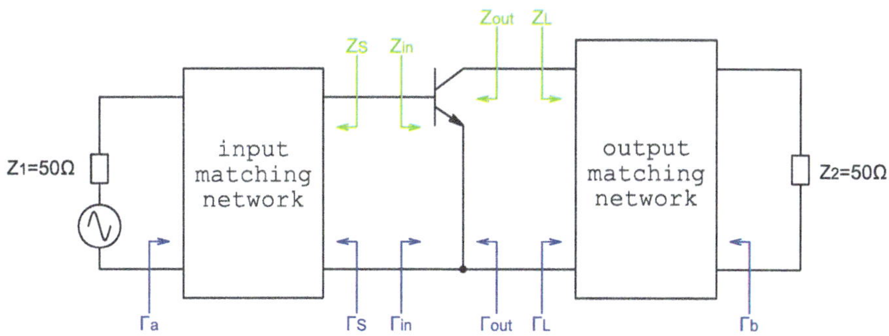

Fig. 4.6 Reflections within the subcircuits of an amplifier

$$P_{avs} = P_{in}|_{\Gamma_{in}=\Gamma_{S}*} = \frac{\frac{1}{2}|b_S|^2}{1-|\Gamma_S|^2} \tag{4.5}$$

P_{in} is the power delivered by the input matching network to the amplifier and is equal to P_{avs} at a conjugate complex input impedance of the matching network.

$$P_{avs} = P_{in}|_{\Gamma_{in}=\Gamma_{S}*} \tag{4.6}$$

Otherwise P_{in} is equal to

$$P_{in} = \frac{1}{2}|b_S|^2 \frac{1-|\Gamma_{in}|^2}{|1-\Gamma_{in}\Gamma_S|^2} \,. \tag{4.7}$$

For the combination of P_{avs} and P_{in} this results in

$$P_{in} = P_{avs} M_S \tag{4.8}$$

with

$$M_S = \frac{(1-|\Gamma_S|^2)(1-|\Gamma_{in}|^2)}{|1-\Gamma_S\Gamma_{in}|^2} \,. \tag{4.9}$$

M_S is the input mismatch factor and a measure of how much power is reflected at the transistor.

On the output side, there are also two power terms P_{avn} and P_L. P_{avn} is the power that can be delivered by the transistor with conjugate complex output matching. P_L is the power that is delivered to the load.

$$P_{avn} = \frac{\frac{1}{2}|b_2|^2}{1-|\Gamma_{out}|^2} \tag{4.10}$$

For conjugate complex output matching, the following applies

$$P_{avn} = P_L|_{\Gamma_L=\Gamma_{out}*} \,. \tag{4.11}$$

In general, P_L is calculated as follows

$$P_L = \frac{1}{2}|b_2|^2 \frac{1-|\Gamma_L|^2}{|1-\Gamma_{out}\Gamma_L|^2} \,. \tag{4.12}$$

P_{avn} and P_L are linked via the output mismatch factor M_L .

$$P_L = P_{avn} M_L \tag{4.13}$$

with

$$M_L = \frac{(1-|\Gamma_L|^2)(1-|\Gamma_{out}|^2)}{|1-\Gamma_L\Gamma_{out}|^2} \,. \tag{4.14}$$

4.1.4 Further Basic Terms

Amplification

Since there are different power concepts, it is obvious that different that there are also different definitions for amplification definitions also exist for amplification [26, 116]. The same restrictions apply here with regard to small and large signal operation as in section 2.2.6.

Firstly, the "transducer gain" G_T, which is defined as

$$G_T = \frac{P_L}{P_{avs}} = \frac{\text{Power, which is emitted at the load}}{\text{Power, which is transmitted from the source}} \,. \tag{4.15}$$

Secondly, the "power gain" G_P, which is defined as follows:

$$G_P = \frac{P_L}{P_{in}} = \frac{\text{Power, which is emitted at the load}}{\text{Power, which is absorbed by the network}} \,, \tag{4.16}$$

and last but not least the "Available Power Gain" G_A, the according to

$$G_A = \frac{P_{avn}}{P_{avs}} = \frac{\text{Power, which can be delivered by the network}}{\text{Power, which is transmitted from the source}} \tag{4.17}$$

is defined. The gains are calculated using the equations

$$G_T = \frac{1-|\Gamma_S|^2}{|1-\Gamma_{in}\Gamma_S|^2}|S_{21}|^2 \frac{1-|\Gamma_L|^2}{|1-S_{22}\Gamma_L|^2} \tag{4.18}$$

or

$$G_T = \frac{1-|\Gamma_S|^2}{|1-S_{11}\Gamma_S|^2}|S_{21}|^2\frac{1-|\Gamma_L|^2}{|1-\Gamma_{out}\Gamma_L|^2} \tag{4.19}$$

resp.

$$G_P = \frac{1}{1-|\Gamma_{in}|^2}|S_{21}|^2\frac{1-|\Gamma_L|^2}{|1-S_{22}\Gamma_L|^2} \tag{4.20}$$

and

$$G_A = \frac{1-|\Gamma_S|^2}{|1-S_{11}\Gamma_S|^2}|S_{21}|^2\frac{1}{1-|\Gamma_{out}|^2}\ . \tag{4.21}$$

Efficiency

There are several definitions in the literature for the efficiency of a amplifier stage. The most common are PAE (Power Added Efficiency) and the general efficiency η. They differ in that in the PAE calculation, in addition to the DC power P_{DC}, the input power P_{in} is also taken into account. which is fed to the amplifier is also taken into account. The efficiency η only takes into account the DC power, which means that the PAE is always worse than η.

$$PAE = \frac{P_{out}-P_{in}}{P_{DC}} \tag{4.22}$$

$$\eta = \frac{P_{out}}{P_{DC}} \tag{4.23}$$

The lower the input power in relation to the output power P_{out}, the lower the difference between the two efficiency terms. As a result, they are sometimes equated.

Maximum Available Power Gain and Maximum Stable Power Gain

The maximum available gain, or MAG for short, is a measure of the maximum gain that can be achieved with a transistor with complex input and output conjugation. achievable amplification. It only applies to stable circuits and can otherwise lead to incorrect results [38]. The MAG is calculated with

$$MAG = \frac{|S_{21}|}{|S_{12}|}\left(k-\sqrt{k^2-1}\right)\ . \tag{4.24}$$

k is the stability factor (see Eq. (4.28)) and is calculated using Eq. (4.24).

The maximum stable gain (MSG) denotes the gain that an amplifier circuit delivers for $k = 1$ and is therefore calculated using

$$MSG = \frac{|S_{21}|}{|S_{12}|} \ . \tag{4.25}$$

One of these terms is usually listed in transistor data sheets and serves as an important criterion when selecting a suitable transistor.

4.1.5 Stability

Stability is an elementary parameter that must be considered when designing amplifiers. Most transistors tend to oscillate without external circuitry and must be stabilized by circuitry measures such as current or voltage feedback.

Absolute stability means that it is not possible to stabilize with any passive generator or load termination ($\Gamma_S < 1$, $\Gamma_L < 1$) to make a circuit oscillate. In order to generate this absolute stability for active two-port the stability factor k over the frequency is considered. Its derivation is already described in numerous books [26], which is why only the formulas and application are presented here. An active two-port is **absolutely stable** if $\underline{k > 1}$ applies and the conditions

$$1 - |S_{11}|^2 > |S_{12}S_{21}| \tag{4.26}$$

and

$$1 - |S_{22}|^2 > |S_{12}S_{21}| \tag{4.27}$$

are fulfilled. The stability factor k is calculated as follows,

$$k = \frac{1 - |S_{11}|^2 - |S_{22}|^2 + |\Delta|^2}{2\,|S_{12}S_{21}|} \tag{4.28}$$

with

$$\Delta = S_{11}S_{22} - S_{12}S_{21} \ . \tag{4.29}$$

Conditional stability means that an amplifier only works stably with certain terminations. This can be represented and monitored by the so-called stability circles in the Smith chart.

All derivations and further explanations or additions to this section can be found in [38, 116] or [26] can be found.

4.1.6 Selection of the Load Impedance

The design of the passive networks for wiring a power amplifier has a very large influence on its available maximum output power and efficiency. These networks

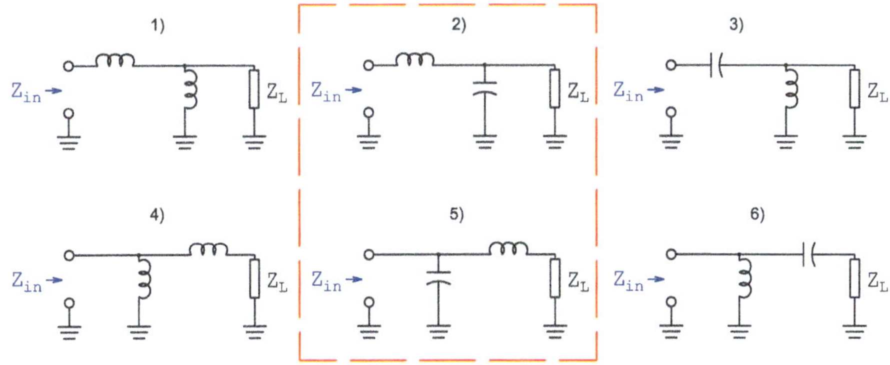

Fig. 4.7 Various constellations for Γ-transformers (LC networks) or L-networks (contain only L or C) with concentrated coils and capacitors

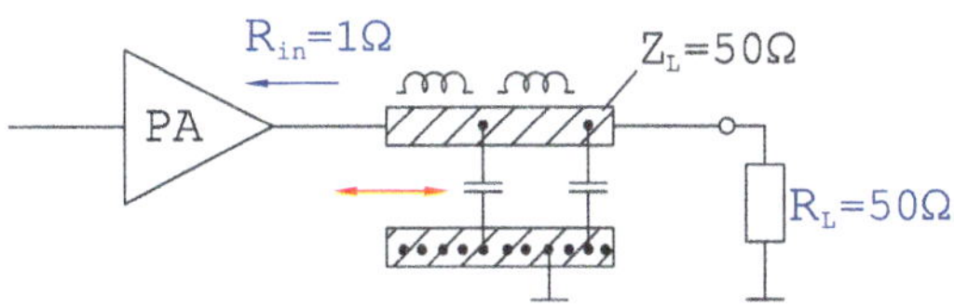

Fig. 4.8 Practical realization of a two-stage Γ-transformer (LC networks) from the easily variable quasi-concentrated line inductances and the shunt capacitors

transform the system impedance ($Z_0 = Z_L$) into an optimum impedance Z_{in} for the transistor. Which input impedance (Z_{in}) is aimed for is explained in the following section.

In this section, only the behavior of the circuit networks at the operating frequency is to be investigated. At the amplifier input, such a transformation network is used to realize the conjugate complex matching. The criteria for matching at the output are discussed in detail below.

Fig. 4.7 shows the simplest networks that can be used for this purpose.

If additional line components are allowed, the number of possible networks increases significantly. The same applies if three components are used instead of two. In [38], the introduction of the Γ transformer showed how the transformation ratio and the usable bandwidth are related. Furthermore the finite qualities of the components also have a reducing influence on the bandwidth. All these design aspects were described in [38] explained in detail. However, for reasons of optimal tuning possibilities, in practice short high-impedance lines (in relation to the transistor output impedance) in the form of 50 Ω-lines and shunt capacitors and thus the networks marked red in Fig. 4.7 are used.

Fig. 4.8 shows the simple implementation form, which allows a lot of tuning via the shifting of the capacitors.

There are two classes of power transistors in high MHz and GHz technology: Low-resistance transistors and transistors with `pre-matching`.

In general, the collector-emitter or drain-source path of the power transistor is very low-resistance when switched through, which is measured as a simple multimeter measurement with the base-emitter or gate-source voltage applied as a simple function test in the DC range.

Typical values for a 10 - 20 W power transistor are:
- Low-resistance transistors: 0.1 - 1 Ω and
- Transistors with `Pre-Matching`: 5 - 20 Ω .

For transistors with pre-matching, the circuit highlighted in Fig. 4.7 is often implemented using a bond wire (as series inductance) and a shunt capacitor (package capacitance).

Power Transistors with `Pre-Matching`

To avoid reflections, a conjugate complex matching is usually generated in high-frequency technology in order to minimize losses, as is generally the case in amplifier circuits. Small signal transistors are adapted in this way, which is described in detail in [38]. Power transistors with pre-matching are used in treated in the same way as small signal transistors for matching the fundamental wave. When matching the harmonics, which are large in power transistors and therefore have a great deal of influence, there are several explanations in the following sections.

These power transistors are already matched by the semiconductor manufacturers with the necessary operating voltage and the maximum achievable power so that the developer does not need to make any further considerations here.

Low-Resistance Power Transistors

In electronics, conjugate complex matching and power matching are often equated. However, this is only true if the physical limits of transistors and DC voltage sources are disregarded. When using conjugate complex matching, maximum amplification is always achieved. It can happen that the maximum output power is limited by the maximum voltage U_{max} of the DC voltage source (Fig. 4.9), although the maximum current I_{max} has not yet been reached.

This means that the potential of the transistor is not optimally utilized. In order to make optimum use of the maximum current and voltage, a smaller load resistance must generally be selected. This load resistance $R_L \simeq R_{opt}$ is referred to as the load line impedance and can be calculated assuming that the internal resistance of the source R_G is much greater $R_G >> R_{opt}$, can be calculated using the simple equation

$$R_{opt} = \frac{U_{max}}{I_{max}} \,. \tag{4.30}$$

If R_G must be taken into account in the calculation, then the above equation changes as follows due to the parallel connection of R_G and R_L

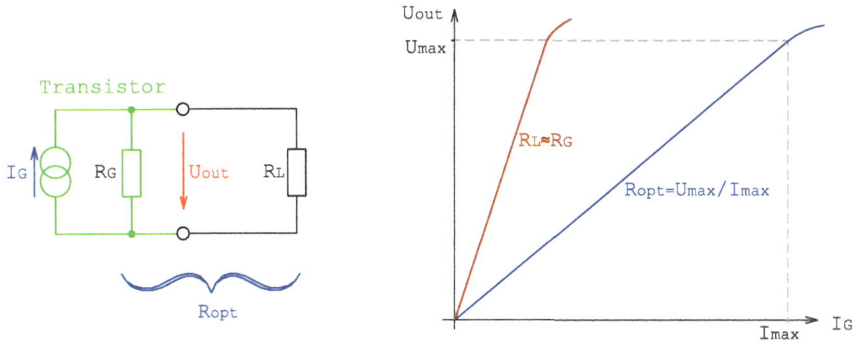

Fig. 4.9 Current-voltage characteristic curve with conjugate complex matching network and with optimal power matching network for low impedance power transistors (R_G is high impedance) with the loadline impedance R_{opt}

$$R_{opt} = \frac{R_L \cdot R_G}{R_L + R_G} = \frac{U_{max}}{I_{max}} . \tag{4.31}$$

Often R_G is a complex impedance Z_G. In practice, however, the imaginary part is often quite small. In this case, it is often sufficient to use the estimate shown above for the initial values of the simulation. Otherwise, the imaginary part of Z_G can also be compensated in a first step.

This raises the question of whether a network optimized in this way does not cause problems due to the "mismatch" and the reflections that occur. The standing waves caused by reflections can only occur if there is no wave returning from the antenna in a transmission system. The antenna can be adapted by means of an isolator or a special antenna construction according to [75]. Another alternative presented below is a balanced amplifier concept.

In high-frequency technology, it is common to abstract current and voltage and only consider the input and output power. The effect described above can also be demonstrated here. Fig. 4.10 shows the output power over the input power of a transistor with two different output matches. The solid graph belongs to a conjugate complex matching network and the dashed graph to an optimized so-called power matching network. The points A.i denote the maximum linear output power and the points B.i the 1 dB compression point. With conjugate complex matching, the gain is 1 - 2 dB higher, but the maximum linear output power is 1 - 4 dB lower. The extent of the increase depends on the technology and the component, but the effect can be seen with all transistors.

Here it was clearly shown that the conjugate complex matching is not equal to the optimal power matching.

In radio frequency measurement technology, the so-called load-pull measurement is used to set the optimum output matching. A commercial measurement setup can be seen in Fig. 4.11.

The input matching is always kept at a value that is as close as possible to the conjugate complex matching, whereby the output matching is tuned by the output

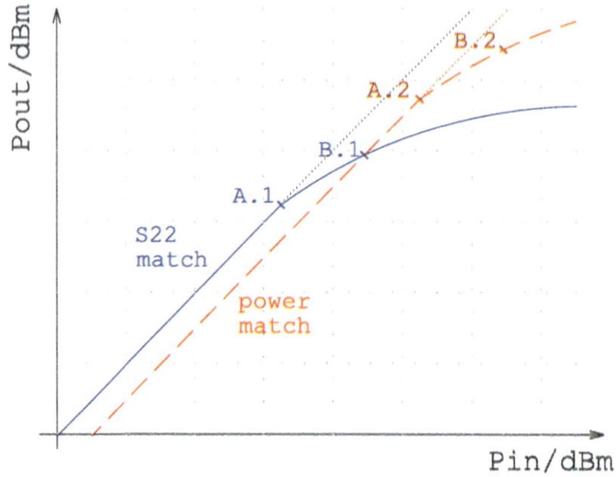

Fig. 4.10 Pout over pin for a conjugate complex matching network (in blue) and a power matching network (load line in red)

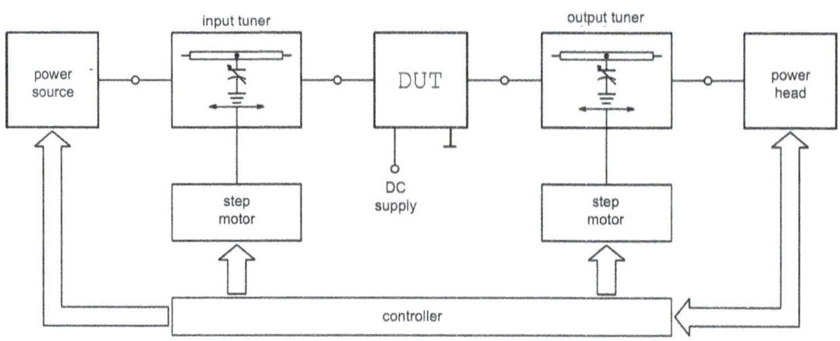

Fig. 4.11 Block diagram of a professional load-pull measuring arrangement in scalar version only

tuner. If the measurement results are plotted in a Smith chart, closed lines of constant output power appear around the point of maximum linear output power, which have an oval shape (Fig. 4.12).

This measurement data gives the RF designer a simple target in the Smith chart to create a suitable matching network. These automatic measurement setups are quite expensive. Therefore, manual tuners are often used, which are relatively easy to build yourself with suitable mechanical manufacturing capabilities.

Load-pull measuring stations are also available in

- vectorial version and/or
- with harmonic tuner and/or
- with electronic tuner.

Load-pull measurements can be simulated with current circuit simulators such as ADS. Predictions can load-pull lines can also be predicted using the load-pull theory [11].

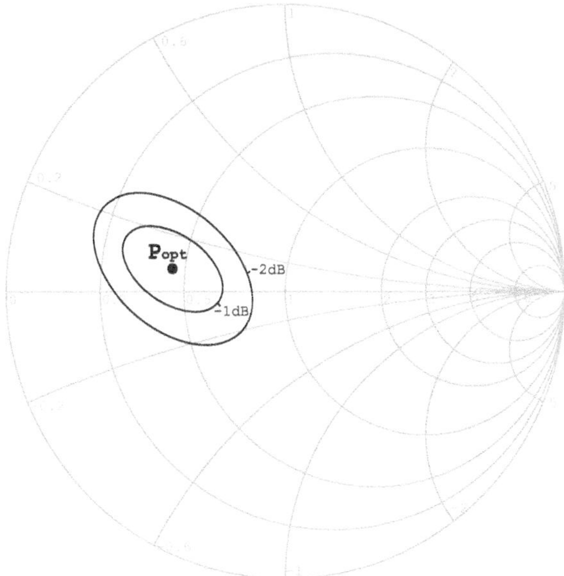

Fig. 4.12 Load-pull lines in the Smith chart for the output power of $P_{opt-1dB}$ and $P_{opt-2dB}$

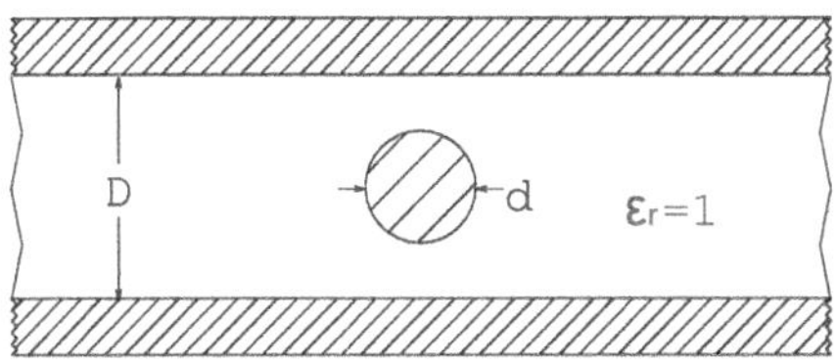

Fig. 4.13 Sectional view of the tuning cable system

4.1.7 Design and Application of a Manual Tuner

A tuner is a 50 Ω TEM line in which one or more adjustable capacitances can be shifted.

First, we will briefly discuss the most commonly used line system. The line system used for tuners as shown in Fig. 4.13 is called "slabline", [112]. It consists of two infinitely extended ground planes with the distance D and a round conductor with the same distance to both ground planes with the diameter d.

The characteristic impedance of this cable system is calculated as follows:

$$Z_0 = \frac{60\,\Omega}{\sqrt{\varepsilon_r}} \ln\left(\frac{4D}{\pi d}\right). \tag{4.32}$$

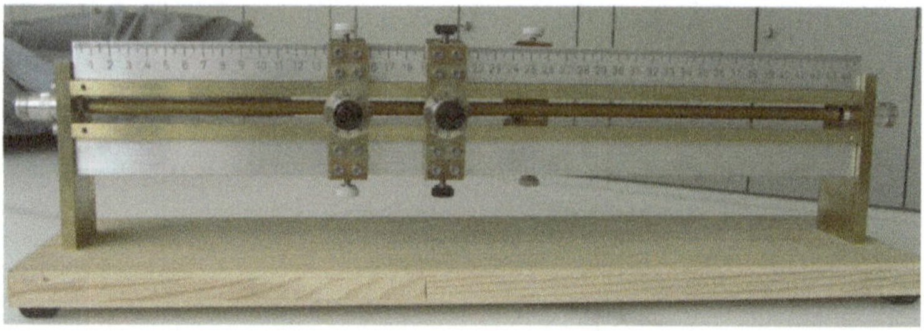

Fig. 4.14 Side view of a self-made tuning transmission line

Fig. 4.15 Optimized transition to the inner conductor with a diameter of 8 mm

This calculation has a deviation of less than 1 % for $\frac{D}{d} \geq 2$. For a more precise calculation, [112] also offers a more exact solution in addition to the calculations shown here.

The transmission line should have a characteristic impedance of 50 Ω and at the base frequency of the amplifier (e. g. 800 MHz) still have an electrical length of a good 400° so that at least two tunable capacitors can be used. All parts relevant to the electrical behavior are often made of brass, as it is a good conductive and easy to machine material.

Fig. 4.14 shows a side view of a self-made tuning lead with three variable setting capacitances. The inner conductor is clearly visible here.

The length of the cable was set at 450 mm, which corresponds to an electrical length of 432° at 800 MHz. The ground surfaces were made torsion-resistant using aluminum U-profiles. These profiles also serve as guides for the variable capacitances. A ruler is engraved on the U-profile to make the positions of the variable capacitances reproducible.

N-adapters have been used as connections. They are screwed into the head plates and connected with the inner conductor of the N connector (diameter 3 mm) to the 8 mm wide inner conductor of the tuner. In order to minimize the mismatch in this transition, the inner conductor must be adapted based on field simulation results (Fig. 4.15).

The variable capacitances are movable slides in which there is an adjusting screw with a diameter of e.g. 10 mm (Fig. 4.16). As the screw approaches the inner conductor, the capacitance to earth increases and so does the mismatch of the tuning

Fig. 4.16 The guide block and the adjusting screw

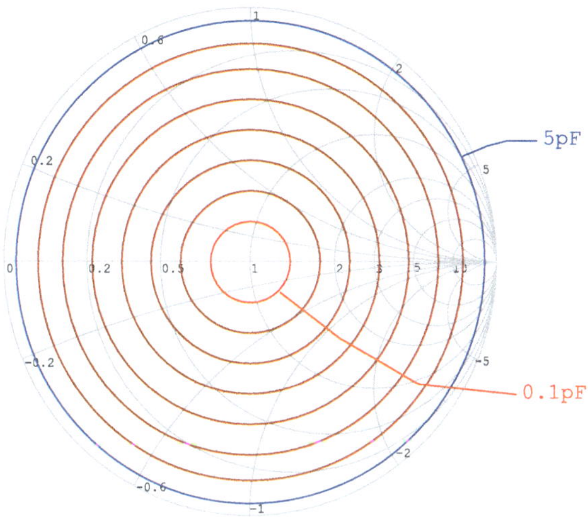

Fig. 4.17 Input impedances of the tuner for various capacitive terminations that are shifted along the tuner line

conductor. If the screw is removed from the cable, the capacitance decreases and the matching improves. At maximum distance, the screw no longer has any influence on the cable. To achieve a better ground connection, the slides are guided by brass blocks that fit precisely between the ground plates (Fig. 4.16). The slides can also be fixed in place with two locking screws.

A large number of input impedances can be realized with such a simple mechanical tuner. If you screw in the adjusting screw to increase the capacitance, you move further and further towards the edge of the Smith chart. The finite attenuation of the line ensures that the values at the edge cannot be covered.

Fig. 4.17 shows which impedance values can be set if the tuning screws for discrete setting values are moved over the entire line length.

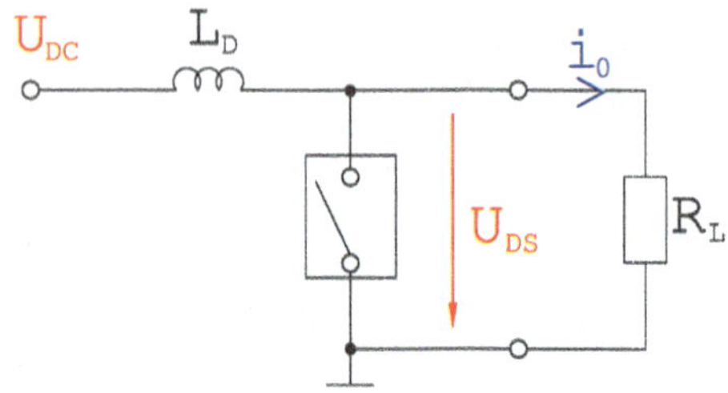

Fig. 4.18 Simplified equivalent circuit diagram for the current path of the DC voltage supply via the drain-source path, which is shown as a switch

These manual tuners are usually only used at the output. If these tuners are used behind an amplifier without an output network the input impedance of the optimum tuner setting corresponds to the input impedance of the network to be developed. Often this input impedance still needs to be de-embedded, [38].

If, on the other hand, the tuner is used behind an amplifier with a network for output matching, it is possible to recognize whether there are output impedances that allow an even greater power yield by quickly setting a very large number of impedances manually.

So far, only broadband tuners have been presented. There are also so-called harmonic tuners (also harmonic tuners). These can only be used at a fixed frequency and are optimized in such a way that a harmonic wave is reflected as strongly as possible and the fundamental wave is not reflected at all. In mechanical designs, the reflection factors of the harmonics cannot usually be adjusted. The reflection phase can only be influenced by the position of the harmonic tuners. However, this aid is very important in the practice of power amplifier development.

4.1.8 Selecting the Drain Coil

Power transistors in the high MHz and GHz range block in the fraction of a nanosecond. A high current flows through the drain coil while the transistor has a low resistance. Fig. 4.18 shows the ECD of the power supply via the drain coil up to the 50 Ω load resistor.

This circuit is similar to a DC-DC converter (boost converter). If the transistor is low resistance, a relatively large current builds up via this direct path to ground. If the transistor opens, the drain coil initially wants to maintain the current. However, as the load resistor is significantly higher impedance than the previously low impedance drain-source path, a significantly higher voltage builds up at the load resistor than is applied with U_{DC}. In practice, a doubling of the voltage is typically achieved with a well-tuned drain coil.

If the inductance value of the coil is too small, there is no isolation between the RF and DC circuits and no sufficient magnetic field for voltage boosting. If the inductance is too high, no noticeable current builds up in the short low-impedance switching period of the drain-source path and hardly any energy is coupled into the RF circuit.

The best way to design this inductor is to optimize the value in a non-linear simulation, which, if possible, includes other parasitic components occurring in the later design in addition to the selected transistor.

This book often uses the operating voltage $U_0 \geq U_{DC}$ is often used in this book to refer to this very important voltage boost in practice.

4.2 The Classic Operating Modes A, AB, B, and C

In this chapter, only the classic operating modes A, AB, B, and C will be discussed. The operating modes classify transistor circuits based on the position of the operating point.

Overview

In A mode, the operating point for maximum modulation is placed in the middle of the transistor's output characteristic so that that $U_{AP} = \frac{U_{DC}}{2}$. U_{AP} is the collector-emitter voltage at the operating point and U_{DC} is the supply voltage. In general, $U_{AP} \geq \frac{\hat{u}_{Amax}}{2}$.

The quantity $\hat{u}_{Amax}$ denotes the maximum required peak output voltage. A mode is the most linear operating mode but also the most inefficient with a theoretical efficiency of maximum 50 % at full scale. In practice, the efficiency is tends to be between 10 % and 35 %. This is why A mode is mainly used in small signal or preamplifier circuits.

Fig. 4.19 shows the output characteristic of a power transistor in A mode with the operating point setting (OP) at around half the operating voltage U_{Cq} as quiescent voltage at the collector and I_{Cq} as quiescent current at the collector. The destruction power is entered with P_{tot} and the breakdown voltage with U_{Db}.

The operating point of an amplifier in B mode is placed as precisely as possible in the inflection point of the input characteristic curve, so that only one half-wave of the input signal is amplified. This achieves a higher efficiency of max. 78.5 %. With suitable filtering, B mode can be used for the construction of linear power amplifiers. However, the generation of harmonics is relatively high. For this class B operation, the fundamental wave is raised linearly and with 6 dB less gain compared to A mode.

A compromise between the very high gain of the A mode and the high efficiency of the B mode is the AB mode. The operating point is placed between those of A and B mode, so that small signals are amplified in A mode and a half-wave is clipped for large signals. The generation of harmonics is not as pronounced as in B mode. Depending on the requirements of the power amplifier, filtering is also necessary

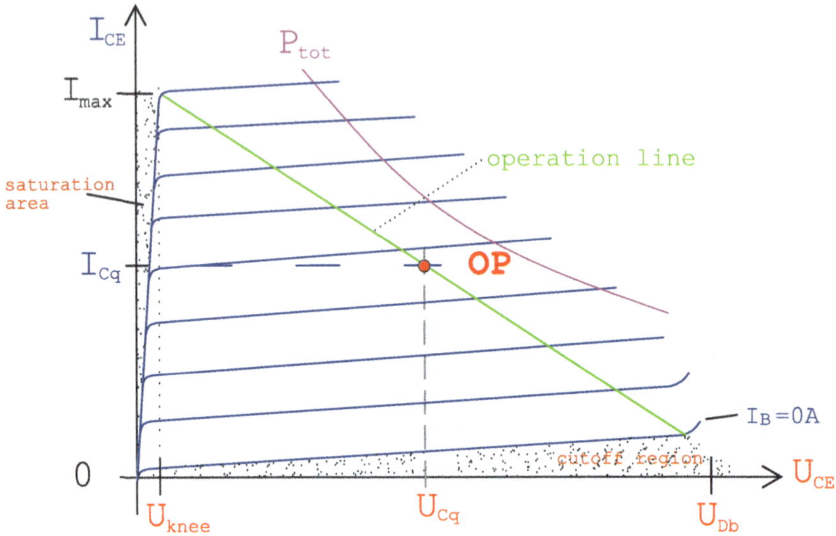

Fig. 4.19 Output characteristic for the transistor in A mode

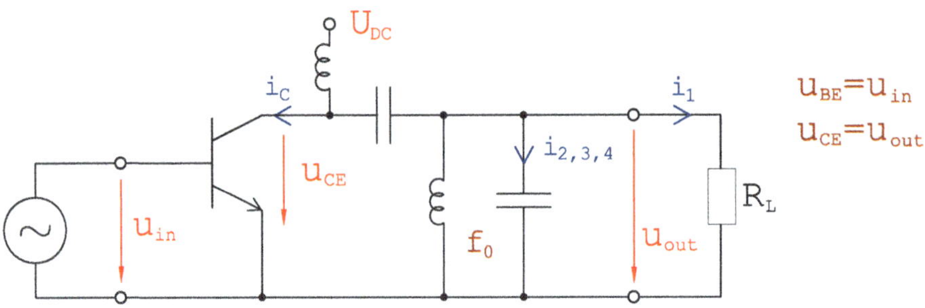

Fig. 4.20 Simplified circuit diagram of a narrowband power amplifier for A, AB, B, or C operation

here. In practice, efficiencies of up to 50 % can be achieved in a low AB mode. However, this operating mode is non-linear.

C mode is a non-linear mode in which the transistor receives no bias voltage at all or a negative bias voltage, so that only the peak of a half-wave is amplified. so that only the peak of a half-wave is amplified. Transistor circuits in C mode are used, among other things, as used as auxiliary amplifiers in Doherty power amplifier circuits to improve efficiency, [11].

Analytic Description

Fig. 4.20 is intended to provide a further description of an amplifier for the classic operating modes.

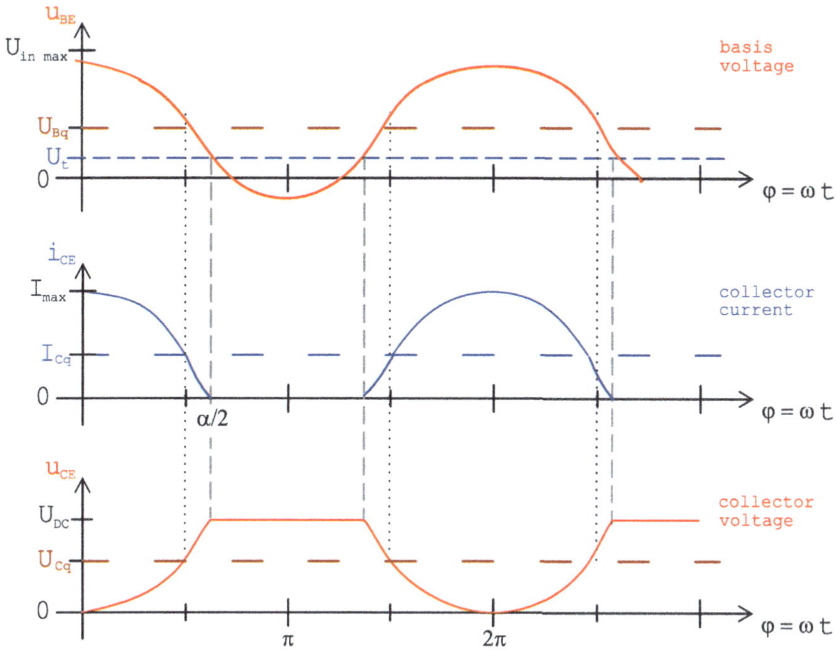

Fig. 4.21 Current and voltages at the transistor with large signal modulation

The applied alternating signals with a DC signal component are labeled with lower case letters. There is a parallel resonant circuit with high Q at the output. This should serve as a bandpass. This means that the fundamental signal with the center frequency f_0 remains unaffected and the harmonics are terminated with a short circuit. The index q indicates the quiescent voltages and currents at the base and collector.

The voltages and collector current applied to the transistor in this amplifier circuit are shown over time in Fig. 4.21.

The contained labels [2] are:

U_{Bq}: Quiescent voltage at the base,	U_t: Threshold voltage,
I_{max}: Saturation current,	I_{Cq}: Quiescent current at the collector,
α: Conductivity angle,	U_0: Operating voltage.

The threshold voltage U_t for a bipolar transistor is around 0.7 V. The saturation current and the maximum voltage or saturation voltage are specified in the data sheet. The operating voltage U_0 is slightly higher than U_{DC} for many amplifiers, as the choke coil provides a voltage boost. The book often uses $U_0 = U_{DC}$.

[2] All equal parts are marked with capital letters

Tab. 4.1 Display of the operating points

Class	$(U_{Bq} - U_t)/U_{inmax}$	I_{Cq}/I_{max}	α
A	0.5	0.5	2π
AB	0 - 0.5	0 - 0.5	$\pi - 2\pi$
B	0	0	π
C	<0	0	$<\pi$

The difference between the various operating modes lies in the choice of line angle α. For a choice of $\alpha = 2\pi$, $u_{BE} > U_t$ applies and therefore the sine signal is not clipped. The class A operation is present here. The other examples are described in Tab. 4.1.

In the following, the efficiency, gain and output power are calculated for these four operating modes.

Calculation of Power and Efficiency

As an unknown quantity for specifying the output power of the fundamental wave or the 1st harmonic must be calculated according to

$$P_1 = \frac{U_{DC}}{\sqrt{2}} \cdot \frac{I_1}{\sqrt{2}} \tag{4.33}$$

the corresponding current I_1 can be calculated. Here, U_{DC} corresponds to the peak value of the AC voltage, as the parallel resonant circuit (at f_0) in the Fig. 4.21 contributes to the generation of the negative half-wave.

The power consumption is calculated from

$$P_{DC} = U_{DC} \cdot I_{DC} \,. \tag{4.34}$$

If I_{DC} is known, this can be specified.

The efficiency η has already been introduced as:

$$\eta = \frac{P_1}{P_{DC}} \,. \tag{4.35}$$

The *PAE* (Power Added Efficiency) was also introduced:

$$PAE = \frac{P_1 - P_{in}}{P_{DC}} \,. \tag{4.36}$$

The formulas show that if the two currents I_1 and I_{DC} are calculated with the time curves as shown in Fig. 4.21, all statements regarding power and efficiency can be made.

Calculation of the Currents I_1 and I_{DC}

For a conduction angle of $-\alpha/2 \leq \varphi \leq \alpha/2$, the following applies to the collector current as a function of the phase φ:

$$i_{C(\varphi)} = I_{Cq} + (I_{max} - I_{Cq}) \cdot \cos(\varphi) \tag{4.37}$$

resp.

$$i_{C(\varphi)} = I_{max} \cdot \cos(\varphi) + I_{Cq}(1 - \cos(\varphi)) \,. \tag{4.38}$$

For a conduction angle of $-\pi \leq \varphi \leq -\alpha/2$ and $\alpha/2 \leq \varphi \leq \pi$ applies to the collector current: $i_{C(\varphi)} = 0A$.

Since there is a continuous transition, the current i_C must be zero at the phase $\varphi = \alpha/2$. Inserting this into the Eq. (4.38) results in

$$I_{Cq} = I_{max} \cdot \frac{\cos(\alpha/2)}{\cos(\alpha/2) - 1} \,. \tag{4.39}$$

The result for the quiescent current I_{Cq} can now be used in the Eq. (4.38):

$$i_{C(\varphi)} = I_{max} \cdot \left(\cos(\varphi) + \frac{\cos(\alpha/2)(1 - \cos(\varphi))}{\cos(\alpha/2) - 1} \right) . \tag{4.40}$$

After a short calculation you get:

$$i_{C(\varphi)} = I_{max} \cdot \frac{\cos(\alpha/2) - \cos(\varphi)}{\cos(\alpha/2) - 1} \,. \tag{4.41}$$

Based on this equation for the collector current in the time domain, all current components (DC component and harmonics) can be calculated using the Fourier transformation.

The DC component can be specified using the following calculation.

$$I_{DC} = \frac{1}{2\pi} \int_{-\alpha/2}^{\alpha/2} i_{C(\varphi)} \, d\varphi = \frac{I_{max}}{2\pi} \int_{-\alpha/2}^{\alpha/2} \frac{\cos(\alpha/2) - \cos(\varphi)}{\cos(\alpha/2) - 1} \, d\varphi \tag{4.42}$$

$$I_{DC} = \frac{I_{max}}{2\pi(\cos(\alpha/2) - 1)} \left[\varphi \cos(\alpha/2) - \sin(\varphi) \right] \Big|_{-\alpha/2}^{\alpha/2} \tag{4.43}$$

$$I_{DC} = \frac{I_{max}}{2\pi(\cos(\alpha/2) - 1)} \left[\frac{\alpha}{2} \cos\left(\frac{\alpha}{2}\right) + \frac{\alpha}{2} \cos\left(\frac{\alpha}{2}\right) - \sin\left(\frac{\alpha}{2}\right) + \sin\left(-\frac{\alpha}{2}\right) \right] \tag{4.44}$$

$$\underline{I_{DC} = I_{max} \frac{\alpha \cos(\alpha/2) - 2\sin(\alpha/2)}{2\pi(\cos(\alpha/2) - 1)}} \tag{4.45}$$

In general, the following applies to the calculation of the n. harmonics:

$$I_n = \frac{1}{\pi} \int_{-\alpha/2}^{\alpha/2} i_{C(\varphi)} \cdot \cos(n\varphi)\, d\varphi \,. \tag{4.46}$$

For the calculation of the 1st harmonic, Eq. (4.41):

$$I_1 = \frac{I_{max}}{\pi(\cos(\alpha/2) - 1)} \int_{-\alpha/2}^{\alpha/2} (\cos(\alpha/2)\cos(\varphi) - \cos^2(\varphi))\, d\varphi \,. \tag{4.47}$$

With the context $\cos^2(\varphi) = 1/2 + 1/2\cos(2\varphi)$ follows

$$I_1 = \frac{I_{max}}{\pi(\cos(\alpha/2) - 1)} \left[\cos(\alpha/2)\sin(\varphi) - \frac{\varphi}{2} - \frac{\sin(2\varphi)}{4}\right]\Bigg|_{-\alpha/2}^{\alpha/2} , \tag{4.48}$$

$$I_1 = \frac{I_{max}}{\pi(\cos(\alpha/2) - 1)} \left[\underbrace{2\cos(\alpha/2)\sin(\alpha/2)}_{=\sin(\alpha)} - 2\frac{\alpha}{4} - 2\frac{\sin(\alpha)}{4}\right] , \tag{4.49}$$

$$\underline{I_1 = I_{max} \frac{\sin(\alpha) - \alpha}{2\pi(\cos(\alpha/2) - 1)}} \,. \tag{4.50}$$

Analysis of the Calculation of I_1 and I_{DC}

The calculation of I_{DC} for class A and B operation provides the following direct current components:

$$I_{DC\,(Class\,A)} = I_{max}/2 \,, \qquad I_{DC\,(Class\,B)} = \frac{I_{max}}{\pi} \,.$$

The calculation of I_1 provides the fundamental wave components for class A and B operation:

$$I_{1\,(Class\,A)} = I_{max}/2 \,, \qquad I_{1\,(Class\,B)} = \frac{I_{max}}{2} \,.$$

The current values of the DC component and the first to fourth harmonics are given over the conduction angle α and thus also for all operating modes in Fig. 4.22.

The four operating modes can be clearly presented in tabular form if the following information or standardizations are introduced:

For the threshold or knee voltage,

the following should be idealized: $U_t = U_k = 0\,\mathrm{V}$,

Normalized voltage values: $U_{qnorm} = U_q/U_0$,

Normalized power values, input: $P_{innorm}^{dB} = P_{in}^{dBm} - P_{inmax}^{dBm}$,

Normalized power values, output: $P_{outnorm}^{dB} = P_1^{dBm} - P_{1\,Class\,A}^{dBm}$,

Normalized load resistance: $R_{Lnorm} = R_L/R_{opt}$,

Normalized power gain: $G_{norm}^{dB} = G_1^{dB} - G_{1\,Class\,A}^{dB}$.

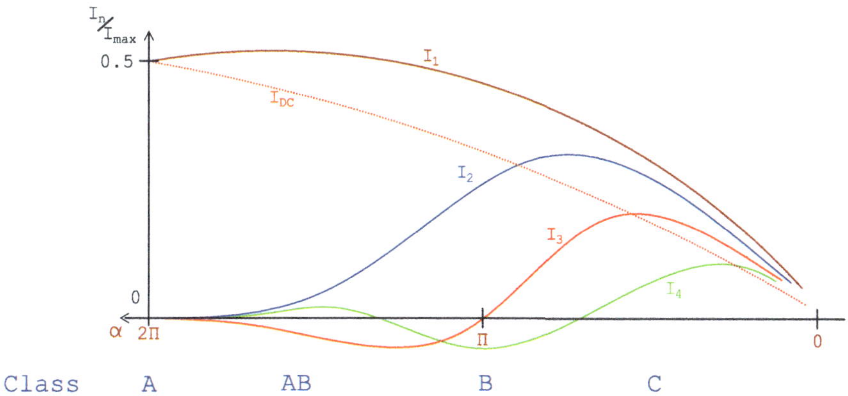

Fig. 4.22 Current values of the direct current component and the first four harmonics via the conduction angle α or the operating modes A to C

Tab. 4.2 Representation of the maximum output power, efficiency and linearity of the four operating modes A-C

P_{innorm}^{dB}		class A	class AB	class B	class C
	U_{qnorm}	0.5	0.25	0	−0.5
	R_{Lnorm}	1.0	0.94	1.0	1.14
0 dB	$P_{outnorm}^{dB}$	0 dB	0.25 dB	0 dB	−0.6 dB
−3 dB	$P_{outnorm}^{dB}$	−3 dB	−1.8 dB	−3 dB	−6.2 dB
0 dB	G_{norm}^{dB}	0 dB		−6 dB	
−3 dB	G_{norm}^{dB}	0 dB		−6 dB	
0 dB	η	50 %	70 %	78.5 %	80 %
−3 dB	η	25 %	53 %	55 %	45 %
	Type	linear	nonlinear	linear	nonlinear

It should be noted that operating mode B requires 6 dB more input power compared to operating mode A is required. Consequently, operating mode A has a 6 dB greater amplification.

Operating mode C has an even lower gain than operating mode B. Please note that the following applies:

Class B: Linear amplification but harmonic generation!

In principle, the class B amplifiers amplify the fundamental wave linearly. In practice, class B power amplifiers exhibit nonlinear behavior due to the crossover distortion that occurs when transitioning between the conduction regions of the two transistors in the push-pull configuration. This distortion results in signal distortion and generation of harmonics, impacting the amplifier's linearity and introducing spectral regrowth, which can degrade signal quality in communication systems.

4.3 Circuit Concepts of Power Amplifiers

In addition, various circuit concepts are presented that can be used to improve the efficiency of power amplifiers in particular. A very high efficiency RF power amplifier offers significant benefits in various applications, particularly in wireless communication systems. Firstly, it reduces power consumption, leading to longer battery life in portable devices and lower operating costs in infrastructure. Secondly, it minimizes heat dissipation, which enhances system reliability and simplifies thermal management. Thirdly, it enables smaller form factors and lighter weight designs, contributing to compact and lightweight devices, especially important in space-constrained applications like mobile phones and IoT (Internet of Things) devices. Lastly, high efficiency amplifiers reduce electromagnetic interference (EMI), improving spectral purity and reducing the likelihood of interference with nearby electronic devices.

4.3.1 Class H Mode

Tab. 4.2 shows that the efficiency for all operating modes is greatly degraded if the amplifier is not fully driven. Mobile radio systems in particular require very different output powers, the values of which depend on the distance.

The basic idea of H mode is that the operating voltage is reduced if the power amplifier is only to deliver a low output power. The circuit concept is shown in Fig. 4.23.

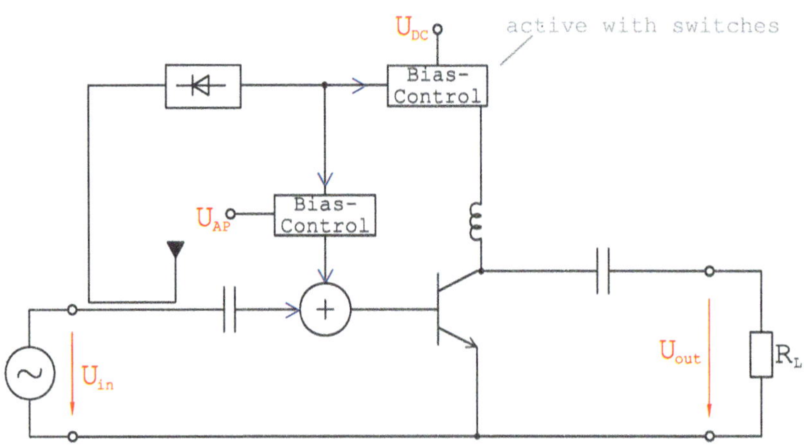

Fig. 4.23 Simplified circuit for an amplifier in H mode with a controlled controlled operating point setting of the A mode

The coupler is used to extract some of the energy from the signal to be amplified and feed it to a detector. Its output signal drives the control units.

If this circuit concept is applied to A mode, the operating point must be set at the base as shown in Fig. 4.23. This setting can be omitted for B mode.

One disadvantage of this concept is the technical complexity of the voltage regulation at the collector. Here, switching transistors are used that are the same size as the amplifier transistor, and on the other hand the contamination of the signals by the necessary switching regulators in the drain or collector branch.

The necessary circuit hardware has fallen immensely in price in recent years. The contamination of the reduced operating voltage can be reduced by switching regulator topologies with the so-called ΣΔ modulators, which are described in detail in the synthesizer chapter.

For amplifiers with RF transistors that are "normally-on" (that means, that the drain-source path for a gate voltage of 0 V is low impedance), class H operation does not involve any major additional effort, as an AF transistor[3] is required as a drain switch anyway.

Up to now, class H amplifiers are commonly used in audio applications where high efficiency and low power dissipation are required, such as in professional sound reinforcement systems and high-quality home audio setups. They are also utilized in mobile devices and automotive audio systems, where energy efficiency and compact design are essential considerations.

4.3.2 Class F Mode

The circuit concept that is most frequently used after the classic operating modes in the high-frequency range is F mode. In this operating mode, the harmonics are reflected in such a way that the efficiency ideally improves to 100 %, as either i_C or u_{CE} are zero. A possible circuit realization for this operating mode can be seen in Fig. 4.24.

The two reflections r_x and r_y with the following properties are generated by the short-circuited line for high-frequency signals at the end and the series resonant circuit:

$$r_x\text{: Open circuit for all harmonics;}$$
$$r_y\text{: Short circuit for even harmonics } (H_2,\ H_4,\ \ldots).$$

As the terminations for the harmonics are to be utilized, A mode cannot be used as the operating point setting.

The short circuit for even harmonics and the open circuit for odd harmonics results in an output voltage at the collector that resembles a square wave signal, Fig. 4.25.

[3] AF: Audio Frequency

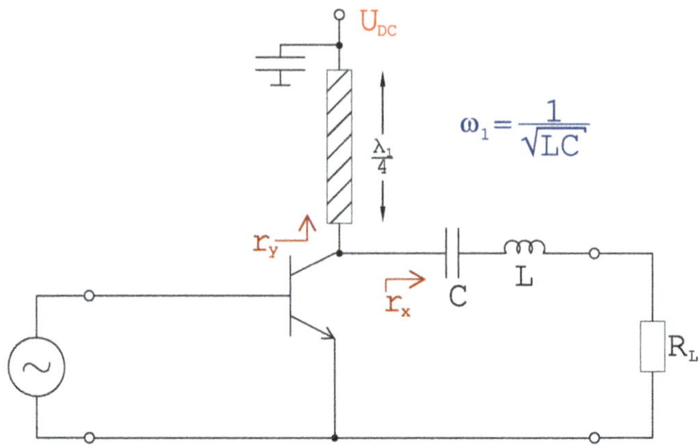

Fig. 4.24 Basic circuit for an amplifier in F mode using a $\lambda/4$ long line

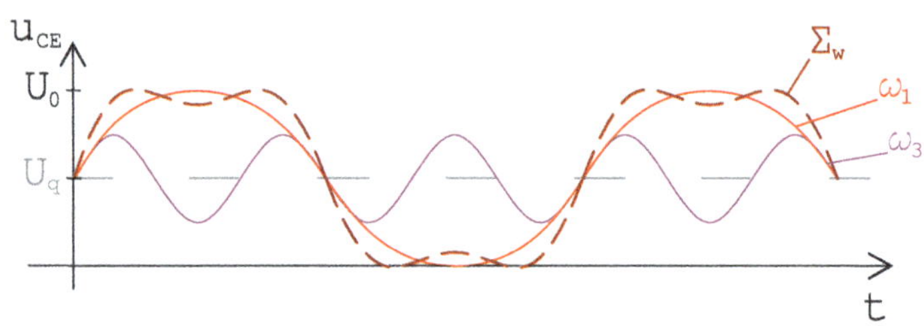

Fig. 4.25 Voltage curves of the fundamental and 3rd harmonics at the collector for the F operation mode

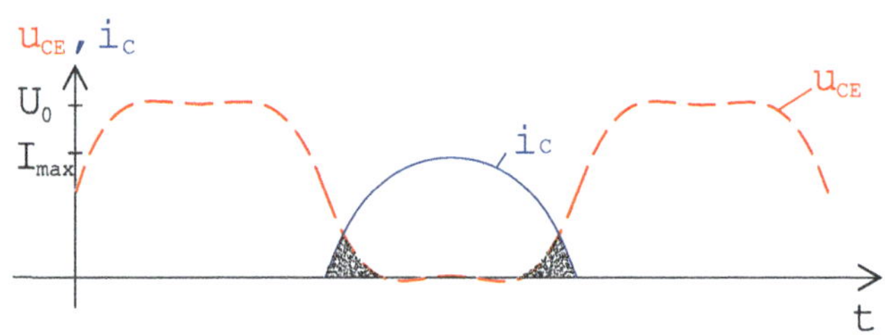

Fig. 4.26 Voltage and current curve at the collector for F operation with an operating point setting according to B mode

In B mode, only one half-wave of the current flows through the collector-emitter path. Only during this period can a power loss occur at the transistor. If the harmonic termination introduced ensures that the voltage applied to the transistor is zero for the period of the current flow, no power can be dissipated at the transistor during this period either. In practice, it is not possible to achieve a perfectly rectangular voltage curve. The area of the overlap region is then proportional to the losses, see Fig. 4.26.

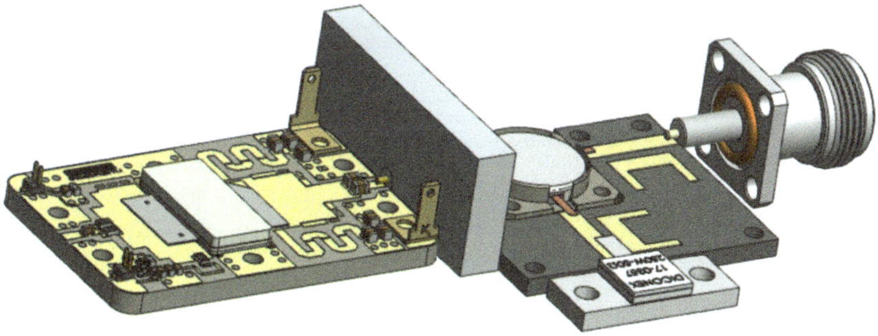

Fig. 4.27 Power amplifier output stage with $\lambda/4$ feed lines for the drain voltages, image from Ampleon

If the amplifier is in AB mode, you must also ensure that the current i_C is as rectangular as possible. This operating mode is also the most common in mobile GSM handsets. For standards with high linearity requirements such as UMTS, a weak AB mode with slight improvements in efficiency is selected and for standards with lower linearity requirements, a strong AB mode with higher efficiency is used. In such devices, the use of $\lambda/4$ long lines is avoided wherever possible.

Fig. 4.24 shows a possible circuit realization. The harmonic terminations required for F operation – especially for the first and second harmonics – can be achieved with a huge range of other circuits.

A possible hardware realization form is illustrated in the drawing of Fig. 4.27 including the protection circuit of an isolator (consisting of the circulator and a 50 Ω power termination).

Class F amplifiers are primarily utilized in RF applications, such as in radar systems, wireless communication systems, and RF power amplifiers for broadcasting. Due to their ability to achieve high efficiency and high output power levels, class F amplifiers are particularly well-suited for applications requiring strong RF signals with minimal power loss. They are also used in applications where harmonic suppression is critical, as Class F amplifiers can efficiently suppress higher-order harmonics, ensuring cleaner output signals. Additionally, class F amplifiers find application in laboratory instrumentation and scientific research where precise RF signal generation is necessary.

4.3.3 Class D Mode

In D mode, the transistor is used as a "quasi-ideal" switch. This approximation can only be used for applications up to the MHz range. The circuit concept of D mode is shown in Fig. 4.28.

The input signal not shown in this image opens and closes the switches. Consequently, this operating mode generates many harmonics. Only signal components

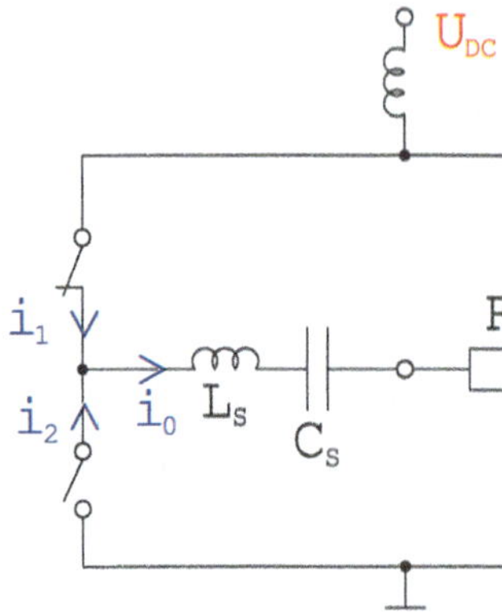

Fig. 4.28 Circuit concept with ideal switches for D mode in the form of a full bridge

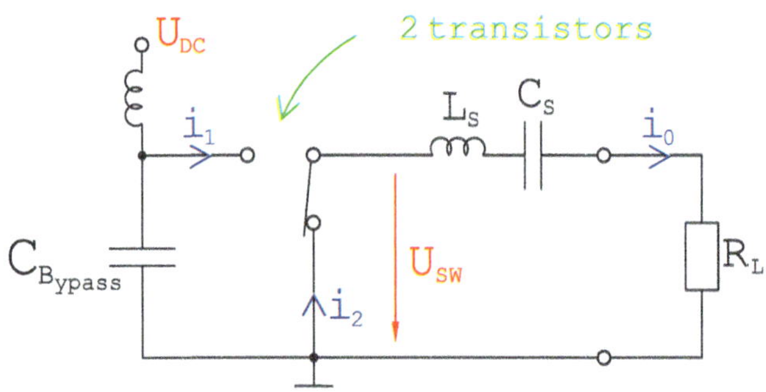

Fig. 4.29 Circuit concept with ideal switches for D operation in the form of the unbalanced version

around the resonant frequency should reach the load R_L through the series resonant circuit. A symmetrical output is available for the load R_L.

As the full bridge according to Fig. 4.28 requires four transistors, this circuit is too expensive for many applications. A cheaper alternative that is suitable for higher frequencies is a version with only two transistors and an unbalanced output, as shown in Fig. 4.29.

The operation of this amplifier can be explained using the diagram in Fig. 4.30.

The capacitor C_{Bypass} only has the function of a supporting capacitor, so that a current flow can start after the upper transistor is closed. This capacitor is therefore also much larger than C_S.

To amplify a pure sinusoidal signal, the switch is switched periodically. It should be open half the time and closed the other half. This is only possible if the input amplitude has a fixed value. This periodic input signal ensures that the rectangular voltage U_{SW} is applied to the changeover switch. During the time in which U_{SW} corresponds to the supply voltage U_{DC}, the current i_1 flows. The two reactive components of the series resonant circuit are designed for the center frequency. This ensures that the current returns to zero at the end of the period. After the switchover,

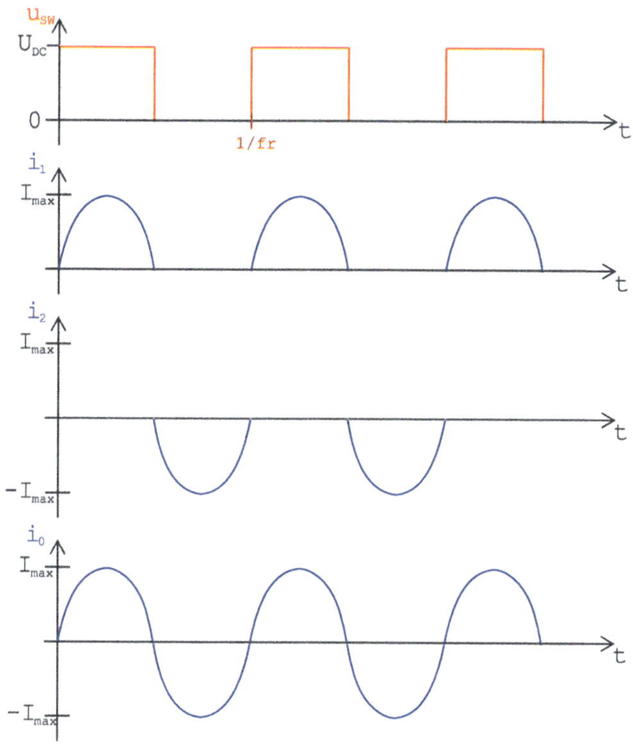

Fig. 4.30 Voltage and current curves for an amplifier in D mode

the voltage U_{SW} is zero and the current i_2 is driven by the energy stored in the series resonant circuit. The two currents i_1 and i_2 result in the sinusoidal total current i_0.

For lossless switches, an efficiency of 100 % is achieved with D mode. In practice, however, time synchronization is very difficult to achieve for signals >10 MHz. Furthermore, very high demands are placed on the component qualities of the series resonant circuit, which is increasingly difficult to fulfill at higher frequencies. For this reason, this switch operation is not found at higher frequencies.

Class D amplifiers are commonly used in audio frequency (AF) applications, such as in portable speakers, home theater systems, and car audio systems, due to their high efficiency and compact design. In RF applications, Class D amplifiers are utilized in RF power amplification for wireless communication systems and in RF transmitters for applications like cellular base stations and satellite communications, providing efficient amplification with minimal power loss and heat generation.

4.3.4 Class E Mode

E-mode is the only switch mode that is also used in high-frequency applications is used, [100].

In practice, the class E amplifier is used for many special implementations up to the GHz range. It is very useful that this operation includes the finite transistor characteristic of a very large collector-emitter capacitance (C_P). Another advantage is that there are no time synchronization problems.

Fig. 4.31 shows this special switching amplifier.

The transistor, which works as a switch, is also shown in this way. The (parasitic) capacitance C_P takes over the current i_R driven by the resonant circuit when the transistor has a high impedance. For precise design, there is only one solution for the component values, which can be calculated using a complex derivation. Numerical solution methods are also used here. The components and the associated voltages (which are no longer freely selectable here) are calculated according to the following dependencies.

$Q_L \geq 1.79$	Free parameter/Q_L: Güte von L_S
$U_{CE} = B \cdot U_{CEV} \times$ Security factor	
$U_{DC} = f(U_{CE})$	Not free choice!
$R = f(P, U_{DC}, Q_C)$	P: Output power /Q_C: Quality value of C_S
$L_S = \frac{Q_L \cdot R}{\omega}$	ω : Operation frequency
$C_S = f(R, f, Q_C)$	
$L_D \geq 30 \cdot \frac{1}{\omega^2} \cdot C_P$	Iterative process
$C_P = f(R, \omega, Q_C, L_D)$	Iterative process

The function of the class E amplifier will be explained in more detail with the help of Fig. 4.32.

The DC voltage U_{DC} and the DC current I_{C0} are fed in via the choke coil L_D. The series resonant circuit forces a harmonic sinusoidal oscillation with the current i_R (see a)). The voltage u_b shown in b) controls the switch. In the phase interval between Θ_1 and Θ_2, the transistor has a high impedance and blocks. If the transistor is low-resistance and conducts, the current i_C shown under c) flows through it. In

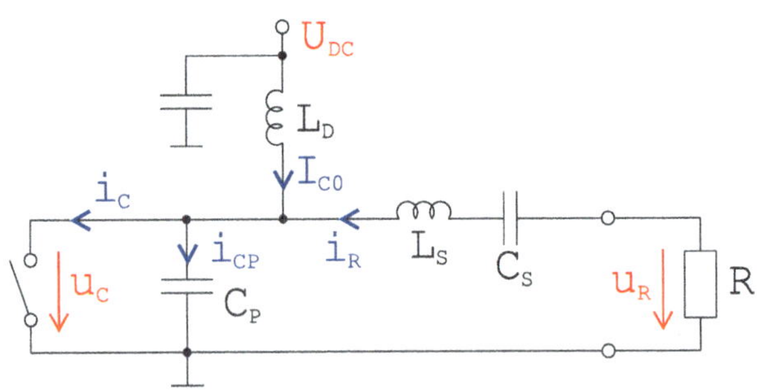

Fig. 4.31 Schematic of the class E amplifier and associated currents and voltages

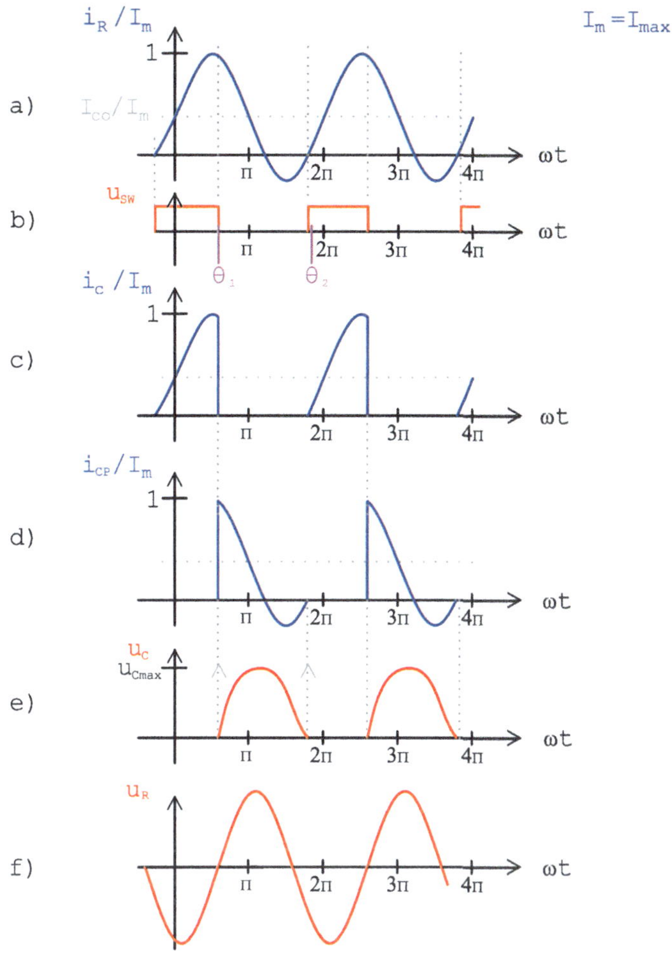

Fig. 4.32 Voltage and current curves for an amplifier in electric mode

the blocking case, the current i_{CP} flows via the capacitance C_P, as shown in d). During this time, the capacitor charges and discharges again. With correctly designed network, the transistor switches if the capacitor is discharged. The voltage at the capacitor or the transistor the voltage u_C according to e). The harmonic component H1 of u_C can pass through the resonant circuit. In practice, a phase shift is also taken into account for this, so that between the current i_R a) and the voltage u_R (~f)) there is no phase shift.

As in F mode, there is never a current and a voltage at the collector at the same time in E mode: Either i_C or u_C is zero!

The following applies to the theoretical efficiency: $\eta = 100\,\%$.

In practice, the best values at 5 GHz are 81 % for an output power of 0.6 W.

Class E amplifiers are commonly employed in radio frequency power applications, including RF transmitters, induction heating systems, and wireless power transfer systems. Their high efficiency and ability to operate at high frequencies make them suitable for a wide range of RF power generation tasks. Class E amplifiers are frequently utilized in applications where power efficiency and compact design are critical factors, such as in portable electronic devices and energy-efficient RF communication systems. Additionally, they find use in medical devices, industrial equipment, semiconductor production, and aerospace applications where reliable RF power amplification is required while minimizing power consumption and heat dissipation.

4.3.5 Class J and Other Amplifiers with Harmonic Matching

In addition to the classes F and E that have been introduced, there are many other classes that use harmonic degrees. For class F, among others, the inverse class F was introduced, in which the current and voltage curve is inverted for switching behavior runs.

Class J Mode

One of the latest developments is the class J amplifier. Class J amplifiers only use H1 and H2. Ideally, H3 does not occur at all. Nowadays, this is a realistic solution for LDMOS transistors due to the relatively low transit frequency and and the relatively large parasitic effects, this is also a realistic approximation. This is why it is not possible to use these transistors to build a class F design with these transistors.

A basic article by Steve Cripps and others on the class J amplifier [12] shows the best current and voltage signals at the transistor in the time domain, which can be achieved by optimally selecting the magnitude and phase ratio between H1 and H2 (Fig. 4.33).

It can be seen that the complete "switching off" of current and voltage is no longer achieved. However, these curves are very similar to those of class E and F.

The following termination impedances were calculated to obtain these time signals:

$$Z^{H1} = R_L + j \cdot R_L \qquad \text{and} \qquad Z^{H2} = -j\frac{3\pi}{8} \cdot R_L. \tag{4.51}$$

With R_L the previously introduced loadline impedance (see also R_{opt}) is used in Eq. (4.51). The operating point for the operating voltage is in low AB mode and thus almost in B mode, which minimizes the H3 generation, among other things.

At 1.8 GHz and 10 W power as well as the hardware from Fig. 4.34 efficiencies of over 80 % were achieved, [12].

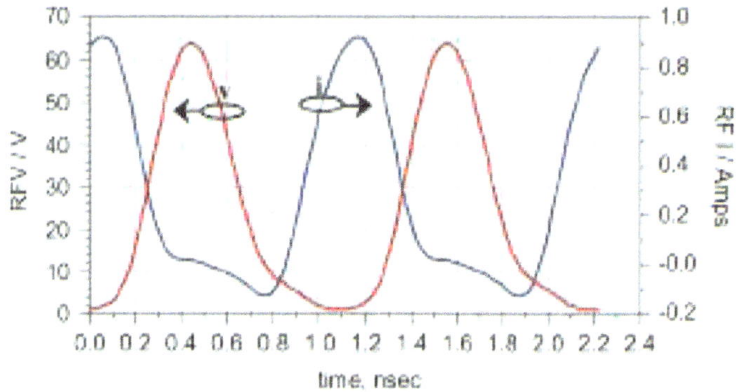

Fig. 4.33 Measured current and voltage curves on the transistor of a class J amplifier. [12]

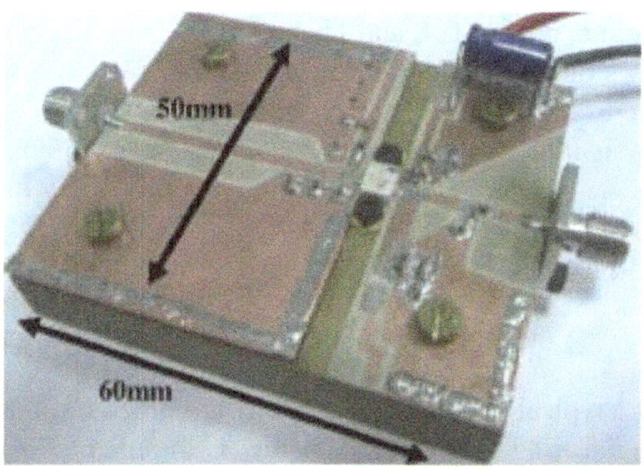

Fig. 4.34 Hardware design of a class J amplifier [12]

For broadband applications with relative bandwidths of over 50 %, efficiencies of over 60 % were even achieved.

Class J amplifiers are primarily used in high-frequency RF power applications where efficiency is crucial, such as in wireless communication systems, radar transmitters, and microwave links. Their unique design allows for high efficiency operation, making them suitable for applications where power consumption and heat dissipation need to be minimized while maintaining high output power levels.

Wave Form Engineering

In modern PA development, the optimization of amplifiers with harmonic matching is now referred to as "wave form engineering". The waveforms (especially at the drain) are considered both in the circuit simulation and in the measurement technology and the matching is changed so that the optimum waveform results. Here, the optimum is always found when either the voltage or the current at the drain is almost

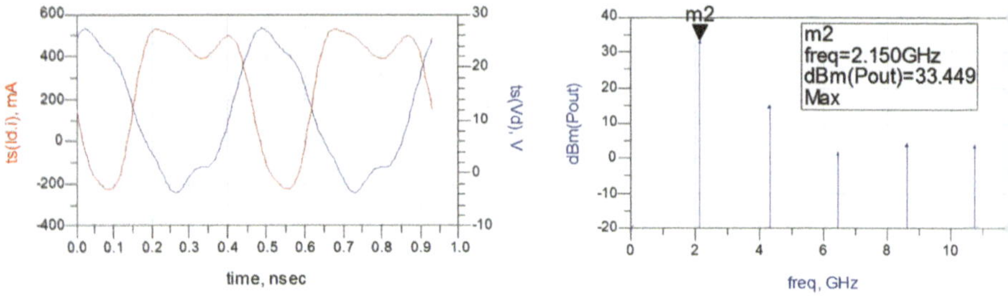

Fig. 4.35 Display of a drain current and voltage optimized in the simulator and the associated associated variables such as spectral power (efficiency is 72 %)

zero. This has already been explained for the class F amplifier using the Fig. 4.26. The operating form F, inverse F, E and J have this property in common.

Fig. 4.33 shows that the condition $I = 0$ A in particular was only satisfactorily achieved for the current.

The efficiency is almost 100 % when the transistor is wired without losses, if applies,

that – the signal at the load is a pure sine wave and
– the drain current or drain voltage is zero and
– the output network is lossless.

The results of a typical optimization of a 2 W amplifier are shown in Fig. 4.35.

Waveform engineering is essential in power amplifiers for shaping the output signal to meet specific requirements, such as reducing distortion or improving efficiency, especially in applications like wireless communication systems and radar. By carefully manipulating the waveform characteristics, power amplifiers can achieve higher linearity, lower distortion, and improved spectral efficiency, contributing to better overall system performance.

4.3.6 Class S Mode

The class S amplifier is based on pulse width modulation (PWM). Depending on, one PNP and one NPN transistor is controlled via an input signal, Fig. 4.36.

As a first approximation, the PNP and NPN transistors switch through at slightly above and slightly below 0 V. This results in the time signals according to Fig. 4.37. The PWM can be clearly seen in the time curve of U_C.

The modulation frequency must be $>> \omega_1$. There are currently no HF PNP transistors and therefore this technology is used in the lower frequency range.

Efficiencies of up to 100 % are possible at low frequencies.

Class S amplifiers are commonly used in AF applications, such as high-fidelity audio systems, due to their ability to provide high efficiency and low distortion audio amplification. In RF applications in the high MHz-range, class S amplifiers find use in power amplification for wireless communication systems, radar transmitters, and satellite communications, providing efficient amplification with minimal power loss and heat generation.

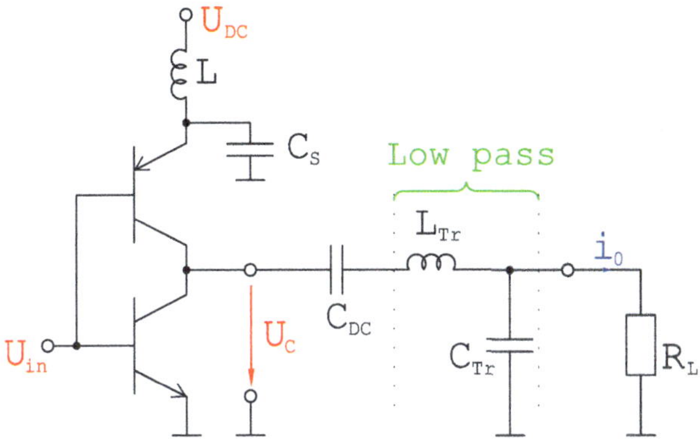

Fig. 4.36 Schematic of the class S amplifier

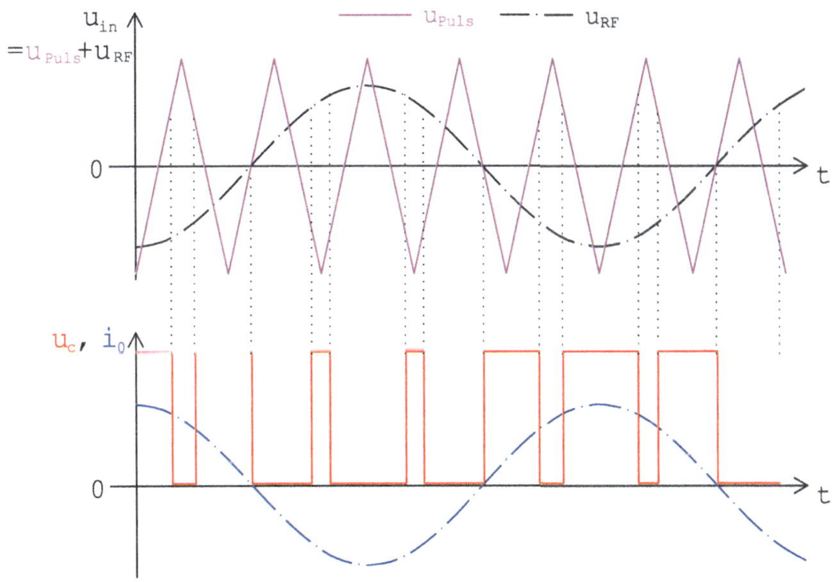

Fig. 4.37 Signal curve on the class S amplifier with PWM behavior

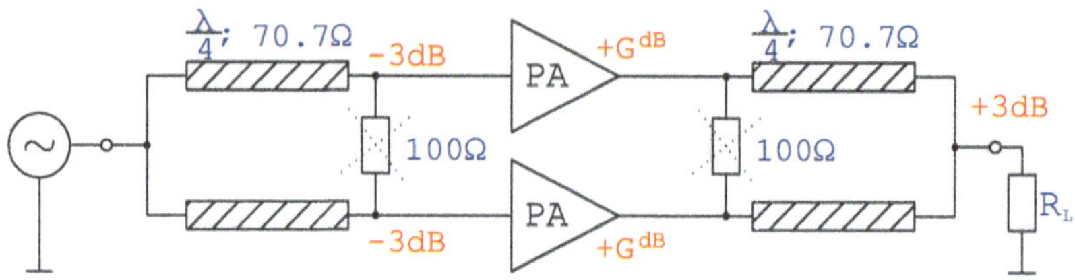

Fig. 4.38 Construction of an output stage using the 0°-coupler power combiner

4.4 Combining of Power Amplifiers

In many applications, several PA stages (usually short for power amplifier stages) are interconnected. Many different interconnection concepts using different power combiners have been established.

These different concepts serve the:

- Reduction of harmonics and/or
- Increasing power and/or
- Stability improvement and/or noise reduction and/or
- Signal splitting and/or heat loss distribution and/or
- Gain and cut-off frequency enhancement and/or
- Matching improvement and/or efficiency optimization.

The simplest type of power combination is presented first.

4.4.1 0°-Couple Power Combiner

In the 0°-coupler power combiner, a Wilkinson coupler is used before and after each of the two identical PA stages, Fig. 4.38.

The advantages of this simple, very compact and very low-loss circuit are

⇒ Generates twice the output power:
e.g. 33 dBm is achieved with two 30 dBm amplifiers.
⇒ Very broadband can be realized using
2-stage Wilkinson power divider.

The disadvantage of this concept is that both output stages are operated in common mode. Thereby both stages draw the maximum current at the same time.

The circuit diagram of the two-stage Wilkinson coupler for broadband applications is shown in Fig. 4.39.

Fig. 4.40 shows the adaptations of the single-stage and two-stage Wilkinson coupler in comparison.

Resistors are not so easy to use for power applications and are not loss-free. This is why leads, coils and capacitors are preferred.

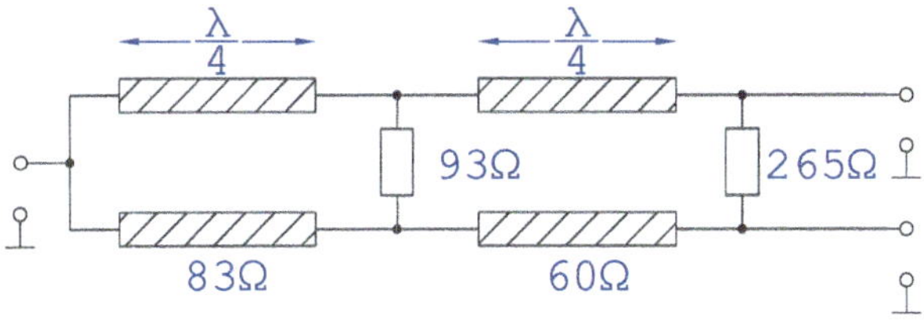

Fig. 4.39 0°-coupler for broadband applications

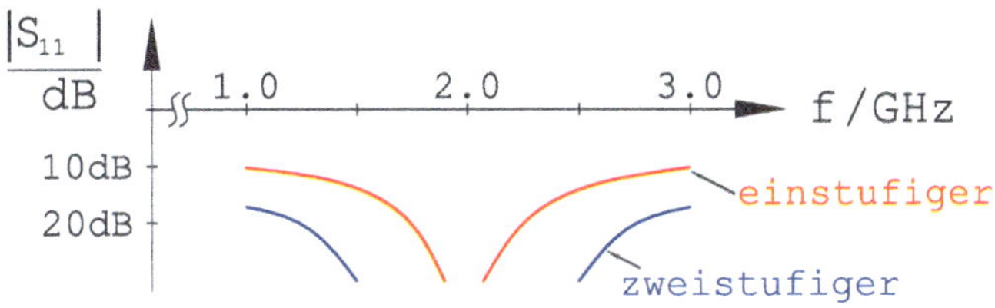

Fig. 4.40 Adaptation of the one- and two-stage 0°-coupler

Fig. 4.41 Design of an amplifier output stage circuit with 0°-couplers, figure from Macom

Practitioners also omit the cross resistance before and after the amplifiers. As a pure common mode wave is generated here and the PAs at the input and output also have the same reflection values, the resistors shown in Fig. 4.38 and Fig. 4.40 can be omitted.

The resistors are only required for unbalanced operation or push-pull operation.

An example of a hardware realization for an operating frequency around 900 MHz for an output power of 1100 W is presented in Fig. 4.41.

4.4.2 The Doherty Amplifier

This circuit concept, introduced by its namesake as early as 1936, uses a main amplifier in A or B mode for the amplification of small output signals and, if required, an auxiliary amplifier in (low) C mode for large input powers.

The circuit concept is illustrated in Fig. 4.42.

The 90°-lines ensure that T1 and T2 have current peaks at different times, which has a positive effect on the voltage supply and the non-linear behavior of the circuit. The 90°-line at the output transforms the lower output impedance of the transistor T2 to the higher value of the load resistor R_L. T1 is predominantly high-impedance.

An additional detector at the input can be used to reduce the main amplifier T2 at high input powers in order to prevent it from saturating.

However, there are also circuits in which the power detection is used to ensure that a controllable attenuator upstream of the auxiliary amplifier T1 causes less loss at higher input powers and thus brings this amplifier into use.

Typically, the auxiliary amplifier circuit is designed so that T1 provides 6 dB more output power than the main amplifier.

This Doherty circuit is often used in cellular base stations and often in combination with linearization circuits, which will be presented below. With this Doherty technology, it is possible to keep the power consumption of a base station for modern linear operating modes to 2-3 kW. Doherty amplifiers find also extensive use in modern wireless communication systems, such as broadcast transmitters, due to their high efficiency and linearity. Their unique architecture allows for efficient power amplification, particularly in scenarios with varying signal levels, contributing to improved spectral efficiency and reduced power consumption in RF power amplifiers.

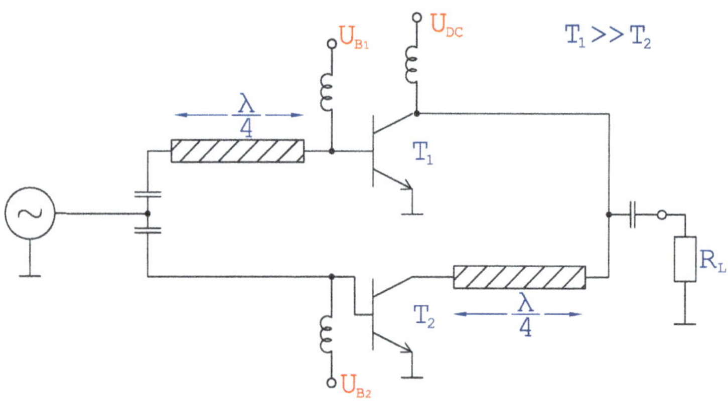

Fig. 4.42 Simplified circuit concept of the Doherty amplifier with the auxiliary amplifier T1 (large transistor) in C mode and the main amplifier T2 (with small transistor) in A or B mode

4.4.3 The Push-Pull Amplifier

In highly integrated circuit technology, the push-pull amplifier is used as the standard amplifier due to the widespread use of differential circuit technology (Fig. 4.43).

The advantages of this amplifier combination concept are:

⇒ Generates power doubling and
⇒ Reduces even harmonics and
⇒ Does not require a GND connection on the amplifier.

The reduction of the even-numbered harmonics only applies to a broadband balun. In practice, it is particularly important fulfilled for the first harmonic that the balun has the same transit times (double phase rotations) as for the fundamental. A simple example of this is the balun with $\lambda/2$ redirection.

Fig. 4.44 shows that the 2nd harmonic is then "phased out".

Due to the virtual GND point, the cut-off frequency of these amplifiers is much higher than with unbalanced setups, as the current feedback of the parasitic GND inductance is eliminated.

In addition, push-pull circuits offer much more stability, which is described in detail in the chapter on mixed-mode circuit technology is explained in detail in [38]. Push-pull amplifiers are widely utilized in audio applications, such as hi-fi stereo systems and guitar amplifiers, due to their ability to deliver high power output with low distortion. They are also employed in RF applications, including wireless communication systems and radio transmitters, where efficient amplification of signals is necessary while minimizing distortion and power consumption.

4.4.4 The Balanced Amplifier

The basic structure of a balanced amplifier is shown in Fig. 4.45.

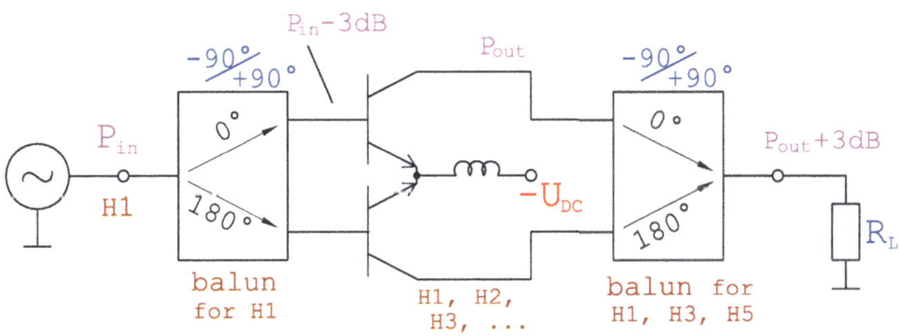

Fig. 4.43 Push-pull amplifier with suppression of the even harmonics

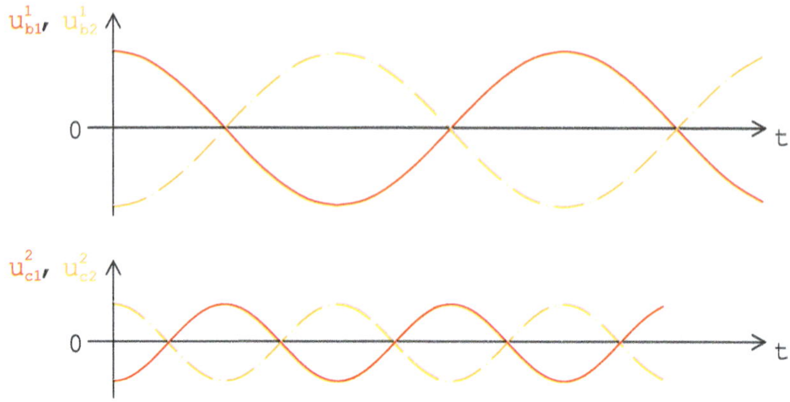

Fig. 4.44 H1 signals at the inputs of the transistors and the corresponding H2 signals at the outputs

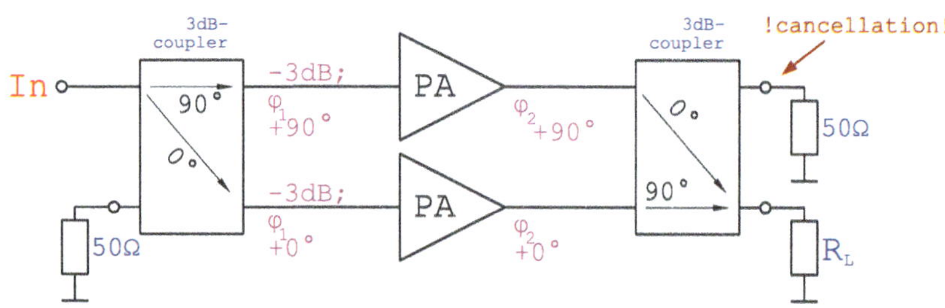

Fig. 4.45 Balanced amplifier with best output matching for signals at the input (In)

This balanced amplifier has the following advantages:

⇒ Generates power doubling
⇒ Eliminates even harmonics through the couplers
⇒ Realizes broadband input and output matching
⇒ Eliminates correlated noise interferers
⇒ Improves stability (k-factor)
⇒ Mismatched amplifier sees broadband conjugate complex matching

As this amplifier thus eliminates the need for complex circulators, it is particularly popular and is therefore frequently used. This broadband input and output matching can be explained using Fig. 4.45 as follows explained as follows: A wave reflected at R_L travels 50 % each on the lower PA (with 90° phase rotation) and the upper path (with 0° phase rotation). These two waves are partially reflected in the same way. However, these two reflected waves arrive in antiphase at R_L and in phase at the upper resistor with the indication "triggering". This means that no energy is reflected back at R_L no energy is reflected back.

Fig. 4.46 shows a setup with amplifier modules from the company Triquint, which was implemented at FH Aachen.

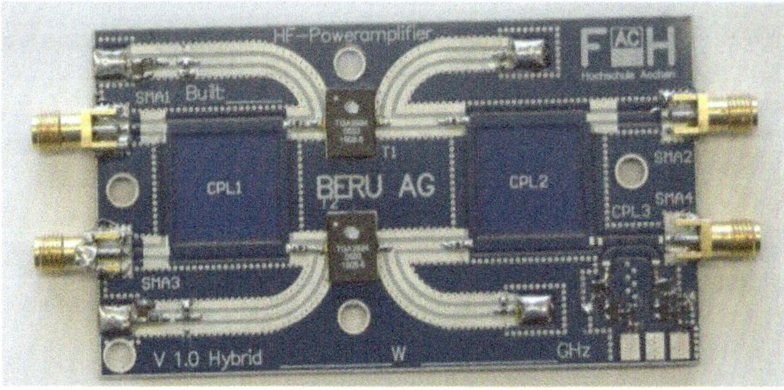

Fig. 4.46 Balanced amplifier at 2.45 GHz with 20 W output power

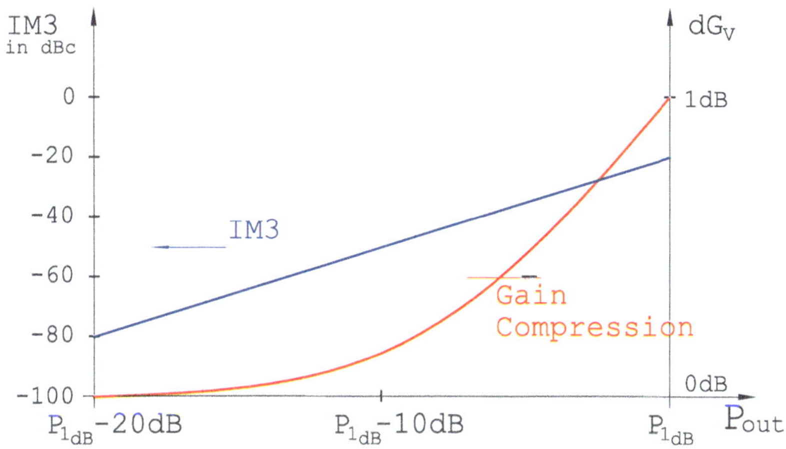

Fig. 4.47 Linearization Techniques

Balanced amplifiers are commonly used in radio frequency (RF) applications where high linearity and improved noise performance are critical, such as in wireless communication systems and radar systems. Their balanced architecture helps suppress even-order distortion and common-mode noise, making them ideal for applications requiring high-fidelity signal amplification and transmission.

4.5 Linearization Techniques for Power Amplifiers

The problem of a class A amplifier for linear use is illustrated in Fig. 4.47. The amplifier becomes nonlinear well below the 1 dB compression point!

The following variables are used in the Fig. 4.47 and in linearization technology:

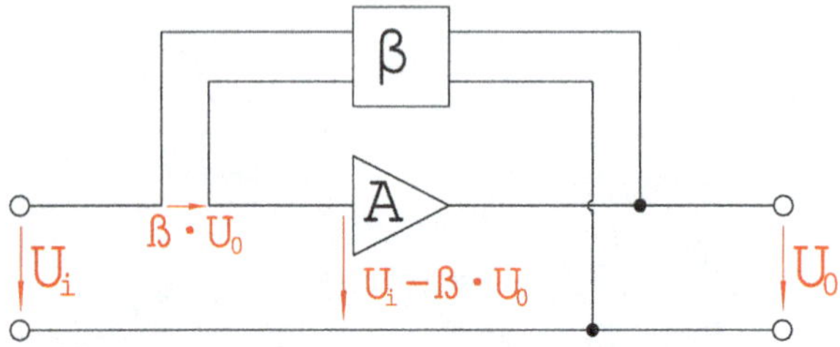

Fig. 4.48 Direct feedback

G_V: Small signal gain and
dG_V: Amplification dip.

$$dG_V = G_V - (P_{out}^{dBm} - P_{in}^{dBm})$$

The CDMA requirement is < -60 dBc for the ACPR values[4].

These requirements and the "early" compression require circuitry measures for linearization, which are explained below.

4.5.1 Direct Feedback

Direct feedback is known from electronics, as illustrated in Fig. 4.48.

From $U_0 = A \cdot (U_i - \beta \cdot U_0)$ & $G = \frac{U_0}{U_i}$ follows:

$$G = \frac{A}{1 + \beta A} \,. \tag{4.52}$$

This direct feedback is only used in the HF range for "small" single transistor stages! Especially in semiconductors, this feedback can be realized very well for transistors with lower power (and therefore smaller size). "Long" structures for feedback have a very large and constantly increasing phase rotation over the frequency, since $\beta = f(\omega)$.

The pure 180° rotation, as known from the electronics sector, is generally not possible.

[4] ACPR: Adjacent Channel Power Ratio – Adjacent channel signals generated by intermodulation processes

4.5.2 Predistortion

Predistortion is essential for power RF amplifiers as it helps mitigate nonlinearities that can lead to distortion in the amplified signal. By applying predistortion techniques, such as digital signal processing algorithms, the amplifier can compensate for nonlinearities in the system, resulting in improved linearity and reduced distortion in the output signal. This is particularly crucial in high-power RF applications, such as wireless communication systems and radar transmitters, where maintaining signal fidelity and spectral purity is paramount for reliable and efficient operation.

A simple solution for linearization via "Predistortion" with regard to the hardware effort is also shown in Fig. 4.49.

Both in magnitude and in phase, a simple nonlinear network (NL) is designed so that the $A_{(P_{in})}$ is compensated.

The associated transfer function for the magnitude typically has the characteristic shown in Fig. 4.51 (Fig. 4.50).

An example of a hardware solution for implementing the NL is shown in Fig. 4.51.

The passive LC network in the T-circuit is used for easy coupling of the diodes, whose behavior changes over the input power. This network should have more attenuation at low input power and less attenuation at higher input power. The different attenuation is achieved by changing reflection behavior. In addition, the transmission phase may only be changed slightly. The simplest low-pass or high-pass filters can be used for this purpose.

The following brief description of the advantages and disadvantages illustrates why this circuit technology is only found in mobile radio devices and not in base station technology.

Advantages:	– Simple, stable
	– Can also be realized digitally
Disadvantages:	– Just sufficiently precise
	– Does not react to mismatch at the output

In contrast, the more complex linearization techniques presented below are used in the power amplifiers of the base stations.

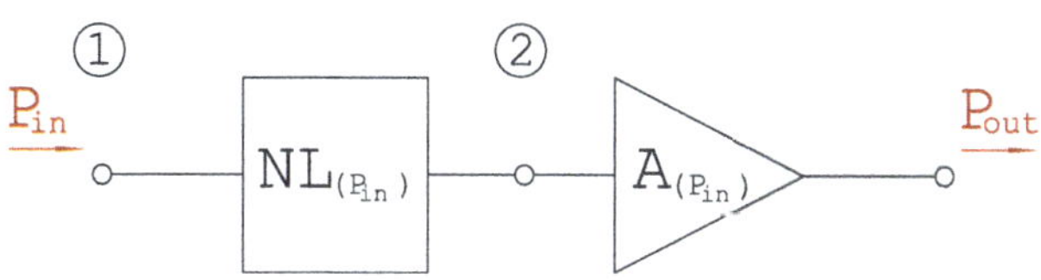

Fig. 4.49 Block diagram of a predistortion network

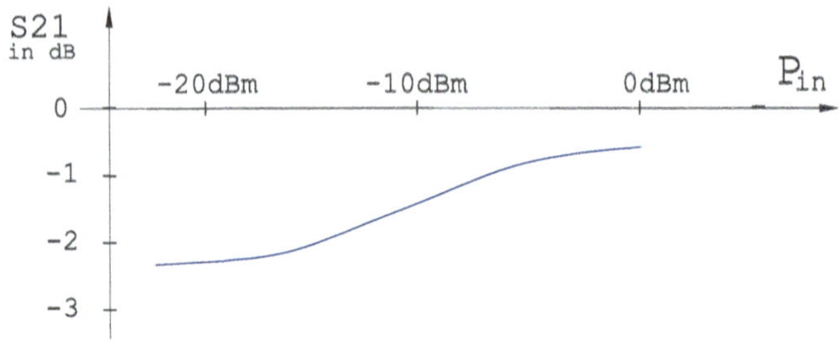

Fig. 4.50 Transmission behavior of a nonlinear network for predistortion such as the following of the T-circuit with nonlinear network (NL) of two diodes

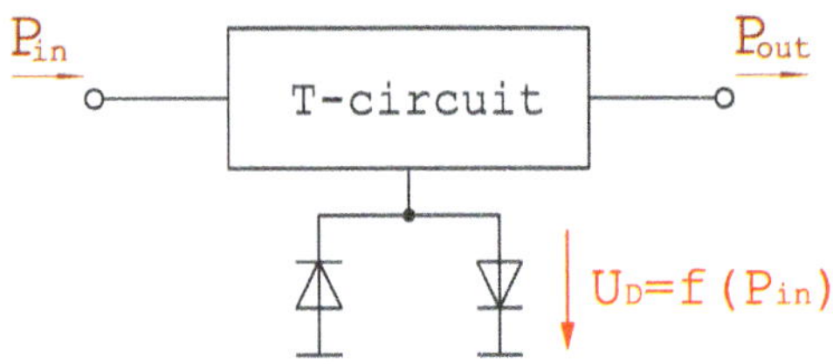

Fig. 4.51 T-circuit (contains reactive components) with nonlinear network of two diodes

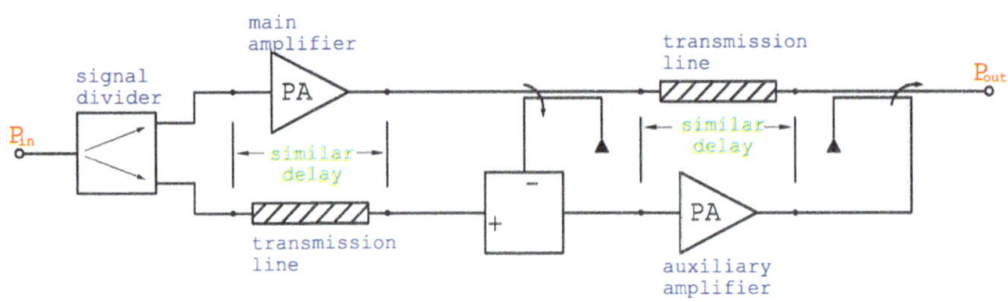

Fig. 4.52 Schematic diagram of an amplifier with feedforward correction

4.5.3 Feedforward

In a "Feedforward" system, errors (e.g. in the output signal) are detected and corrected by a bypass via an additional feed of correction information.

The feedforward circuit is designed in such a way that the auxiliary amplifier shown in Fig. 4.52 makes almost no contribution for small input signals and only adds the missing power for large input signals (where the main amplifier goes into compression).

A feedforward structure has the following properties:

- Technically complex, but stable.
- Corrects amplitude & phase response.

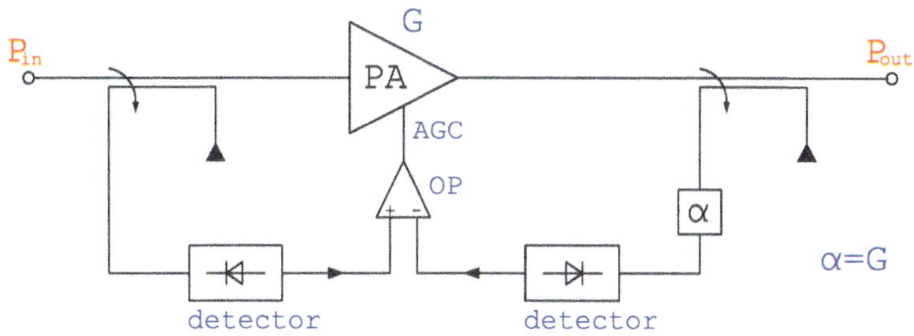

Fig. 4.53 Block diagram of a indirect feedback circuit

- Auxiliary amplifier and coupling loss are 6 dB lower than the output power of the main amplifier.
- A lot of losses in the last coupler and the associated resistive termination.

As the last coupler is typically designed as a 6 dB coupler, the auxiliary amplifier must be similar in size to the main amplifier.

4.5.4 Indirect Feedback

The "Indirect Feedback" circuit controls the gain via a scalar AGC circuit[5] and keeps it constant. The following critical points must be observed with indirect feedback:

- Only correction of the amplitude response!
- Stability problems!

The good efficiency of the easy-to-combine control for H operation is very advantageous. In simplified terms only the output coupler is added to H mode. In modern practice, the functionality of the OP shown in Fig. 4.53 is performed by a microprocessor, which allows many more fine optimizations.

Both, feedforward as well as indirect feedback RF circuits are commonly used in various RF applications, including communication systems, radar systems, and electronic warfare systems. The frequency of their usage depends on the specific requirements of the application, such as desired gain, linearity, bandwidth, and power efficiency. In many RF systems, feedforward as well as indirect feedback circuits are employed to achieve stability, improve linearity, enhance gain, and minimize distortion, making them a fundamental component in RF circuit design.

[5] AGC: Automatic Gain Control.

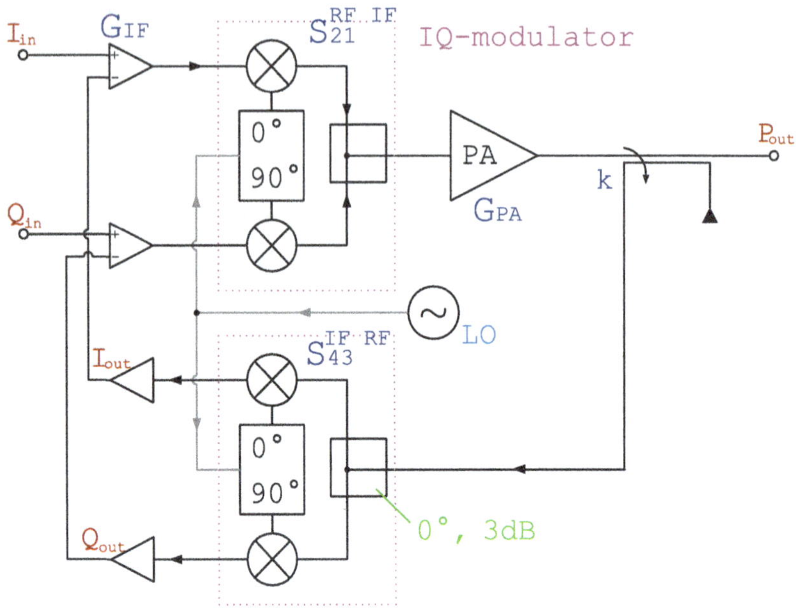

Fig. 4.54 Cartesian loop for linearization of the PA

4.5.5 Cartesian Loop

To the understanding of the Cartesian loop, it is helpful to understand IQ modulators (these are presented in more detail in Chap. 6) and/or homodyne systems is helpful ([38], chapter 10).

In very simplified terms, the IQ modulator shown in Fig. 4.54 converts the vector IF signal (given as I and Q signals) into an RF signal.

As with direct feedback, the gain of the power amplifier results from

$G_{ZF} = S_{21}^{HF\,ZF} \cdot G_{PA} \cdot k \cdot S_{43}^{ZF\,HF}$

with the fixed values $G_{ZF}, S_{21}^{HF\,ZF}, S_{43}^{ZF\,HF}, k$ to

$$G_{PA} = \frac{G_{ZF}}{k \cdot S_{21}^{HF\,ZF} \cdot S_{43}^{ZF\,HF}}\,. \tag{4.53}$$

The value of G_{ZF} is very stable as it is a feedback operational amplifier circuit. The value for G_{PA} is selected slightly (e.g. 1 dB) below the value of G_{PAmax}.

The Cartesian loop has the following properties:

- Broadband but stability problems.
- Magnitude & angle - compensation.
- Interesting for integrated solutions.
- A must for IQ synthesizer sources for measurement technology.

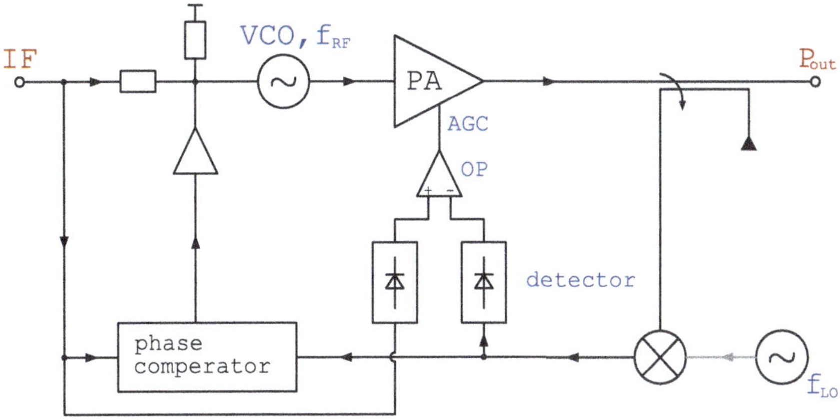

Fig. 4.55 Block diagram of the polar loop

4.5.6 Polar Loop

A heterodyne system has two RF oscillators. The polar loop is controlled by magnitude and phase (Fig. 4.55). With regard to the properties, the same discussion applies as for the cartesian loop. In addition, very good efficiencies can be achieved with the polar loop, as the power amplifier runs in H mode.

Possible phase comparator circuits that can be used are presented in connection with the synthesizers.

The IQ modulator architecture has become established in handsets due to the elimination of an HF oscillator.

4.6 Frequency Multiplier

Above 10 GHz, frequency multipliers are often used to generate RF signals.

For example, a push-pull amplifier is used as a harmonic source for H3 generation. H1 and H2 are completely mismatched at the output so that as much energy as possible is converted from H1 (at the input) to H3, Fig. 4.56.

In this way, a signal source for the IMS band at 24 GHz can be developed with inexpensive amplifiers that operate at 8 GHz and can be optimally stabilized there via a PLL. A class F amplifier naturally provides a very good basis here.

There are already research projects with 65 nm CMOS semiconductor circuits −4 dBm at 288 GHz and with SiGe technology −3 dBm at 325 GHz and −29 dBm at 825 GHz.

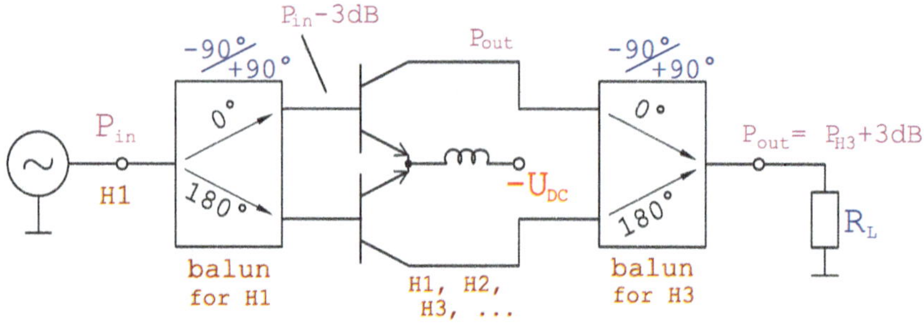

Fig. 4.56 Principle circuit diagram of a tripler

Chapter 5
Oscillators

If active high-frequency circuits oscillate, then at least one mono-frequency output signal is produced. This can often already be observed when designing an RF amplifier. Unfortunately, such unwanted oscillators do not oscillate at a desired frequency and the other properties of the oscillation are also inadequate.

This rather short, easily accessible and concise chapter on oscillators for the GHz range is the first to present the basics derived purely from scattering parameters. The dual-mode technique for improving the phase noise is also innovative here. The basics of noise influences conclude this chapter.

A good HF oscillator consists of an amplifying unit and a resonator with the highest possible Q factor. If the frequency of this resonator can be changed via at least one varactor diode, it is referred to as a voltage-controlled oscillator (VCO). A corresponding ideal signal is shown in Fig. 5.1 above the background noise of the measuring device in the frequency range.

An oscillator exhibits fluctuations in the output amplitude over time (AM noise) on the one hand and (in practice even more serious) fluctuations in the phase around the target frequency on the other. This phase noise is usually represented for the upper sideband with the difference between the noise value and the maximum oscillator signal (which is not shown), as can be seen in Fig. 5.2.

Of the large number of parameters to be considered when using an oscillator, this phase noise often has the greatest influence on the system characteristics. For example, if you want to mix down a signal at a distance of 1 MHz from the LO

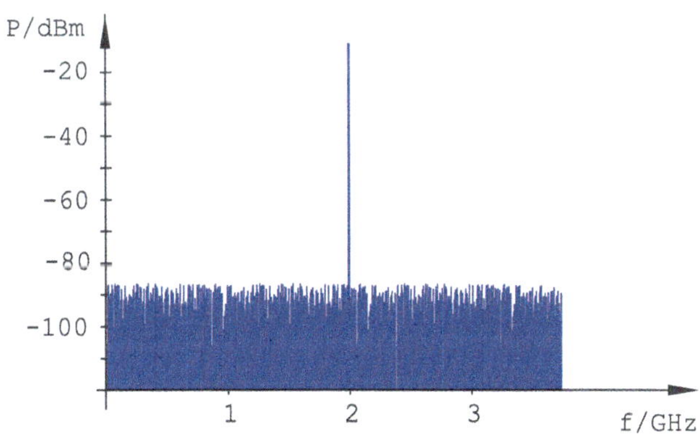

Fig. 5.1 Pure sinusoidal signal of a stabilized oscillator, measured with a spectrum analyzer

H. Heuermann, *Microwave Technology*,
https://doi.org/10.1007/978-3-658-45686-3_5

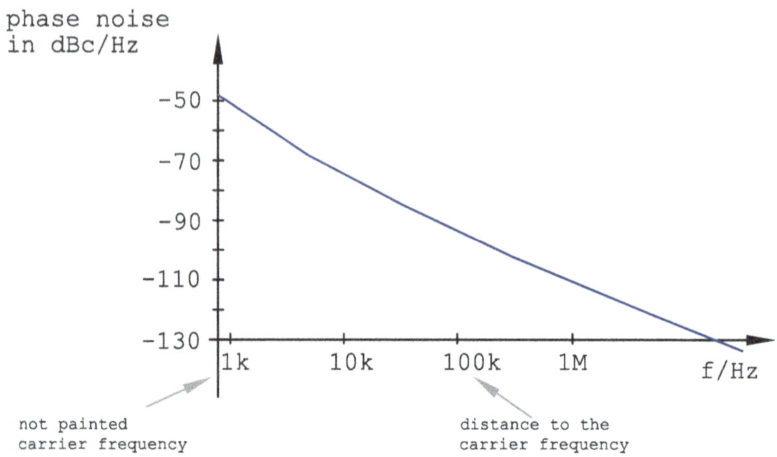

Fig. 5.2 Typical phase noise of an unstabilized oscillator with LC resonator

signal and consider the mixer as a multiplier, the mixer also mixes down the phase noise signal of the LO oscillator. Consequently, the phase noise should be below the expected signal level at the receiver.

In addition to the phase noise, there are a number of points to consider when assessing and using an oscillator:

- Output power and its frequency response for the VCO,
- Power consumption,
- Generation of the harmonics with a typical distance of 20 dBc from the fundamental and, if applicable, the subharmonics (half the frequency of the fundamental),
- *Frequency Pushing:* A variable that characterizes how much the frequency changes with fluctuations in the supply voltage,
- *Frequency Pulling:* Specifies how much the frequency changes when a mismatch occurs at the output,
- Temperature response or drift of the output power and the frequency,
- Tuning characteristic in MHz/V of a VCO and its linearity and maximum changeability (speed).

In the following, oscillation conditions for various oscillators of interest in RF electronics and their circuit technology are presented. In addition, many other oscillators for generating RF signals are known, particularly with other RF technologies (terahertz technology, waveguides, tubes, etc.), which cannot be discussed further here.

5.1 Two-Port Oszillators

Two-port oscillators are also referred to as oscillators with backward amplifiers or four-port oscillators. They have a transmission amplifier and a feedback network.

5.1.1 The Oscillation Condition

An ideal amplifier for voltages with a feedback network is shown in Fig. 5.3.

The following should apply to the transfer function of the amplifier:

$$G(\mathrm{j}\omega) = \frac{u_{out}}{u_s} \ . \tag{5.1}$$

$G(\mathrm{j}\omega)$: Transfer function of the ideal voltage voltage amplifier with the boundary conditions: $R_{in} \to \infty\,\Omega$, $R_{out} \to 0\,\Omega$ and no internal feedback.

The following applies to the feedback network:

$$H(\mathrm{j}\omega) = \frac{u_r}{u_{out}} \ . \tag{5.2}$$

$H(\mathrm{j}\omega)$: Voltage transfer function of an arbitrary feedback network The ideal voltage adder shown in Fig. 5.3 should apply:

$$u_s = u_{in} + u_r \ . \tag{5.3}$$

Eliminating u_s and u_r in Eq. (5.3) by means of (5.1) and (5.2) results in the closed loop voltage gain

$$V_u(\mathrm{j}\omega) = \frac{u_{aus}}{u_{in}} = \frac{G(\mathrm{j}\omega)}{1 - G(\mathrm{j}\omega)H(\mathrm{j}\omega)} \ . \tag{5.4}$$

The loop voltage gain of the open loop is introduced:

$$V_o(\mathrm{j}\omega) = G(\mathrm{j}\omega)H(\mathrm{j}\omega) \ . \tag{5.5}$$

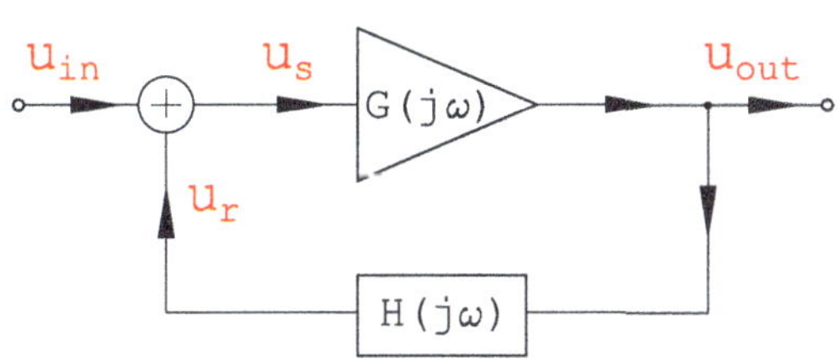

Fig. 5.3 Feedback amplifier

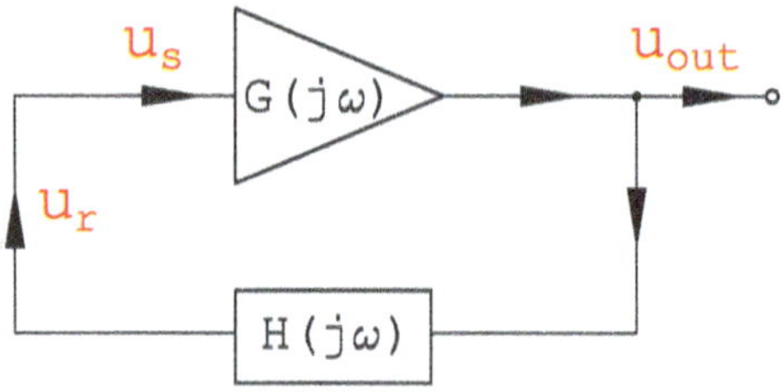

Fig. 5.4 Oscillator with backward amplifier

This feedback voltage amplifier should exhibit oscillation without an input voltage being present:

$$u_{in} = 0V \,. \tag{5.6}$$

However, only a finite output voltage u_{out} can be specified for Eq. (5.4) if the following applies:

$$1 - G(\mathrm{j}\omega) \cdot H(\mathrm{j}\omega) = 0 \,. \tag{5.7}$$

In this case, the ideal adder can be removed and the circuit illustrates an oscillator (Fig. 5.4).

From the condition

$$1 = G(\mathrm{j}\omega_0) \cdot H(\mathrm{j}\omega_0) \tag{5.8}$$

for the oscillation frequency ω_0, the known "Nyquist" criteria result

for the magnitude

$$|G(\mathrm{j}\omega_0) \cdot H(\mathrm{j}\omega_0)| = 1 \tag{5.9}$$

and for the phase

$$\angle(G(\mathrm{j}\omega_0) \cdot H(\mathrm{j}\omega_0)) = 0^\circ \,. \tag{5.10}$$

Brakhausen established similar criteria. These two conditions in (5.9) and (5.10) represent the limit value above which a feedback amplifier oscillates. In practice, for safe oscillation, the conditions

$$|G(\mathrm{j}\omega_0) \cdot H(\mathrm{j}\omega_0)| > 1 \tag{5.11}$$

so that with the real phase condition

$$\angle(G(\mathrm{j}\omega_0) \cdot H(\mathrm{j}\omega_0)) \approx 0^\circ \tag{5.12}$$

the oscillator oscillates reliably.

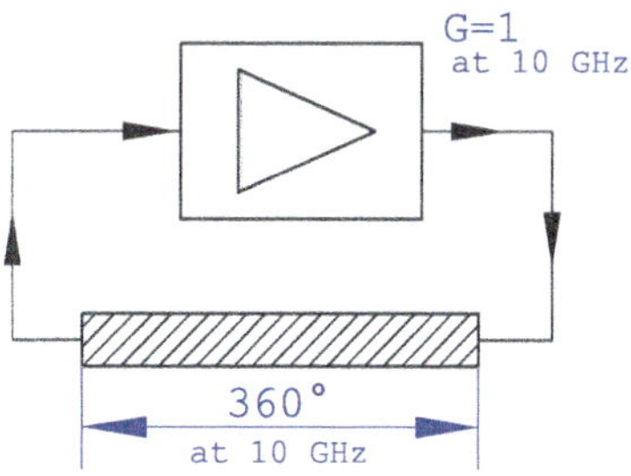

Fig. 5.5 Oscillator example: Amplifier with 360°-transmission line in the feedback loop

In practice, the frequency is set at which the two conditions (5.11) and (5.12) are fulfilled as well as possible (Fig. 5.5):

$$1 < \mathrm{Re}\{G(\mathrm{j}\omega_0) \cdot H(\mathrm{j}\omega_0)\} = V^{\mathrm{Re}}_{o(\mathrm{j}\omega_0)} \tag{5.13}$$

Example

Given: The following applies: Gain = 1 at the cut-off frequency of 10 GHz; phase=360° at 10 GHz; no attenuation above 1 GHz in the feedback (below isolation). The gain and the feedback network have the following frequency dependencies:

$$G(\mathrm{j}\omega) = 100\,\frac{0.1\,GHz}{f}, \tag{5.14}$$

$$\angle H(\mathrm{j}\omega) = 360^\circ \frac{f}{10\,GHz} \mathrel{\hat{=}} 2\pi\,\frac{f}{10\,GHz} \quad \text{resp.} \tag{5.15}$$

$$H(\mathrm{j}\omega) = e^{-\mathrm{j}2\pi\frac{f}{10\,GHz}}. \tag{5.16}$$

Calculation: With these two conditions, the Nyquist criterion is met exactly for 10 GHz. However, the amplifier has a higher gain for lower frequencies.

$$V^{\mathrm{Re}}_{o(\mathrm{j}\omega)} = \mathrm{Re}\left\{\frac{10\,GHz}{f}\, e^{-\mathrm{j}2\pi\frac{f}{10\,GHz}}\right\} \tag{5.17}$$

$$V^{\mathrm{Re}}_{o(\mathrm{j}\omega)} = \frac{20\pi\,GHz}{\omega}\,\cos\left(\frac{\omega}{10\,GHz}\right) \tag{5.18}$$

Determining the maximum value by derivation

$$\frac{dV^{\mathrm{Re}}_{o(\mathrm{j}\omega)}}{d\omega} = 0 \stackrel{!}{=} -\frac{20\pi\,GHz}{\omega^2}\cos\left(\frac{\omega}{10\,GHz}\right) - \frac{20\pi\,GHz}{\omega}\,\frac{\sin\left(\frac{\omega}{10\,GHz}\right)}{10\,GHz} \tag{5.19}$$

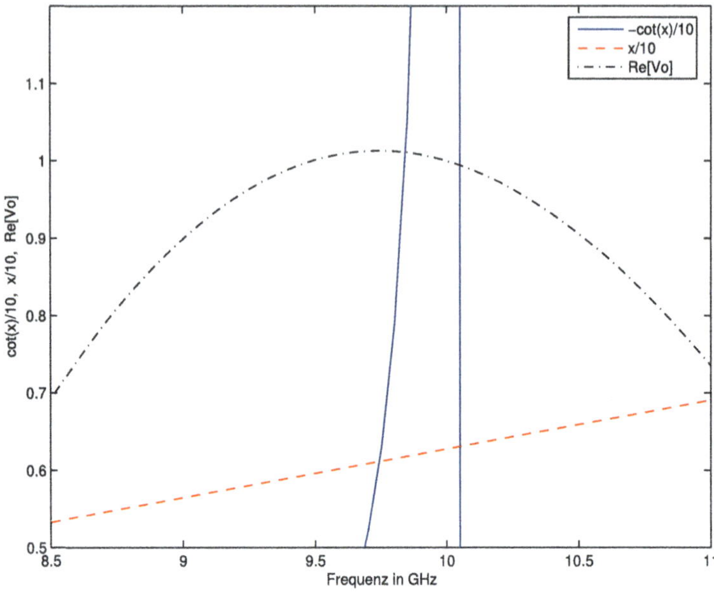

Fig. 5.6 Graphical method for determining the oscillation frequency from Eq. (5.22)

$$0 \stackrel{!}{=} \frac{1}{\omega}\cos\left(\frac{\omega}{10\,GHz}\right) + \frac{1}{10\,GHz}\sin\left(\frac{\omega}{10\,GHz}\right) \tag{5.20}$$

$$0 = \frac{10\,GHz}{\omega} + \tan\left(\frac{\omega}{10\,GHz}\right) \tag{5.21}$$

results in with $x = \frac{\omega}{10\,GHz}$*:*

$$x = -\cot(x). \tag{5.22}$$

The solution is given by the intersection of the functions $x/10$ *and* $\cot(x)/10$[1] *in Fig. 5.6 . Consequently, the resonant frequency is not 10 GHz, but around 9.7 GHz, as the higher gain compensates for the imperfect phase condition. Furthermore, in Fig. 5.6 the result of* Eq. (5.18) *is shown. As expected, this maximum is around 9.7 GHz.*

In order to achieve an optimally low phase noise of the oscillator, there is even an amplification of around 6 dB.

[1] The two expressions have only been multiplied by $\frac{1}{10}$ for better visualization

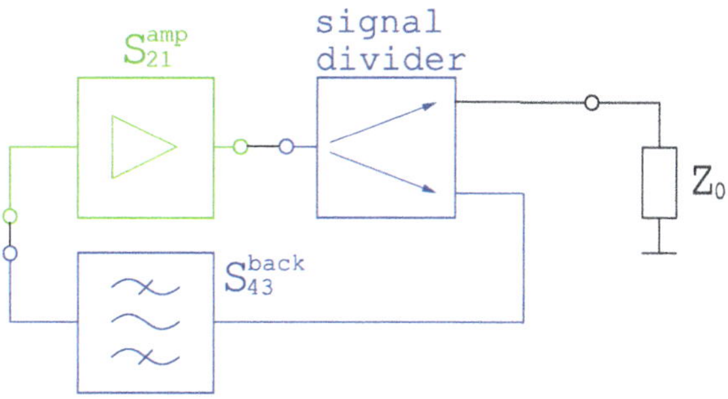

Fig. 5.7 Two-port oscillator with transmission resonator

5.2 Construction of a Backward RF Oscillator

The two transfer functions $G_{(j\omega)}$ and $H_{(j\omega)}$ can be applied in general and thus also replaced by the scattering parameters S_{21}^{Amp} and S_{43}^{Back}. The optimization condition (5.13) does not change:

$$1 < \mathrm{Re}\left\{S_{21}^{Amp} S_{43}^{Back}\right\}. \tag{5.23}$$

It does not matter which phase rotation the amplifier makes. The gain of the open loop must be maximum and the phase rotation must be as close as possible to $n \cdot 360°$.

The generated signal should be decoupled from an oscillator. The general associated block diagram is shown in Fig. 5.7.

This oscillator is also referred to as a second oscillator.

Details to optimize the transmission resonator are explained in [38]. The sparse balun from ([38], Chap. 5) can be used as the simplest signal divider. In the following analysis, the losses and phase rotation of the lower sub-branch are assigned to the feedback network and thus to S_{43}^{Back}.

The magnitude of the transfer function for large output powers of an amplifier can be calculated using the empirically determined correlation

$$\left|S_{21}^{Amp}\right| = \sqrt{\frac{P_{sat}}{P_{in}}\left[1 - e^{-\frac{G_o P_{in}}{P_{sat}}}\right]} \tag{5.24}$$

This means that this function can also be modeled well. Included is:

P_{in} : Power at the input of the amplifier,
P_{sat} : Saturation power of the amplifier,
G_o : Small signal gain.

The amplification of an amplifier with the values $P_{sat} = 100\,\mathrm{mW}$ and $G_o = 100$ is shown above the input power in Fig. 5.8.

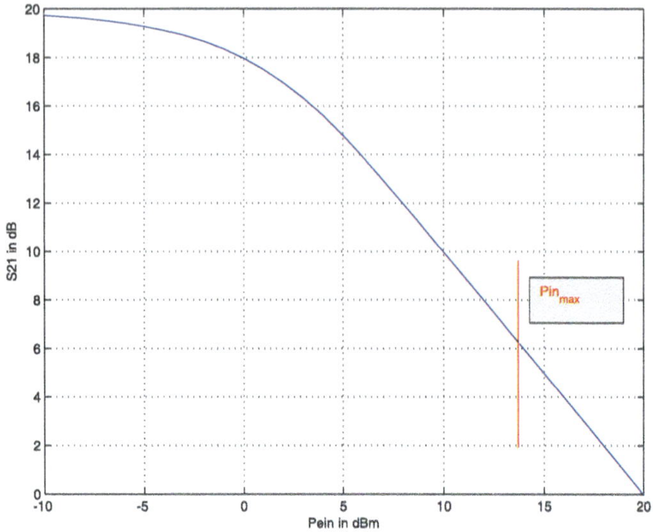

Fig. 5.8 Magnitude of the gain of a small signal amplifier in compression

With the gain decreases as the input power increases. The signal divider and the transmission filter exhibit noticeable losses, which must be well below the small signal gain.

If a two-port oscillator is activated, the output power increases up to the value at which the amplifier can only compensate for the losses of the feedback network.

$$\left|S_{21}^{\text{Amp}}\right| = \left|S_{43}^{\text{Back}}\right| \tag{5.25}$$

The output power is therefore calculated from

$$P_{out}^{dBm} = S_{43}^{\text{Back}^{dB}} + P_{in}^{dBm} . \tag{5.26}$$

Since the usual attenuations for the feedback path are 4 dB to 10 dB, it can be seen from Fig. 5.8 that the output power is in the saturation range in practice:

$$P_{out}^{dBm} = P_{sat}^{dBm} . \tag{5.27}$$

5.2.1 Cross-Coupled and Colpitts Oscillator

A in particular important oscillator for understanding (intentional and unintentional) oscillation processes is the cross-coupled oscillator shown in Fig. 5.9.

In the oscillator illustrated in Fig. 5.9, the necessary circuits for the operating point settings at the gate, the decoupling network and the matching circuits of the transistors are not shown for a clearer representation. In the other diagrams, coupling

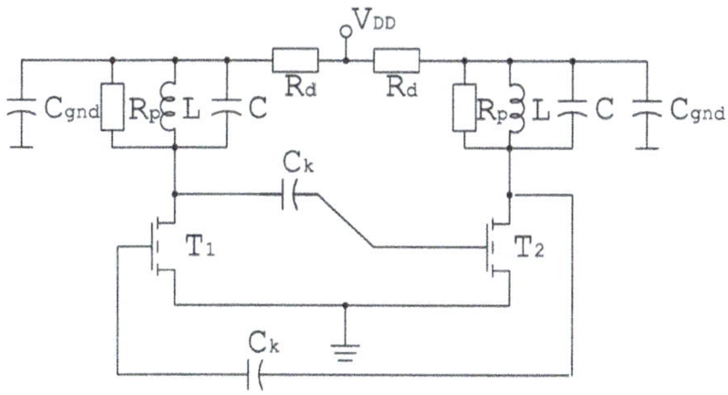

Fig. 5.9 Simplified circuit diagram of a cross-coupled oscillator

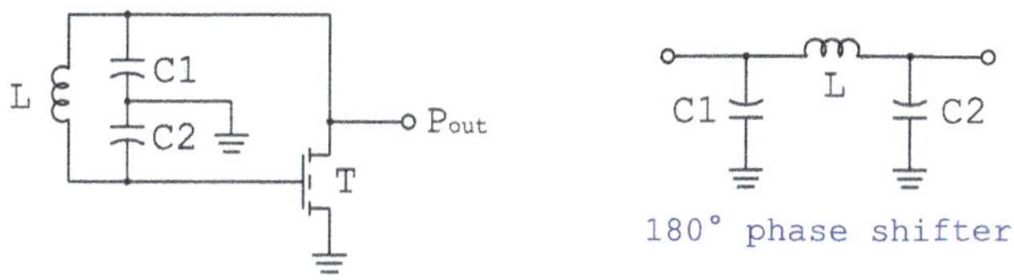

Fig. 5.10 Colpitts oscillator in source or emitter basic circuit and individually displayed phase shifter

capacitors, loss resistors (here R_p) and capacitors that realize the RF ground for a given DC supply are also not shown (as is standard in the international literature).

The two CMOS transistor amplifiers shown in the source circuit should each have a transmission phase rotation of 180°. This oscillator could therefore oscillate at any frequency with a feedback of 1. However, the supply voltage V_{DD} for the RF signal must be regarded as ground. Therefore, these two oscillating circuits behave like bandpass filters.

This circuit, which is very simple in terms of the phase condition, has the practical disadvantage that two transistors and two resonant circuits (which may need to be tuned simultaneously) are required. The advantage is that signals that are differential can be coupled out at T_1 and T_2 respectively.

In semiconductors, the drain resistors are omitted and are replaced by a transistor operated in saturation, which serves as a current source.

If you only want to use a transistor and an LC resonant circuit, you can use a feedback network in a source or emitter amplifier according to the procedure described above, which on the one hand makes 180° phase rotation and on the other hand counter-couples the amplifier via a parallel resonant circuit outside the resonant frequency.

The circuit shown in Fig. 5.10 can only oscillate at the frequency at which there is the 180° phase rotation. Only at this frequency is there optimal positive feedback.

The phase shifter contained in Fig. 5.10 has a low-pass character. Consequently, the feedback network has no bandpass behavior! The included low-pass filter reduces the oscillation tendencies at lower frequencies.

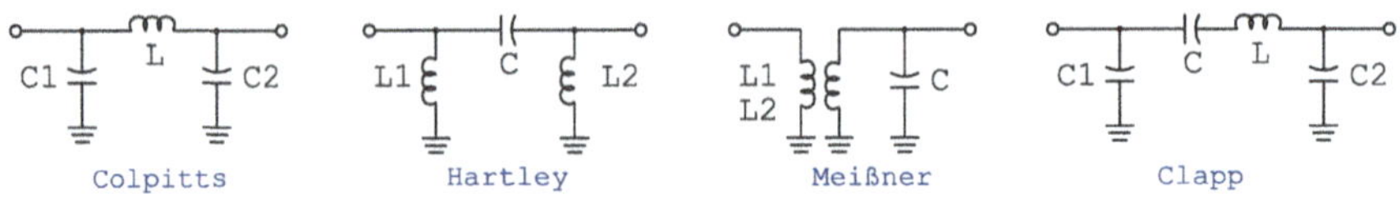

Fig. 5.11 Feedback networks and their names for the associated oscillators

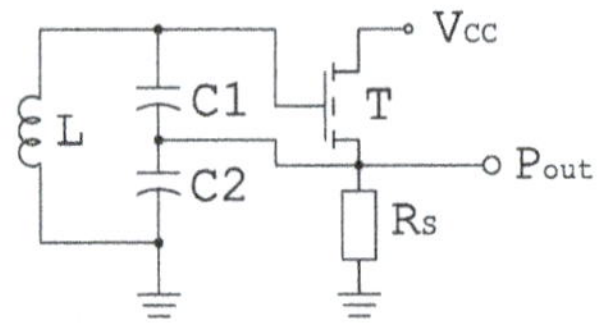

Fig. 5.12 Colpitts oscillator in basic drain circuit with resonator at the gate

The advantage of the Colpitts oscillator is that the feedback network requires only one coil and has no DC short circuit (and therefore requires no additional coupling capacitors). The majority of the capacitors used have a higher Q factor than the coil. This is also the reason why this design is often used in semiconductors.

In the same way as the Colpitts oscillator, a phase shifter can be constructed from the two shunt elements *L*1 and *L*2 and the series element *C*. This phase shifter has a high-pass character. The associated circuit is called a `Hartley oscillator`. If you replace the coil of the parallel resonant circuit with a transformer, you get the `Meissner oscillator`.

The advantages of these two circuits lie in the tuning of the phase shift via only one electronically variable capacitor and in the low oscillation tendency at lower frequencies.

An overview of the various feedback networks is shown in Fig. 5.11.

In practice, the networks are not symmetrical, as an impedance transformation is usually necessary.

These oscillators can also be implemented in the two other basic circuits (base circuit and collector circuit). These two circuits have 0° phase rotation for the transmission signal. Consequently, the feedback network does not need to have a phase rotation or should have 360°.

This 0° phase rotation can also be achieved using a simple resonator. A parallel resonant circuit with a coupling between the capacitors is often used. This circuit is also classified as a Colpitts oscillator.

The parallel resonant circuit can be implemented on the gate side, Fig. 5.12.

The resonant frequency is calculated as usual from *L* and *C*, whereby *C* is calculated from the series connection of C_1 and C_2. If the value of C_2 is set very high, then the parallel resonant circuit is only very weakly coupled and the quality is mainly determined by the losses of the components.

The DC voltage at the gate (the associated network is not shown) of the transistor T is set so that only a small current flows due to the current feedback with R_S. At the resonance frequency, R_S is short-circuited by *C*2. At this frequency, the transistor amplifier has a high gain value and oscillates.

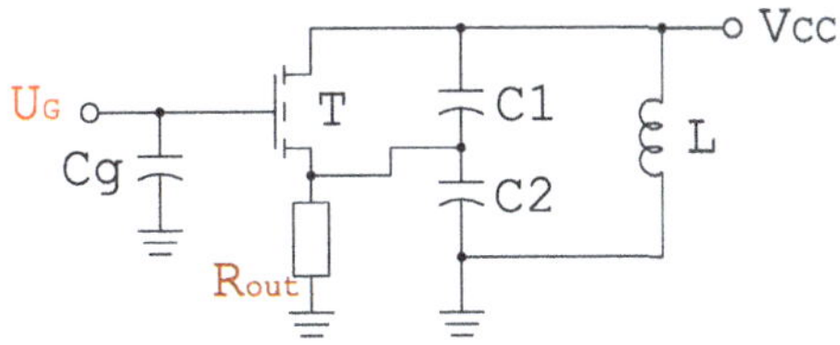

Fig. 5.13 Colpitts oscillator in basic or gate basic circuit with resonator

In semiconductors in particular, R_S is replaced by a constant current source.

Fig. 5.13 shows the gate circuit with parallel resonant circuit.

The Colpitts oscillator is widely used in voltage-controlled oscillators (VCOs) due to several key advantages. Firstly, its simple design allows for easy implementation and integration into various electronic circuits. Secondly, the Colpitts oscillator offers good frequency stability and tunability, making it suitable for applications where precise frequency control is required, such as in radio frequency communication systems. Additionally, its configuration allows for relatively low phase noise, which is critical in applications like frequency modulation (FM) and phase-locked loops (PLLs) where signal fidelity is essential. Lastly, the Colpitts oscillator's ability to operate at high frequencies with low power consumption makes it a preferred choice for VCOs used in modern electronic devices and communication equipment.

5.3 One-Port Oscillators

With two-port oscillators, the transmission behavior (S_{21} of amplifier and feedback) is considered. With one-port oscillators, on the other hand, the reflection behavior plays the decisive role. In the nomenclature of radio frequency technology, an associated amplifier must have a reflection gain and the resonator must have a strong magnitude reflection and the appropriate phase rotation at the desired oscillation frequency.

5.3.1 Oscillation Condition of One-Port Oscillators

These oscillators are particularly preferred in semiconductor circuits. Fig. 5.14 shows the preferred block diagram here.

The impedance transformation upstream of the resonator, which is accomplished with the capacitors C_K, is described in detail in [[38], chapter 5]. The resonator contained in Fig. 5.14 reflects the entire energy in broadband. However, the following applies outside the resonance frequency for the reflection phase: $M_{33}^{-} = -1$. Consequently, this circuit oscillates only via the phase condition:

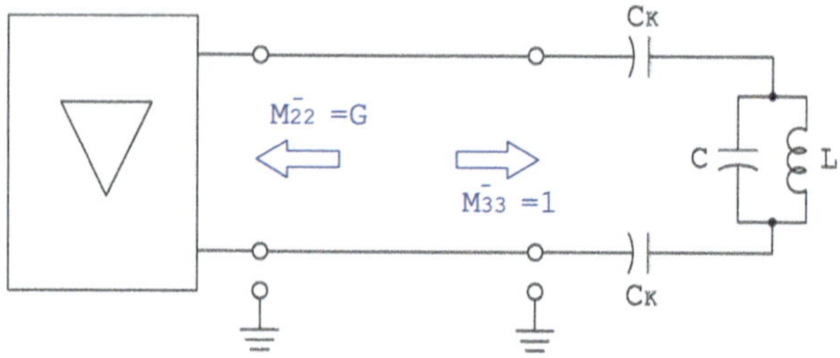

Fig. 5.14 One-port oscillator in symmetrical circuit technology, consisting of reflection amplifier (left) and resonator including impedance transformation network (Γ-transformer)

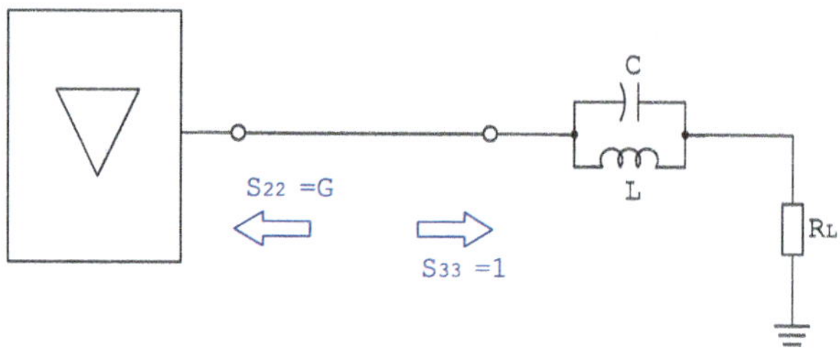

Fig. 5.15 One-port oscillator in unbalanced circuit technology with band-stop filter

$$\angle(G(\mathrm{j}\omega_0)\cdot H(\mathrm{j}\omega_0)) = 0^\circ \quad \text{resp.} \quad \angle\left(M_{22}^{-}(\mathrm{j}\omega_0)\,M_{33}^{-}(\mathrm{j}\omega_0)\right) = 0^\circ\,. \tag{5.28}$$

In addition, the reflection amplification must be at least as large as the reflection attenuation of the resonator:

$$\left|M_{22}^{-}(\mathrm{j}\omega_0)\,M_{33}^{-}(\mathrm{j}\omega_0)\right| \geq 1\,. \tag{5.29}$$

In summary, the oscillation condition for one-port oscillators is obtained:

$$\mathrm{Re}\left\{M_{22}^{-}(\mathrm{j}\omega_0)\,M_{33}^{-}(\mathrm{j}\omega_0)\right\} \geq 1\,. \tag{5.30}$$

In practice, the matching of the resonator at the resonance frequency with magnitude values in the range 2 - 8 dB is better than at other frequencies. Reflection amplifiers with values in the range of 10 - 20 dB are used for safe oscillation.

If you also want to use the amplitude condition for one-port oscillators, you must use a bandstop filter, as is also done for DRO oscillators, [[38], Kap. 5].

As with the transmission oscillator, this oscillator has an oscillation condition via the phase and the magnitude. At resonance, the band-stop filter reflects most of the energy.

It is of course arbitrary whether the definition for symmetrical line systems is formulated in M-parameters (as in Eq. (5.28) and (5.29)) or in S-parameters (as in Fig. 5.15).

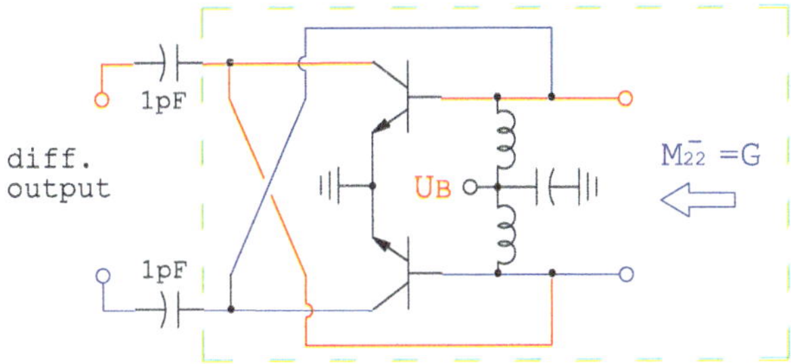

Fig. 5.16 Reflection amplifier for one-port oscillators in symmetrical circuit technology with small decoupling capacitors

5.3.2 Used Reflection Amplifier

The reflection amplifier frequently used in semiconductor technology in particular is a differential amplifier with direct feedback, as shown in Fig. 5.16.

The input signal of the upper red branch is amplified, rotated in phase by 180°, then partially decoupled via a coupling capacitor with a small capacitance (e.g. 1 pF) and made available to the lower blue input. A larger signal is now available at this blue input than just the incident wave component of the symmetrical input signal. Therefore, more energy is reflected than is incident. The same applies to the energy component that is coupled in via the blue branch and coupled out via the red branch.

This amplifier is *not* a linear amplifier, as it goes into compression due to the positive feedback. This is not retroactive for oscillator operation. In practice, this amplifier has a very high input impedance (typically 5 kΩ) and therefore places very little load on the resonator.

Often, and thus also in many other literature sources (e.g. [79]), this reflection amplifier is described as a negative resistance is described.

A further advantage of this amplifier is that the Miller effect can be compensated for and thus a large gain can be achieved at high frequencies, [38].

5.3.3 Setups of Differential Oscillators

The structure of the simplest differential oscillator, which is very frequently used in RF IC technology, is shown in Fig. 5.17.

Since the amplifier is correspondingly high impedance, a small coupling transformer for impedance transformation (by means of Γ-transformer) is not always used.

In [38], in addition to the simple resonators, the dual-mode resonators were also introduced, which, due to the double of the signal in direct and push-pull (differential) mode, they have a significantly higher loaded Q factor than simple resonators.

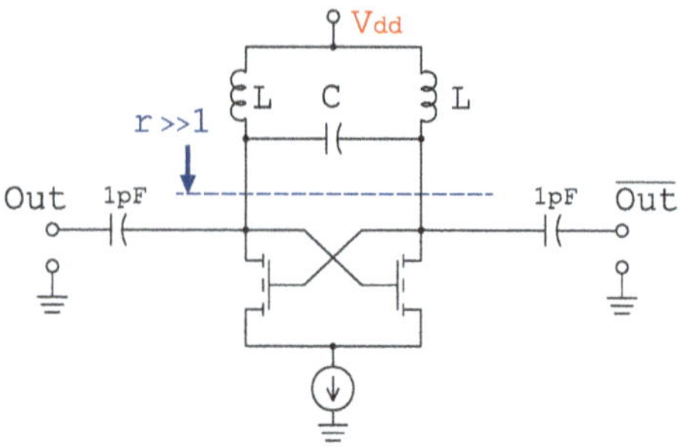

Fig. 5.17 Differential or push-push or cross-coupled oscillator

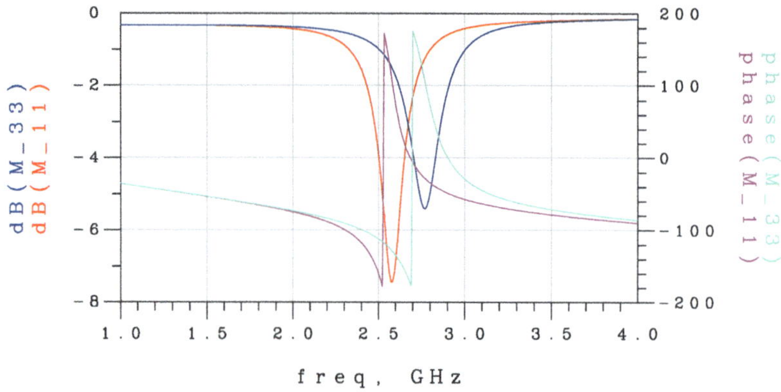

Fig. 5.18 Differential reflection for the classic resonator ($M_{33} = M\_33$) and the dual-mode resonator ($M_{11} = M\_11$)

The achievable increase in the loaded Q factor due to the mixed-mode concept is illustrated in Fig. 5.18.

The values apply to 2.5 GHz and the components: $Q_L = Q_C = 40$, L_{min}: 1 nH, C_{min}: 0.5 pF. A more stable oscillation is achieved due to the steep phase response.

The phase curve passes more steeply through the point of resonant frequency, as the loaded Q is increased. At the same time, the losses in the resonator increase, as the resonant circuit is passed through twice (with common mode and differential mode).

If a reflection oscillator is implemented in this circuit technology, the basic circuit technology is as shown in Fig. 5.19.

The reflection amplifier must now be considered for the common mode and differential mode. At the resonator, the differential mode wave is converted into a common mode wave with 90° phase rotation and reflected. The common mode wave is reflected back from the amplifier with 180° phase rotation and converted back into a differential mode wave with 90° phase rotation at the resonator.

This process of wave reflection and conversion is illustrated in the lower part of Fig. 5.19.

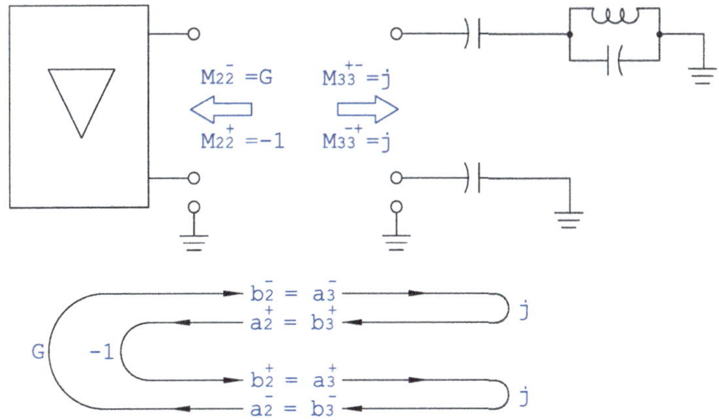

Fig. 5.19 Left: Reflection amplifier; right: dual-mode resonator

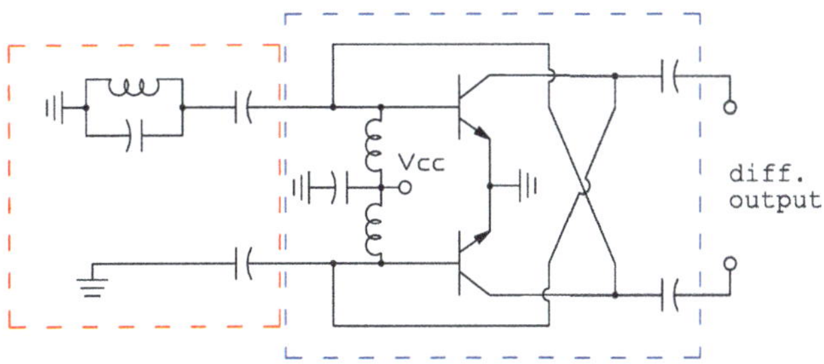

Fig. 5.20 Left: Dual-mode resonator; right: reflection amplifier

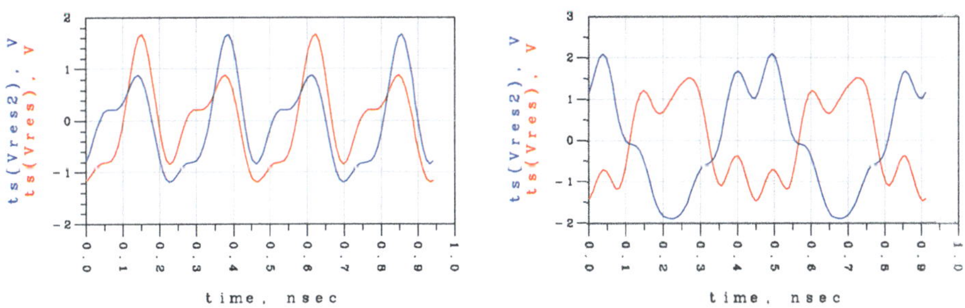

Fig. 5.21 Resonator voltages at base 1 and base 2 in push-pull and dual-mode operation (right)

This oscillator, which is also known as a dual-mode or mixed-mode oscillator, differs very little from the differential oscillator shown in Fig. 5.17: Only one capacitor, which is contacted to ground, needs to be used on the resonator (Fig. 5.20).

The difference in the voltages at the resonators can be seen in Fig. 5.21.

The same oscillator was used for these simulation results of the mixed-mode oscillator and the differential oscillator, as the oscillator mode can be switched via

Fig. 5.22 Image of the realized dual-mode reflection oscillator

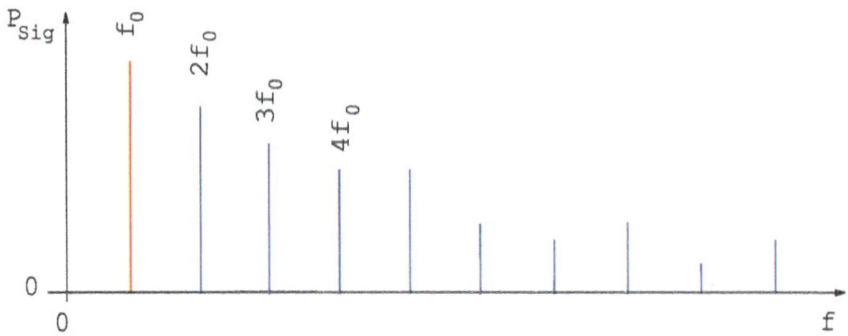

Fig. 5.23 Typical frequency spectrum of semiconductor oscillators

a short circuit. You can clearly see that the oscillator has a higher resonator voltage in dual-mode operation.

The structure of the mixed-mode oscillator with SMD components and SiGe transistors is shown in Fig. 5.22.

In the simulation, a 6 dB better phase noise was calculated at 1 MHz distance. In the measurement, it was even 7 dB better, [49].

5.3.4 Push-Push-Oscillators

It has already been mentioned that the differential oscillator is also sometimes referred to as a push-push oscillator. Here, push-push oscillators only refer to oscillators that oscillate at a fundamental frequency f_0, generate many harmonics and from which only one harmonic is tapped as the output signal.

Fig. 5.23 shows the typical output signal of an oscillator in semiconductor technology. In special circuits, the energy of the harmonics is optimized to produce a spectrum that resembles a comb. In these cases, one speaks of a special comb oscillator or comb generator.

The simplest push-push oscillator is controlled by an oscillator with two outputs. The output signals are each rotated via a phase shifter and then recombined.

This general structure of a push-push oscillator is shown in Fig. 5.24.

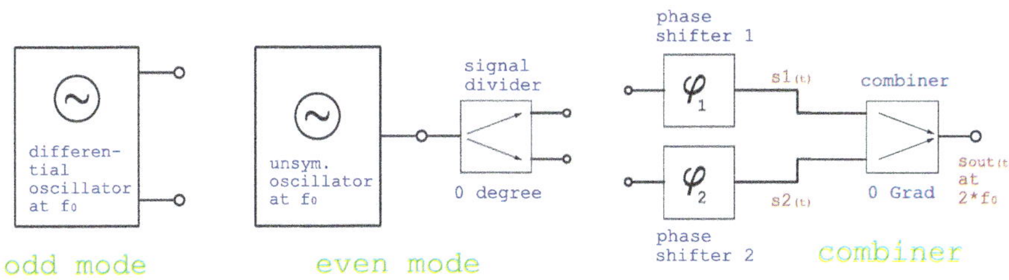

Fig. 5.24 Oscillators and combining network for push-push oscillators

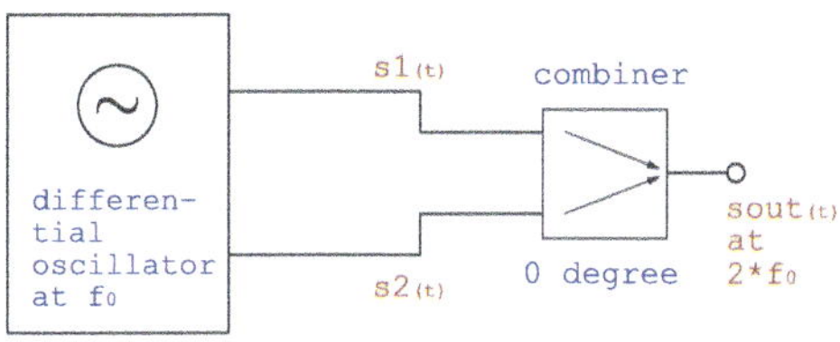

Fig. 5.25 Simplest push-push oscillator

Mathematically, it can be described as follows. The following applies to the signal in the upper path:

$$s_{1(t)} = a\, e^{j\omega_0 t} + b\, e^{j2\omega_0 t} + c\, e^{j3\omega_0 t} + \dots \,. \tag{5.31}$$

In the lower path, the signal is delayed due to an additional phase rotation:

$$s_{2(t)} = a\, e^{j\omega_0 (t-\Delta t)} + b\, e^{j2\omega_0 (t-\Delta t)} + c\, e^{j3\omega_0 (t-\Delta t)} + \dots \,. \tag{5.32}$$

The combiner network adds these two signals together:

$$\begin{aligned} s_{out(t)} = {} & a'\, e^{j\omega_0 t}(1 + e^{-j\omega_0 \Delta t}) + b'\, e^{j2\omega_0 t}(1 + e^{-j2\omega_0 \Delta t}) \\ & + c'\, e^{j3\omega_0 t}(1 + e^{-j3\omega_0 \Delta t}) + \dots \,. \end{aligned} \tag{5.33}$$

With a phase difference of 180° or $\pi = \omega_0 \Delta t$, only the even numbered harmonics remain:

$$s_{out(t)} = 2b'\, e^{j2\omega_0 t} + 2d'\, e^{j4\omega_0 t} + \dots \,. \tag{5.34}$$

The simplest and most frequently used implementation (in semiconductors) is shown in Fig. 5.25.

These oscillators have been described in many publications, particularly for semiconductor circuit technology, in a very large number of variants. The phase shifters can be omitted as the differential oscillator already supplies the necessary push-pull signal.

A more detailed scientific publication on push-push oscillators is printed in [95].

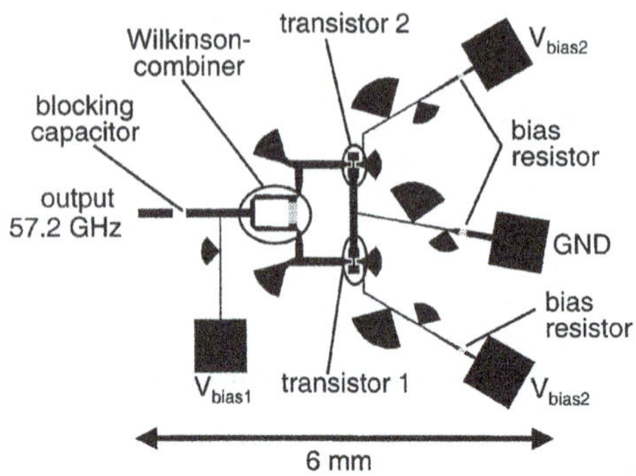

Fig. 5.26 Push-push oscillator at 57 GHz in SiGe-HBT technology from [95]

Fig. 5.26 shows an implementation of two 28.5 GHz oscillators realized on Al_2O_3 substrate in unbalanced circuit technology.

A Wilkinson coupler designed for 57 GHz is used here as a power combiner. The oscillator has very good characteristics with an output power of 1 dBm and a phase noise value of 108 dBc/Hz at 1 MHz spacing.

Others have even developed a 51 GHz push-push oscillator from two 6.4 GHz oscillators using a combiner network for the eighth harmonic, [58]. This solution, created using SMD technology on a Teflon substrate, has almost as good values as the previously presented oscillator.

5.3.5 Electrically Tuneable Oscillators

In practice, quartz oscillators and oscillators with dielectric resonators (DRO) are the only fixed oscillators.

The ground of the HF oscillators use a varactor diode in the oscillating circuit. The varactor diode is related to the PIN diode and is bred so that a blocking voltage can change the blocking capacitance over a very large range. To put it simply:

Component	Function
PIN diode	Electric tuneable resistor
Varactor diode	Electrally tuneable capacitor

This functionality of an electrically tunable oscillator (VCO; voltage controlled oscillator) can be described with this simple function:

$$\omega_{out} = \omega_0 + K_V \cdot U_{control} \; . \tag{5.35}$$

Here, K_V describes the slope of the VCO in $\frac{rad}{Vs}$.

In practice, a varactor diode cannot be tuned linearly. This non-linearity must therefore be compensated for by a non-linear control via a microprocessor or a memory or by a more complex analog varactor diode arrangement.

Modern integrated circuits for synthesis oscillator applications now include a large matrix of switchable capacitors, with which the frequency range can be detuned by several GHz.

5.4 Noise Behavior of Oscillators

Already in the introduction to this chapter, various finite properties and in particular the noise of the oscillator were discussed.

An ideal RF oscillator would generate a pure sinusoidal signal with a fixed frequency. Real oscillators exhibit more or less strong noise around the resonance frequency, Fig. 5.27.

The following values can be taken from this figure:

- Signal power (marker 1): 2 dBm
- Noise power (marker 2): −66 dBm

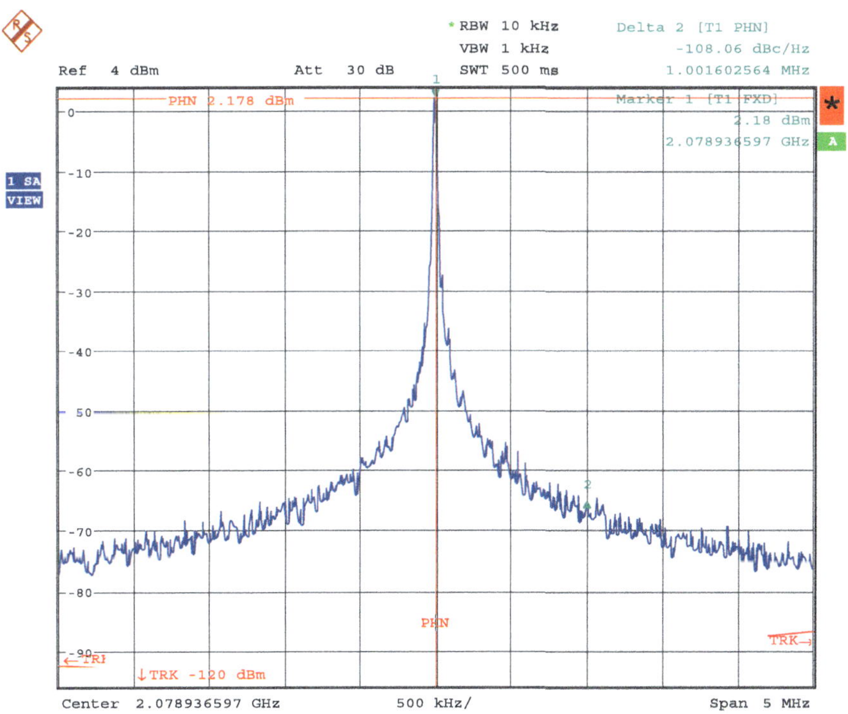

Fig. 5.27 Spectral output signal of a VCO in SMD technology

- Resulting bandwidth (RWB): 10 kHz
- Phase noise value at 1 MHz distance: −108 dBc

A resulting bandwidth of 10 kHz corresponds to 40 dB increased noise compared to a measurement with only 1 Hz bandwidth.

The phase noise value, which is always specified for 1 Hz bandwidth, is calculated as follows in this example:

$$L_{(1\,\mathrm{MHz})} = -66\,\frac{\mathrm{dBm}}{10\,\mathrm{kHz}} - 40\,\mathrm{dB} - 2\,\mathrm{dBm} = -108\,\frac{\mathrm{dBc}}{\mathrm{Hz}}\,. \tag{5.36}$$

This noise around the carrier, which can be limited to a good approximation of phase noise, determines the selectivity and therefore the range in determines the selectivity and therefore the range in most modern radio systems. A great deal of effort is therefore made to minimize phase noise.

As early as 1966, Leeson [65] created a linear invariant model to describe phase noise. This model takes into account various influences on the phase noise. Other scientists refined this model, [27,64,82]. However, the Leeson phase noise model is often considered sufficient and still very clear, [106].

Demnach gilt:

$$L(\Delta\omega) = 10\,log\left[\frac{2FkT}{P_{sig}}\left\{1+\left(\frac{\omega_0}{2Q_L\Delta\omega}\right)^2\right\}\left(1+\frac{\Delta\omega_{1/f^3}}{|\Delta\omega|}\right)\right] \quad \text{with} \tag{5.37}$$

$L(\Delta\omega)$: Single sideband noise power, normalized to P_{sig} for 1 Hz bandwidth
P_{sig}: Signal power of the oscillator
$\Delta\omega$: Circular frequency deviation from ω_0
T: Temperature
$\Delta\omega_{1/f^3}$: Corner frequency of the $1/f$ noise
ω_0: Center frequency of the oscillator
F: Noise figure of the amplifier
k: Boltzmann constant
Q_L: Loaded quality factor of the resonator

The associated noise spectrum is shown in Fig. 5.28.

The cause of the strong sideband noise is in particular the $1/f$ noise present at low frequencies, which is mixed up by nonlinear effects on the oscillator. Up to the limit of $\Delta\omega_{1/f^3}$, the sideband noise a gradient of −30 dB/decade and then −20 dB/decade until it turns into white noise.

The phase noise of an oscillator can be reduced with the following changes:

1. Increase in the output power P_{sig}. Since the thermal noise is fixed, $2kT/P_{sig}$ can be optimized, [64].
2. Increase of the loaded resonator quality Q_L.
3. Reduction of the noise figure F, [77], which is not as efficient, [64].
4. Select semiconductors with a low $1/f$ cut-off frequency. SiGe is chosen here.

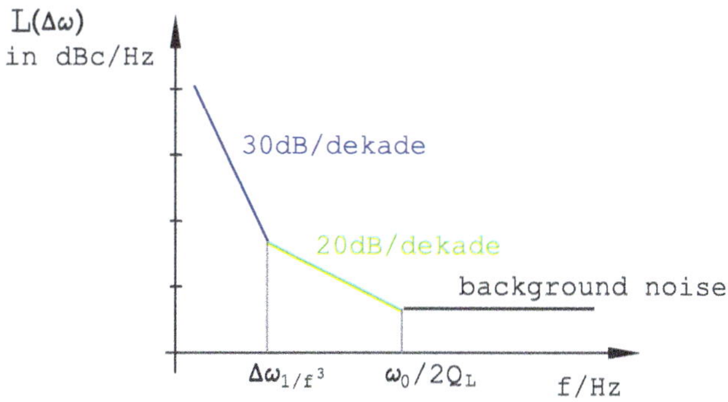

Fig. 5.28 Phase noise of an oscillator according to the Leeson model, [27]

In [38] was shown in the amplifier chapter how to reduce the $1/f$-noise can already be reduced well using circuit technology, thus reducing the phase noise by around 6 dB.

If you want to compare oscillators with each other, you introduce a power factor (for the phase noise) as a so-called FOM value (figure of merit):

$$FOM = L(\Delta\omega) + 10\,log\left[\left(\frac{\Delta\omega}{\omega_0}\right)^2 \frac{P_{VCO}}{1\,mW}\,\frac{V_c}{FTR}\right] \tag{5.38}$$

This FOM value corrects the output power and the relative bandwidth of the VCO.

P_{VCO} describes the power consumed by the oscillator in mW. *FTR* stands for *Frequency Tuning Range* and is given as a percentage. V_c is the tuning voltage range in volts.

International best values for VCOs are −185 dBc to −200 dBc.

Chapter 6
Detectors and Mixers

In the early days of microwave technology, detectors were the only components used to transfer microwave signal information to the DC- and AC-sector. A very simple microwave system can still be set up with an amplitude-modulated signal in the transmitter branch and a detector for frequency conversion or demodulation in the receiver. However, this is no longer state of the art, as a dual-mode radio system that uses a mixer instead of a detector is hardly more complex (detail, in [38]).

A detector can be used to determine the power of an RF signal. This means in detail: The output voltage of the detector is proportional to the microwave power at the input of the detector.

$$\text{Detector:} \quad U_{DCout} \sim P_{RFin}$$

Mixers are used to convert the microwave signal (via conversion losses, typically 6 dB). Minus these losses, a mixer converts a signal from the GHz range into a signal in the electronic range.

$$\text{Mixer:} \quad U_{IFout} \sim U_{RFin} - 6\,\text{dB}$$

For this reason, receivers with mixers are significantly more sensitive than detector circuits. With radio systems, the dynamic difference in dB is almost a factor of 2.

Nowadays, detectors are mainly only used to control the output power.

In contrast, complex circuits in the mixer area have brought about many innovations, which are described in detail below, right up to the IQ receiver. This chapter begins with a brief explanation of detectors and then goes into great detail on the basics of frequency conversion of signals using mixers. Numerous types of mixer circuits through to IQ converters are described.

6.1 Detectors

The wide range of integrated detector circuits with integrated operational amplifiers for the single-digit GHz range means that circuit engineers rarely have to develop detector circuits themselves.

H. Heuermann, *Microwave Technology*,
https://doi.org/10.1007/978-3-658-45686-3_6

Nevertheless, there are still detector diodes on the market for the entire GHz range, as the integrated solutions are not suitable for the desired frequency range, for example, are not available for the desired frequency range, see also [1].

The theory of the detector circuits with the Schottky diode has already been derived in section 3.5.1. It is derived there, how to set the conductance G_S at the diode via the operating point voltage U_0 or the operating point current I_0 can be set.

Two receiver circuits are shown below: In Fig. 6.1 the structure of a pure power detector and in Fig. 6.2 the block diagram of an AM demodulator.

At the input of the power detector, the diode (here without bias voltage) is adapted to the measurement gate (here to Z_0) via a matching network (e.g. Γ transformer, [38]). The series coil L with typically 100 nH blocks the GHz signal with high impedance. The following capacitor C_{IF} with typically 1 nF smoothes the DC voltage signal of the measurement signal.

The AM demodulator in Fig. 6.2 operates with a diode set to the optimum operating point via the bias voltage U_{op}. As described in section 3.5.1, this can be a setting with regard to. the greatest sensitivity or, on the other hand, the greatest bandwidth or a compromise of both. At typically 100 nH, the inductance L is extremely high impedance for the RF signal.

The C_{RF} is a capacitor that allows the GHz signal to be transmitted optimally and at the same time has a high impedance for the IF signal (typically 10 pF). The following low-pass filter must have a high-impedance input impedance for the GHz range, so that only the matching network, the diode and the RF capacitor are circuit elements to be considered for the GHz signal.

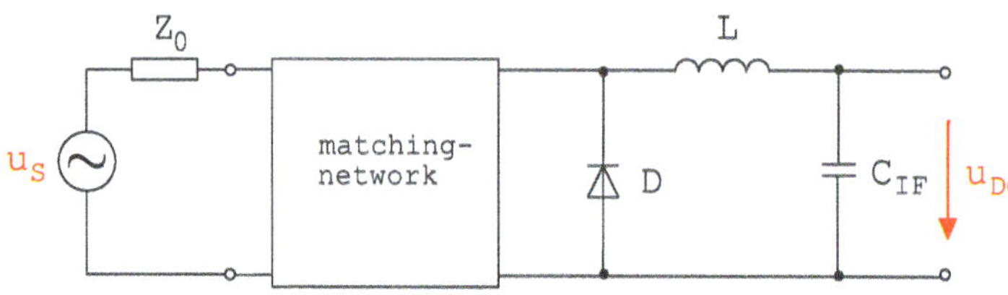

Fig. 6.1 Block diagram of a GHz power detector

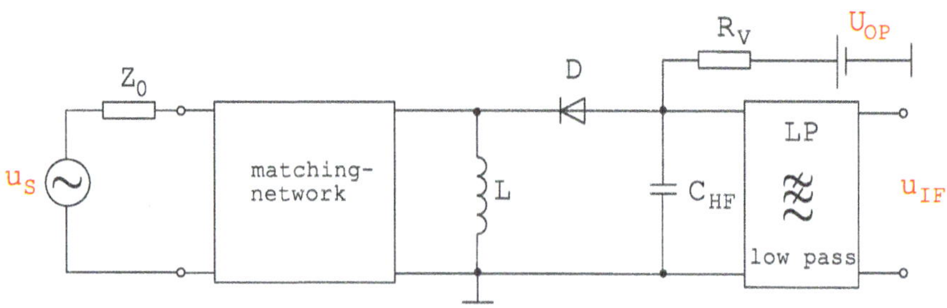

Fig. 6.2 Block diagram of a AM demodulator

6.2 The Superhet or Superheterodyne Principle

In modern communication processes, however, the RF signals are converted directly into the lower frequency range. This conversion preserves the amplitude and phase information. There are three ways to detect these vectorial RF signals:

1. The superheterodyne or heterodyn principle
2. The homodyn principle
3. The sixport principle

Fig. 6.3 represents a heterodyne receiver that feeds the antenna signal with the frequency f_s to a mixer, which is "pumped" by the signal with the frequency f_p and outputs a signal with the IF frequency f_i, which contains the amplitude and phase information of the RF signal with the frequency f_s.

The heterodyne and homodyne stages can also be used as transmitters. These two principles for frequency conversion are based on circuits with mixers.

The six-port principle only has couplers and detectors and can only be used as a receiver.

There is a clear gradation in distribution: heterodyne systems are very common, homodyne stages are rare and six-port receivers are very rare.

6.3 Basics of Frequency Conversion with Mixers

Fig. 6.4 shows the structure of a heterodyne stage with detector receiver with the signal frequency $f_S = f_{RF}$, the local oscillator frequency f_{LO} and the intermediate frequency f_{IF}.

Receivers with mixers have the following properties:

- A signal is converted from one frequency range to another.
- Transmitters of different carrier frequencies are converted into an intermediate frequency signal with a fixed carrier frequency: often f_{IF} is fixed and $f_{RF} = f(f_{LO})$.
- The non-linear properties of components are required for mixing.

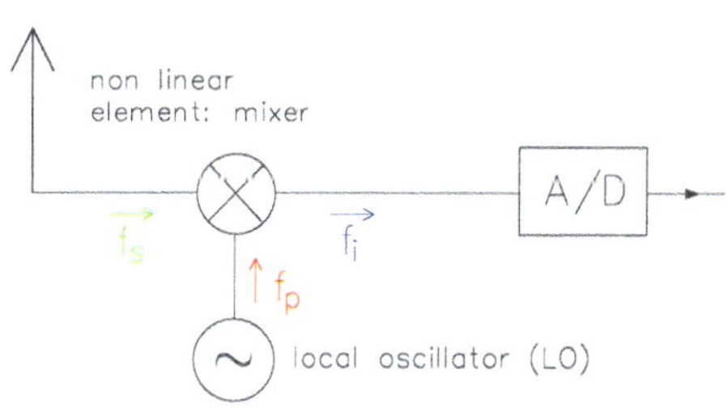

Fig. 6.3 Setup of a heterodyne stage as a receiver

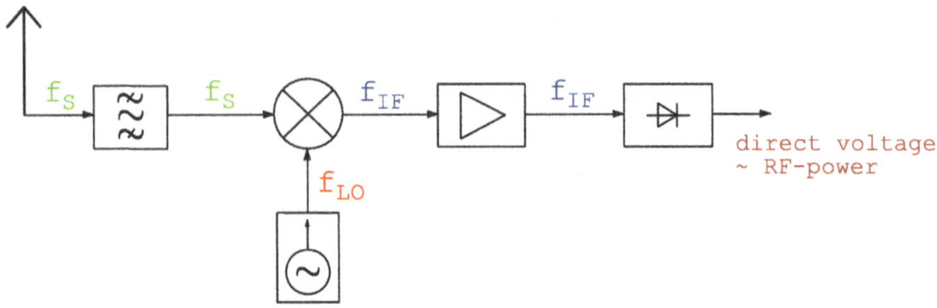

Fig. 6.4 Construction of a heterodyne stage with detector receiver (with $f_S = f_{RF}$)

- So-called Gilbert mixers are often used[1] among others for integrated circuits up to 12 GHz and Schottky diodes for frequencies above 12 GHz.

Fig. 6.5 shows a simplified mixer circuit consisting of the voltage sources of the RF signal U_S of the signal frequency f_S and the pump signal U_{LO} with the local oscillator frequency f_{LO} as well as a non-linear, resistive component D.

Current and voltage characteristics of the mixer element as a Taylor series, as already introduced several times:

$$i_D = c_0 + c_1 u_D + c_2 u_D{}^2 + c_3 u_D{}^3 + c_4 u_D{}^4 + \dots \,. \tag{6.1}$$

Both voltage sources can be summarized as a total source voltage (as already shown in Eq. (3.13)):

$$\text{General:} \quad u_D = U_{LO}\cos(\omega_{LO}t) + U_S\cos(\omega_S t)\,, \tag{6.2}$$

$$\text{For down converter:} \quad u_D = U_{LO}\cos(\omega_{LO}t) - U_{RF}\cos(\omega_{RF} t)\,, \tag{6.3}$$

$$\text{For up converter:} \quad u_D = U_{LO}\cos(\omega_{LO}t) + U_{IF}\cos(\omega_{IF} t)\,. \tag{6.4}$$

The resulting spectral components for a mixer in which all coefficients up to c_4 must be taken into account are illustrated in Fig. 6.6 for the case of upward mixing.

[1] Gilbert mixers are pure semiconductor circuits and are described in detail in [107],

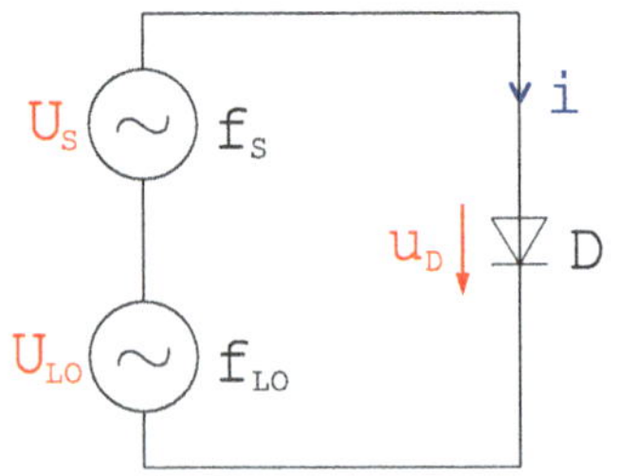

Fig. 6.5 Schematic diagram of a heterodyne stage for upward or downward mixing

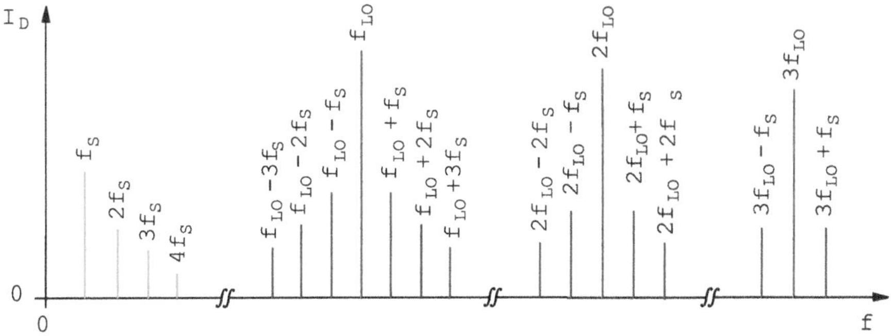

Fig. 6.6 Practically relevant spectral components of a general mixing process for the upward mixing ($f_S = f_{IF}$)

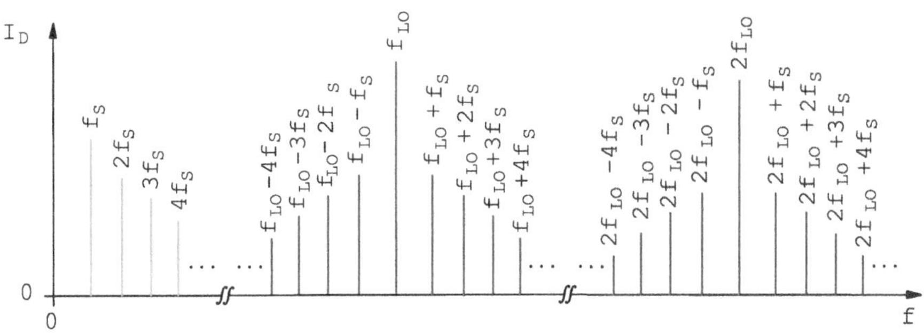

Fig. 6.7 Spectral components of a general mixing process at a high signal level for the upward mixing ($f_S = f_{IF}$)

Theoretically and at very high signal levels, additional spectral lines occur for a mixer in which all coefficients up to c_4 must be taken into account, as shown in Fig. 6.7.

The RF signal f_{RF} and the intermediate frequency, f_{if} or f_i is composed in the `fundamental wave mixer` via

$$f_{IF} = \mp f_{LO} \pm f_{RF} \quad \text{resp.} \quad f_{RF} = f_{LO} \pm f_{IF} \tag{6.5}$$

together. The following therefore applies to the intermediate frequency signal f_{IF}:

$$f_{IF} = f_{LO} - f_{RF} \quad \text{for} \quad f_{LO} \leq f_{RF}\,, \tag{6.6}$$

$$f_{IF} = f_{RF} - f_{LO} \quad \text{for} \quad f_{RF} \leq f_{LO}\,. \tag{6.7}$$

The same applies to mixing with the harmonic n:

$$f_{IF} = \mp n\, f_{LO} \pm f_{RF} \quad \text{resp.} \quad f_{RF} = n\, f_{LO} \pm f_{IF}\,. \tag{6.8}$$

This operation with a strong local oscillator and weak useful signal is normal operation.

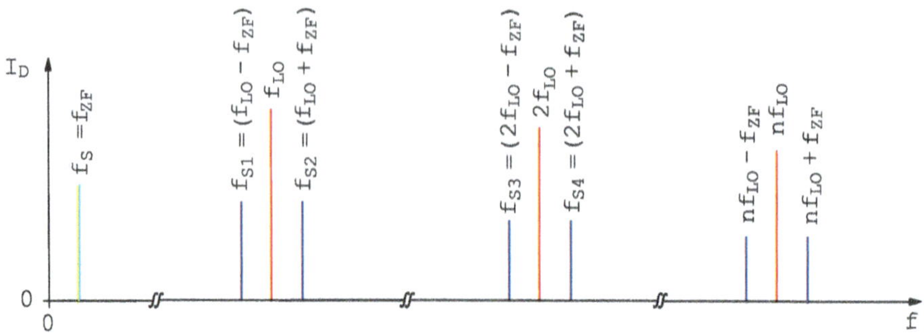

Fig. 6.8 Spectral components of a typical mixing process (at low signal levels) for the upward conversion ($f_S = f_{IF}$)

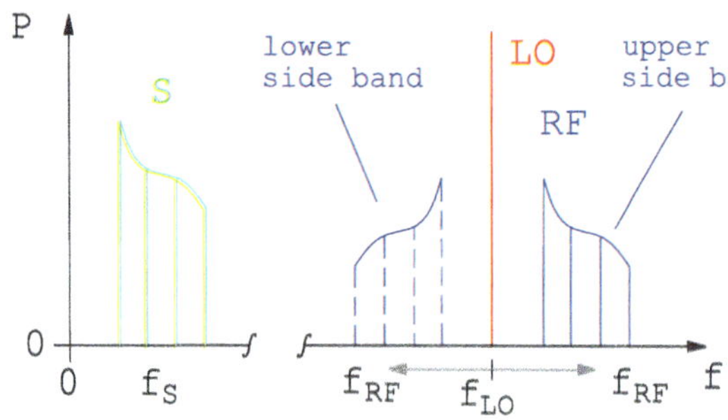

Fig. 6.9 Upconversion products of a broadband signal

For a downward mixer, there are currents in the intermediate frequency range that are proportional to the receive voltage:

$$I_{IF} \sim c_2 U_S \quad \text{for} \quad f_{IF} = \mp f_{LO} \pm f_S\,, \tag{6.9}$$

$$I_{IF} \sim c_3 U_S \quad \text{for} \quad f_{IF} = \mp 2 f_{LO} \pm f_S\,, \tag{6.10}$$

$$\text{ect.} \tag{6.11}$$

These components are illustrated for a sinusoidal signal in Fig. 6.8 and for a broadband frequency spectrum in Fig. 6.9.

Common Terms for Mixed Products

Upward Mixing:
 The input signal is in the intermediate frequency:
 $f_{IF} = f_S$

Downward Mixing:
 The input signal is in the RF frequency range:

$f_{RF} = f_S$

Mirror Frequency:

A 2nd signal frequency, which is a mostly undesirable mixed product of the same order as the desired 1st signal frequency

Fundamental Mixing:

Mixing with the fundamental wave signal of the local oscillator

$f_{RF} = f_{LO} \pm f_{IF}$ or $f_{IF} = \mp f_{LO} \pm f_{RF}$

Harmonic Mixture:

Mixing with a higher harmonic of the local oszillators

$f_{RF} = n\, f_{LO} \pm f_{IF}$ or $f_{IF} = \mp n\, f_{LO} \pm f_{RF}$ for $n \geq 1$

Equal Site:

Or low-side injection: The RF frequency spectrum is shifted on the frequency axis compared to the IF spectrum

$f_{RF} > f_{LO}, \quad f_{mirror} < f_{LO}$

Image Site:

Or high-side injection: the RF frequency spectrum is shifted and mirrored on the frequency axis compared to the IF spectrum

$f_{RF} < f_{LO}, \quad f_{mirror} > f_{LO}$

The equal and inverse position mixing products and other examples are shown in Fig. 6.10.

In RF and microwave technology, the following components are commonly used as mixers for discrete setups:

1. f < 10 GHz: Bipolar and field-effect transistors
2. f > 10 GHz: Schottky diodes

   Mixing with a higher harmonic of the local oscillators:

Transistors are usually used for integrated circuits up to 40 GHz. In the single-digit GHz range, the preferred the Gilbert mixer is used.

6.4 Parametric Calculation

Parametric calculation is a general approach in science and technology for describing weak nonlinear effects. Here, this calculation is performed with mixer parameters.

- Mixer: Clear current-voltage characteristic $i_{(t)} = i(u_{(t)})$
- $u_{p(t)}$: Periodical large signal
- $\Delta u_{(t)} = u_{s(t)} + u_{i(t)}$: Any small signal
- $u_{s(t)}$: Small signal component in the RF range
- $u_{i(t)}$: Small signal component in the lower frequency range (AD or DA converter range)

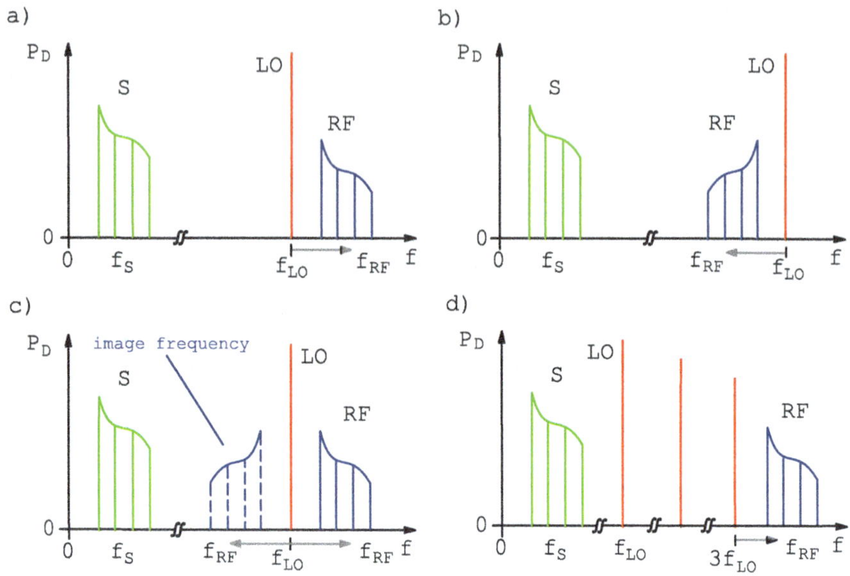

Fig. 6.10 Overview of the mixing processes **a** Up conversion in equal site (position) **b** Up conversion in image site (position) **c** Up conversion into equal and image site **d** Harmonic conversion in equal site

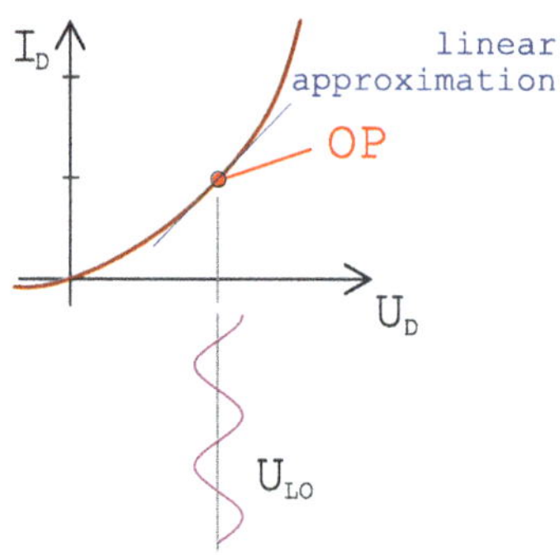

Fig. 6.11 Representation of the diode characteristic with operating point (OP) due to the dc voltage setting and LO signal for modulation in mixer mode

The typical nonlinear characteristic curve of the diode is shown in Fig. 6.11.

The parametric calculation is based on a truncated Taylor series:

$$i(u_{p(t)} + \Delta u_{(t)}) = i(u_{p(t)}) + \left.\frac{di}{du}\right|_{u_{p(t)}} \cdot \Delta u_{(t)} \,, \tag{6.12}$$

$$= i(u_{p(t)}) + \Delta i_{(t)} \,, \text{ with:} \tag{6.13}$$

$$\Delta i_{(t)} = g(u_{p(t)}) \cdot \Delta u_{(t)} \quad \text{and} \quad g(u_{p(t)}) = \left.\frac{di}{du}\right|_{u_{p(t)}} = g_{(t)} \; . \tag{6.14}$$

Caution: It must be ensured that only the pump signal and not the small signal has an influence on the small signal conductance $g_{(t)}$ of the mixer. Mathematically, this means

$$g_{(t)} \sim u_{p(t)} \;\; \text{but} \;\; g_{(t)} \neq f(\Delta u_{(t)}) \; . \tag{6.15}$$

For $\omega_p = 2\pi f_p$ $g_{(t)}$ can be represented as a Fourier series.

Example: *The following non-linear relationship between current and voltage is given with a > b:*

$$i_{(t)} = a \cdot u_{(t)} + b \cdot u^2_{(t)} \qquad \textit{with} \qquad u_{(t)} = \cos(\omega_p t) \; . \tag{6.16}$$

The resulting current has Fourier coefficients greater than zero for three frequencies:

$$i_{(t)} = a \cdot \cos(\omega_p t) + \frac{b}{2} + \frac{b}{2} \cdot \cos(2\,\omega_p t) \; . \tag{6.17}$$

For $g_{(t)}$ applies as a periodically changed small signal conductance

$$g(u_{p(t)}) = \sum_{n=-\infty}^{\infty} G_n \cdot e^{\,jn\omega_p t} \tag{6.18}$$

with:

$$G_n = \frac{1}{2\pi} \int_{-\pi}^{\pi} g(u_{p(t)}) \cdot e^{-jn\omega_p t} \cdot d(\omega_p t) \; . \tag{6.19}$$

Since $g_{(t)}$ is a real function:

$$G_{-n} = G_n^* \, , \qquad G_0 : real \; . \tag{6.20}$$

For $u_{s(t)}$ as a mono-frequency signal with the frequency f_s results for

$$\Delta i_{s(t)} = g_{(t)} \cdot u_{s(t)} \, , \tag{6.21}$$

only signals at $|f_p \pm n \; f_s|$ $n = 0, 1, 2, 3, \cdots$. No harmonics of f_s occur due to small signal approximation.

Thus: $n = 0$, 1 and a linear approximation should be valid in the range of the OP.

$$\Rightarrow \qquad g_{(t)} = G_1^* \cdot e^{-j\omega_p t} + G_0 + G_1 \cdot e^{\,j\omega_p t} \tag{6.22}$$

For others, an alternative to Eq. (6.22)

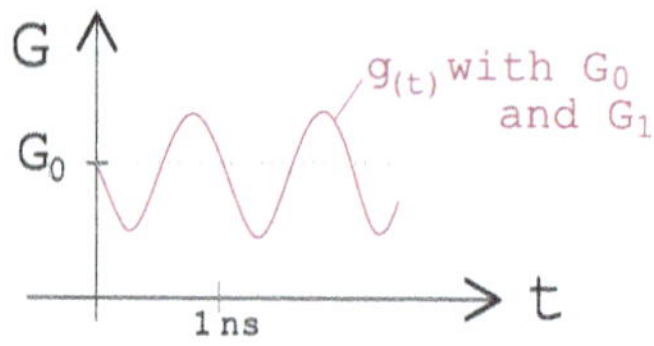

Fig. 6.12 Small signal conductance G_0 for the OP and time-varying small signal conductance $g_{(t)}$ with G_1 at $f_{LO} = 1\,\text{GHz}$

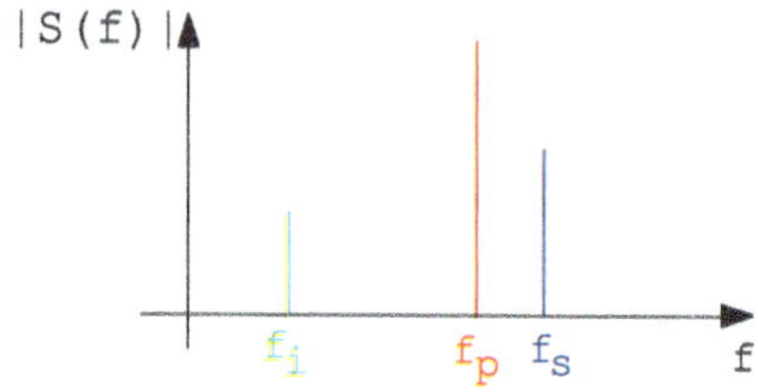

Fig. 6.13 Signal spectrum for a mixing process in equal position

$$g_{(t)} = G_0 + 2 \cdot G_1 \cdot \cos(\omega_p t) , \tag{6.23}$$

which is also clearly shown in Fig. 6.12. Using suitable filters, only the three signals at the frequencies

$$f_s, \qquad f_p, \qquad f_i = f_s - f_p \qquad \text{occur.}$$

The small signal quantities can be calculated with complex pointers as follows:

$$\Delta i_{(t)} = \frac{1}{2}\left\{I_s e^{j\omega_s t} + I_s^* e^{-j\omega_s t} + I_i e^{j\omega_i t} + I_i^* e^{-j\omega_i t}\right\} , \tag{6.24}$$

$$\Delta u_{(t)} = \frac{1}{2}\left\{U_s e^{j\omega_s t} + U_s^* e^{-j\omega_s t} + U_i e^{j\omega_i t} + U_i^* e^{-j\omega_i t}\right\} . \tag{6.25}$$

From

$$\Delta i_{(t)} = \left(G_1^* e^{-j\omega_p t} + G_0 + G_1 e^{j\omega_p t}\right) \cdot \Delta u_{(t)} \tag{6.26}$$

follows by inserting Eqs. (6.24) and (6.25) in Eq. (6.26)

$$I_s = G_0 \cdot U_s + G_1 \cdot U_i \qquad \text{there} \qquad \omega_s = \omega_i + \omega_p \tag{6.27}$$

and in the same way (Fig. 6.13)

$$I_i = G_1^* \cdot U_s + G_0 \cdot U_i . \tag{6.28}$$

The equal position (signal at equal site) should be selected here: $f_s > f_p$.

These results for the equal position can be displayed as a matrix:

$$\begin{pmatrix} I_s \\ I_i \end{pmatrix} = \begin{pmatrix} G_0 & G_1 \\ G_1^* & G_0 \end{pmatrix} \cdot \begin{pmatrix} U_s \\ U_i \end{pmatrix} \quad \text{resp.} \quad \begin{pmatrix} I_s \\ I_i \end{pmatrix} = [\mathbf{G}] \cdot \begin{pmatrix} U_s \\ U_i \end{pmatrix} . \tag{6.29}$$

That means: A mixer accomplishes a linear conversion from one frequency range to another. The amplitude of the pump signal does not appear explicitly.

For the image position, $f_s < f_p$ applies and the result is

$$\begin{pmatrix} I_s \\ I_i^* \end{pmatrix} = \begin{pmatrix} G_0 & G_1 \\ G_1^* & G_0 \end{pmatrix} \cdot \begin{pmatrix} U_s \\ U_i^* \end{pmatrix} \quad \text{resp.} \quad \begin{pmatrix} I_s \\ I_i^* \end{pmatrix} = [\mathbf{G}] \cdot \begin{pmatrix} U_s \\ U_i^* \end{pmatrix} . \tag{6.30}$$

6.5 Conductance Elements of Schottky Diodes

The (nonlinear) equation applies to Schottky diodes:

$$i_{(t)} = I_{ss} \cdot \left(e^{\frac{u_{(t)}}{U_T}} - 1 \right) \quad \text{with} \quad U_T = \frac{nkT}{q} \quad \text{whereby} \tag{6.31}$$

n: Ideality factor 1 ... 1.1 k: Boltzmann constant
I_{ss}: Saturation current q: Elementary charge

The following applies to the variable small-signal conductance (also called differential conductance) of a Schottky diode:

$$g_{(t)} = \left. \frac{di}{du} \right|_{u_{p(t)}} \quad \text{with} \quad u_{p(t)} = U_0 + \hat{U}_p \cos(\omega_p t) \tag{6.32}$$

$$\text{results in} \quad g_{(t)} = \frac{I_{ss}}{U_T} e^{\frac{u_{p(t)}}{U_T}} . \tag{6.33}$$

The corresponding conductance values for the various discrete frequencies can be calculated using Eq. (6.20) as follows:

$$G_n = \frac{I_{ss}}{U_T} e^{\frac{U_0}{U_T}} \cdot \frac{1}{2\pi} \cdot \int_{-\pi}^{\pi} e^{\frac{\hat{U}_p}{U_T} \cos(\omega_p t)} \cdot \cos(n\,\omega_p t) \cdot d(\omega_p t) . \tag{6.34}$$

Eq. (6.34) corresponds to a modified Bessel function.

$$G_n = \frac{I_{ss}}{U_T} e^{\frac{U_0}{U_T}} \cdot J_n \left(\frac{\hat{U}_p}{U_T} \right) \sim \hat{U}_p \tag{6.35}$$

$\Rightarrow G_0 > G_1 > G_2$, all: positive, real.

As an alternative to Eq. (6.18), the small-signal conductance $g_{(t)}$ can also be calculated as follows:

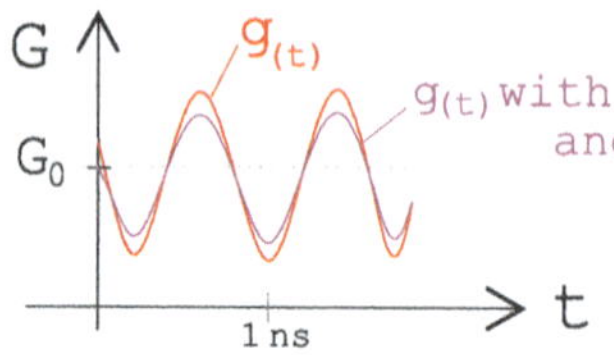

Fig. 6.14 The mixer corresponds to a time-dependent small-signal conductance $g_{(t)}$ and is often approximated by the small-signal conductance coefficients G_0 and G_1

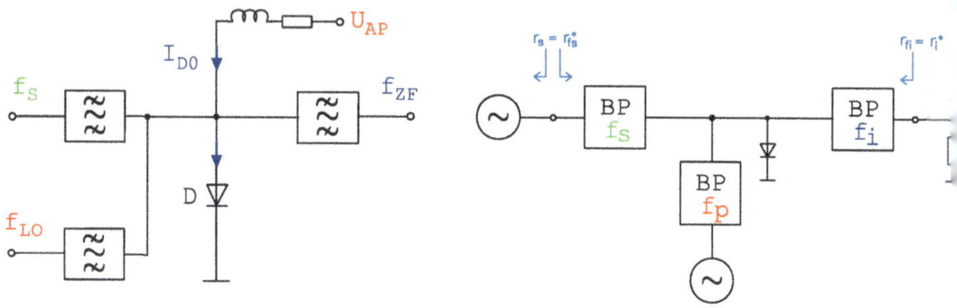

Fig. 6.15 Downward mixer with operating point adjustment of the Schottky diode

$$g_{(t)} = G_0 + 2 \sum_{n=1}^{\infty} G_n \cdot \cos\left(n\,\omega_{LO} t\right) . \tag{6.36}$$

The fact that an approximation with only the two coefficients G_0 and G_1 does not correspond perfectly to the real behavior is illustrated in Fig. 6.14.

6.6 Downward Mixing with Schottky Diode

There are two illustrations of a downward mixer with a Schottky diode in Fig. 6.15.

The gain G is the amount of the frequency-converting transfer function of the mixer (e.g. S_{21}^{is}). The maximum gain is given by G_m.

It is even more common to describe a mixer via the conversion losses:

$$\text{Conversion loss} = (\text{Gain})^{-1} \quad \text{resp.} \quad L = G^{-1} = S_{21}^{-1} = {S_{21}^{is}}^{-1} .$$

The minimum conversion loss is achieved with perfect matching.

The underlying network is derived from the Eq. (6.30). This equation is represented in Y-parameters and can be easily implemented as a PI circuit[2], Fig. 6.16 .

[2] This can be taken directly from auxiliary sheet A.2 of the book [38]

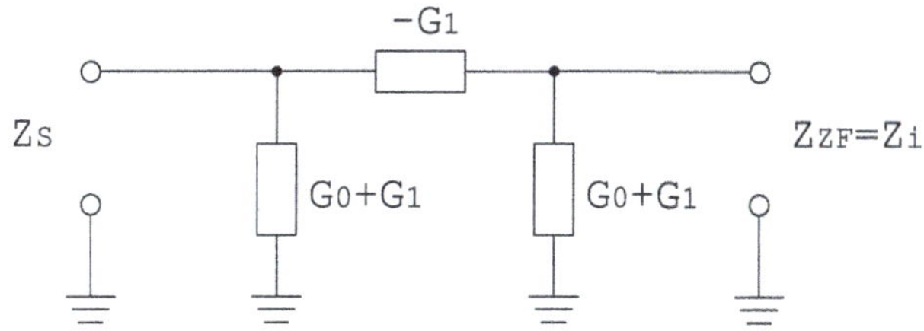

Fig. 6.16 Equivalent circuit diagram of the mixer with G_0 and G_1 according to Eq. (6.30)

In practice, a mixer is often adapted to 50 Ω on both sides. Furthermore, a real input resistance R_{in} or the associated conductance G_{in} for both gates is used.

$$G_{in} = \sqrt{G_0^2 - G_1^2} \tag{6.37}$$

After a little calculation, the calculation equation for the maximum gain (G_m) is obtained:

$$|S_{21}^{is}| = G_m = \left(\frac{G_1}{G_0}\right)^2 \cdot \frac{1}{\left(1 + \sqrt{1 - \frac{G_1^2}{G_0^2}}\right)^2} \tag{6.38}$$

Since: $\frac{G_1}{G_0} < 1 \quad \rightarrow \quad L^{dB} = (5 - 8)\ \text{dB}$.
The minimum conversion loss can be calculated from

$$L_{min} = \left(\frac{G_0}{G_1} + \sqrt{\frac{G_0^2}{G_1^2} - 1}\right)^2 . \tag{6.39}$$

6.7 Noise Behavior of the Mixer

The minimum signal power that can be processed by a receiver is determined by the inherent noise. The noise sources in the receiver are thermal noise, shot noise and up-mixed 1/f noise. To maximize the sensitivity of the receiver, the LNA is placed directly behind the antenna (cascade noise formula), see Fig. 6.17. However, such a receiver is sensitive to other sources of interference that drive the LNA into compression and thus put the receiver out of operation due to excessive power.

The S/N from the input (port 1) to the output (port 2) deteriorates by the conversion loss L. $\Rightarrow F \approx L$

The mixer with the LO oscillator can be seen as a frequency-converting element, Fig. 6.17.

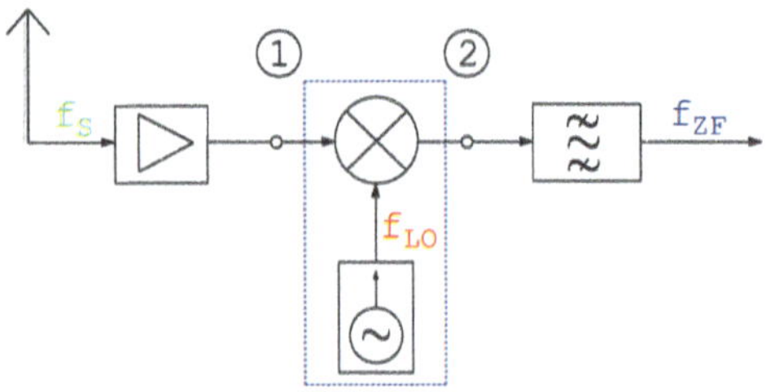

Fig. 6.17 Downward mixer with local oscillator as frequency-converting two-port

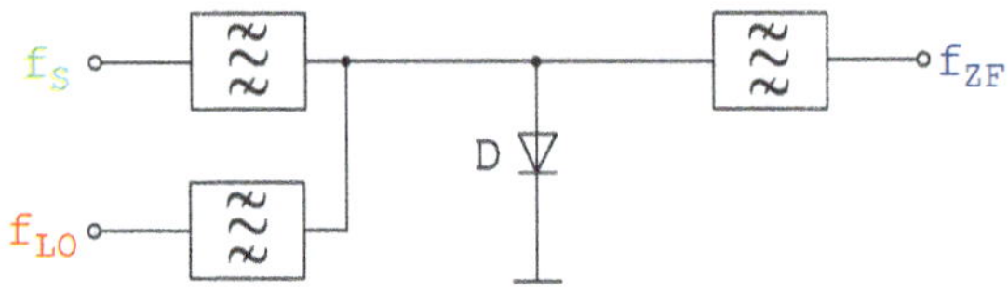

Fig. 6.18 Schematic diagram of a single-diode mixer for receivers with bandpass filters for port isolation

6.8 Performance Parameters of Mixers

The following conditions are imposed on mixers:

- Selection of a desired mixed product, suppression of all other products,
- Isolation between the ports for the radio frequency, local oscillator and intermediate frequency signal and
- High dynamic range: low noise, linearity up to high signal level.

Types of Mixer

Single-Diode Mixer:
These mixers have a single structure.

Fig. 6.18 illustrates a single-diode mixer as applied in the MHz range in particular.

Here there is a limited frequency range due to the isolation of the gates with bandpass filters.

In the GHz range, however, couplers are often used to ensure the isolation of the two GHz signals. The second port A in Fig. 6.19 is intended to represent a matching network.

A slight disadvantage of this solution is that the LO signal is attenuated by e.g. 10 dB. This high LO signal power must be provided for isolation with directional couplers. In the signal path, however, attenuation is irrelevant.

The details for mixer architectures in which the diodes are switched "hard" are also presented, as graphically explained in Fig. 6.20.

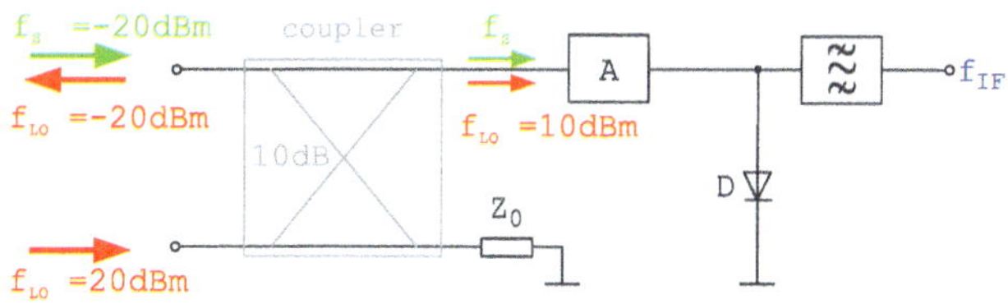

Fig. 6.19 Schematic diagram of a single-diode mixer with coupler for port isolation

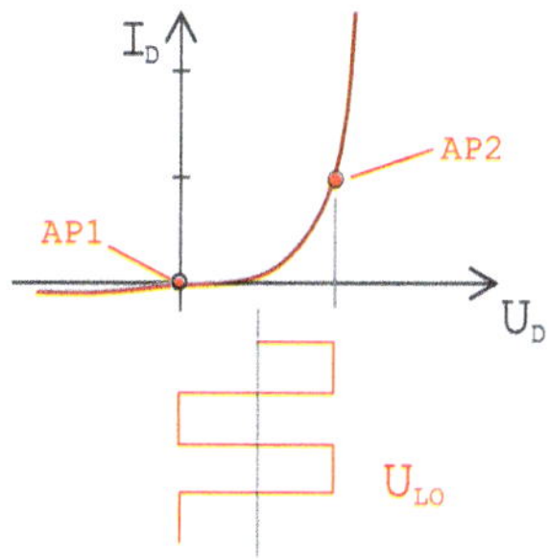

Fig. 6.20 Representation of the two operating points for the operation of a mixer switched by the LO signal operation of a mixer

Fig. 6.21 shows the signals in the time and frequency domain for the case that the LO signal switches the diode hard and the mixer has no conversion losses, so that no distinction is made between upward and downward operation.

Push-Pull/ Differential Mode Mixer:

This suppresses mixed products and ensures well-isolated ports. The frequency conversion is independent of fluctuations in the LO amplitude. Fig. 6.22 shows this kind of mixer with a low pass filter in the IF range.

Push-pull mixers are preferably operated in switching mode. The LO signal is often only a large sinusoidal signal. However, when analyzing the signals, a rectangular LO signal that switches the diodes hard is advantageous, Fig. 6.23.

In differential mode, as with power amplifiers, only odd harmonics occur. This reduces the number of unwanted mixed signals by a factor of 2.

However, it is crucial that the LO signal is suppressed very well in practice.

Ring Mixer:

These are mixers with four diodes arranged in a ring. This behaves largely like an ideal multiplier (Fig. 6.24).

Only very low signals occur around the harmonics of the LO signal. In addition, there are hardly any IF signals on the RF side and vice versa.

Furthermore, it is also very advantageous that the even harmonics of f_S are suppressed and their mixed products are therefore not found in the spectrum.

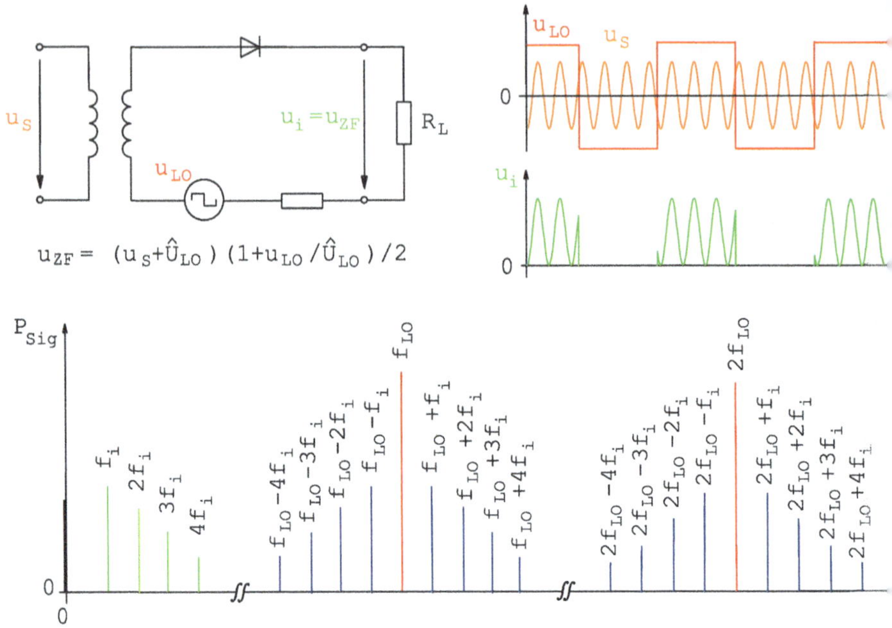

Fig. 6.21 Representation of the signals on a single-diode mixer in switching mode: Time signals for downward mixing and frequency spectrum for upward mixing

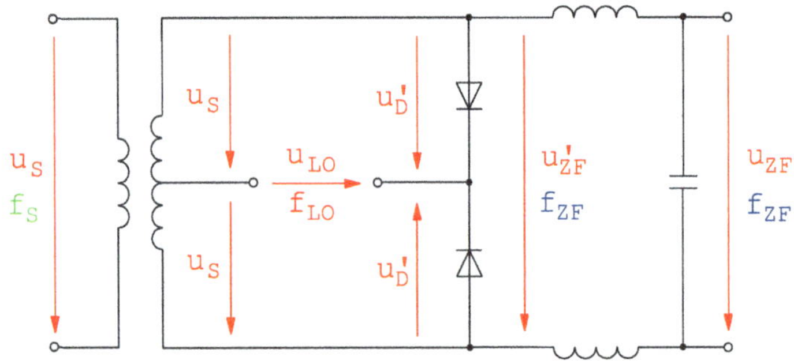

Fig. 6.22 Principle circuit diagram of a push-pull mixer with transformer as 180°-coupler ($u_{ZF} = u_{IF} = u_i$)

6.9 Realization of Mixers with Schottky Diodes

For discrete developments, transformers or couplers are typically used for the frequency ranges

- **f** ≤ 4 GHz ferrite core transformers and
- **f** ≥ 4 GHz couplers made from microstrip, coplanar and slotted line elements are used.

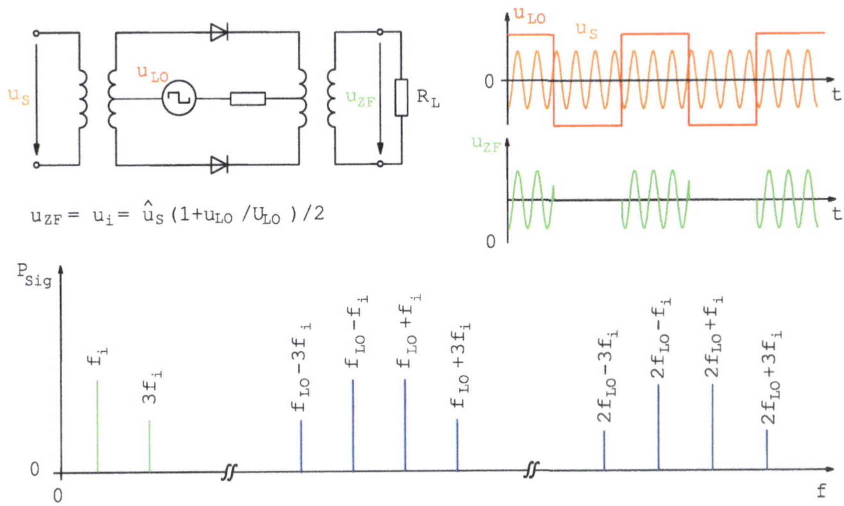

Fig. 6.23 Representation of the signals on a push-pull mixer in switching mode: Time signals for downward mixing and frequency spectrum for upward mixing

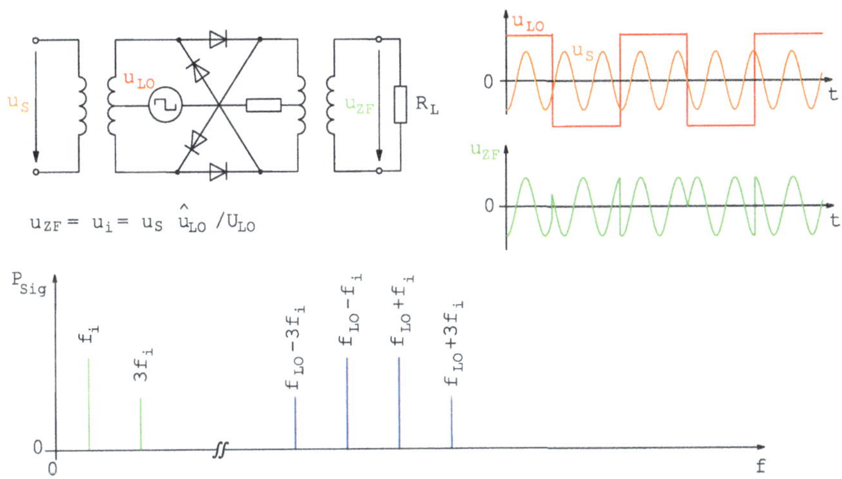

Fig. 6.24 Representation of the signals on a ring mixer in switching mode: Time signals for downward mixing and frequency spectrum for upward mixing

Fig. 6.25 shows the block diagram of a typical mixer setup for planar microstrip conductor technology with SMD components. As this mixer is a balanced or push-pull version, common-mode interference from the power supply or external radiation is suppressed. The corresponding structure in microstrip technology is shown in Fig. 6.26.

With these balanced mixers, the push-pull signal of the input controls the two diodes simultaneously. Here, a common mode signal is generated at the IF output.

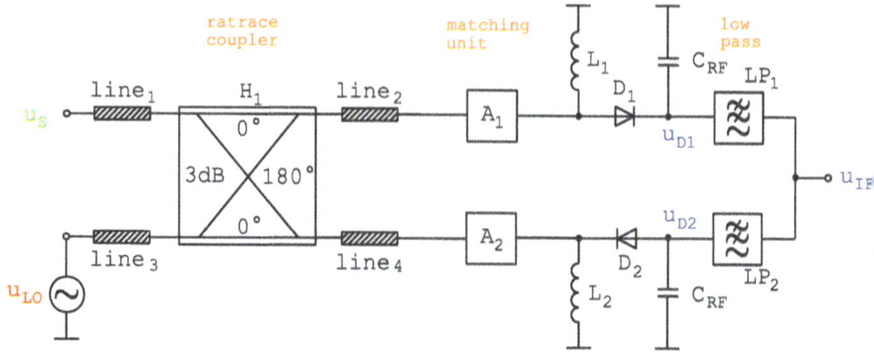

Fig. 6.25 Block diagram of a balanced mixer with ratrace coupler

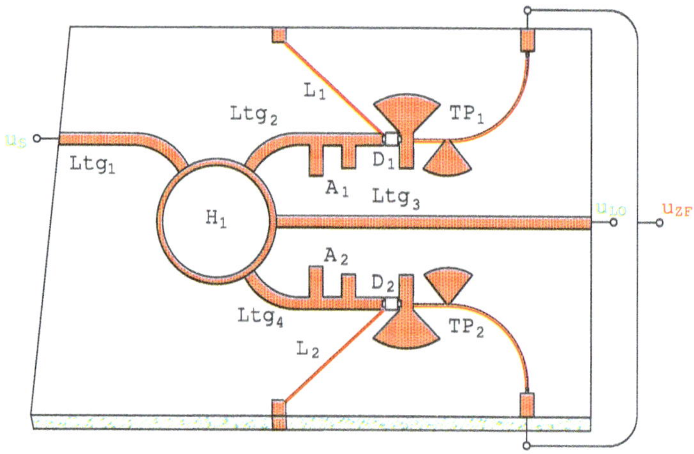

Fig. 6.26 Construction of a balanced mixer with ratrace coupler

In the push-pull mixers of Figs. 6.22 and 6.23, the two diodes are controlled in antiphase. Here, a push-pull signal is generated at the IF output.

The corresponding hardware structure is shown in Fig. 6.26 for a circuit in the high two-digit GHz range. The input signal at connection u_s is fed to the two diodes at 50 % each. No energy arrives at the fourth connection of the coupler (u_{LO}). The LO signal at connection u_{LO} is also fed to the two diodes at 50 % each. No energy is supplied to the fourth connection of the coupler (u_s) either. A downward mix is performed at each of the diodes, whereby the signals are 180 degrees to each other. The downward mixed push-pull signal is applied to u_{ZF}.

For setups in the MHz and single-digit GHz range, the so-called double-balanced mixers can also be used very well, Fig. 6.27.

The associated voltage curves (for the approximation that the LO signal is a square-wave signal) are shown in Fig. 6.28. The voltage u'_{IF} shown is the IF voltage before the low-pass filter. Fig. 6.29 shows the circuit of another push-pull mixer.

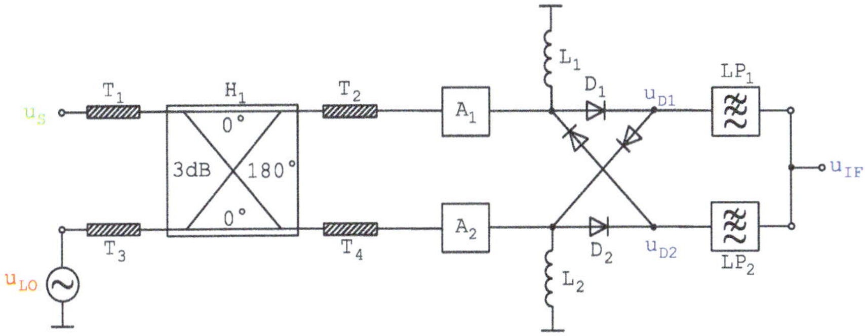

Fig. 6.27 Block diagram of a double balanced mixer with ratrace coupler

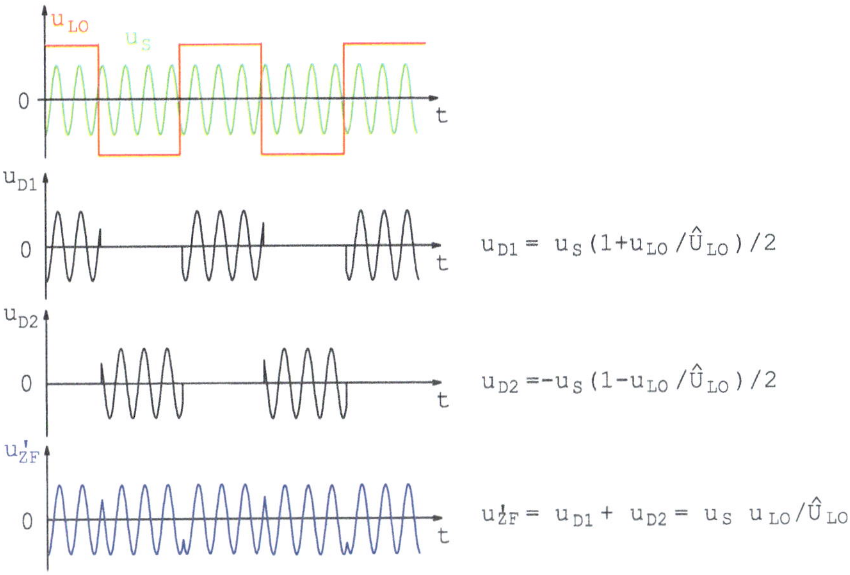

Fig. 6.28 Voltage curves on the double-balanced mixer with ratrace coupler

The line 1 must be a pure two-wire line that only allows one mode (e.g. twisted pair without shielding ground). In this case, the LO source first sees the matched unsymmetrical line 3, then the diodes, which in turn are matched to the system impedance via line 2a and line 2b.

In contrast, thc push-pull signal u_s can flow through the two diodes connected in series.

Typical Data of Diode Mixers:

Tab. 6.1 shows the typical values of a single-diode mixer and Tab. 6.2 shows the values of a push-pull mixer again.

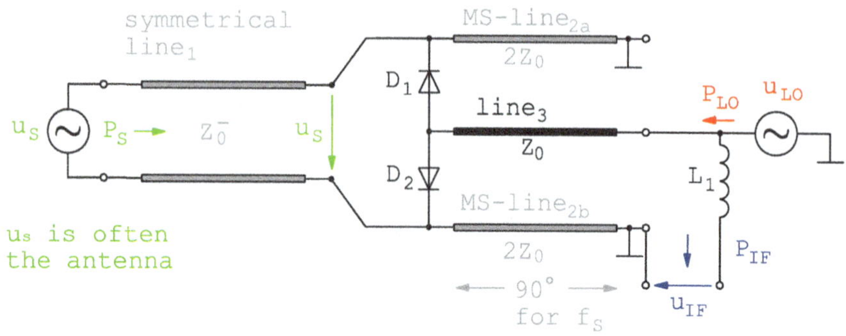

Fig. 6.29 Circuit of a push-pull mixer in balanced line technology

Tab. 6.1 Characteristics of typical single-diode mixers

	Frequency		
Mixer parameters	1 GHz	10 GHz	20 GHz
Conversion loss [dB]	5	7	10
Noise figure F [dB]	6	7	8
LO power [dBm]	5 ... 10		

Tab. 6.2 Characteristics of typical push-pull mixers

	Frequency		
Mixer parameters	1 GHz	10 GHz	20 GHz
Conversion loss [dB]	4	6	9
Noise figure F [dB]	6	7	8
Isolation LO signal	20	20	20
LO power [dBm]	5 ... 10		

6.9.1 Single Sideband Converters and IQ Modulators

To suppress the mirror frequency, single sideband converters, mixers, modulators or are used.

Single sideband converters and the closely related IQ modulators (also vector modulators) are based on interconnected mixers, which are regarded as multipliers. The structure of a single sideband transposer with two mixers and passive components, all of which have already been presented in detail, is shown in Fig. 6.30.

The local oscillator signal

$$u_{LO} = \hat{u}\cos(\omega_{LO}t) \tag{6.40}$$

is divided into the components u_{LOa} and u_{LOb} by means of a 0°/90° coupler and shifted in phase by 90° in the lower branch. The following therefore applies to the two signals reduced by 3 dB:

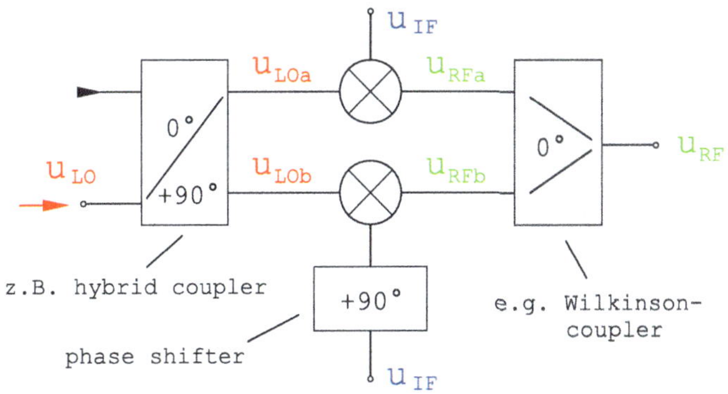

Fig. 6.30 Structure of a single sideband shifter in the image position

$$u_{LOa} = \frac{1}{\sqrt{2}} \hat{u} \cos(\omega_{LO} t) \quad \text{and} \quad u_{LOb} = -\frac{1}{\sqrt{2}} \hat{u} \sin(\omega_{LO} t) \,. \tag{6.41}$$

Assuming the transmission case with an IF transmission signal of the simple form

$$u_{IF} = \hat{u}_{IF} \cos(\omega_{IF} t) \tag{6.42}$$

and the conversion losses L, the multiplication results in the two partially phase-shifted signals with the voltages:

$$u_{RFa} = \frac{\hat{u}_{IF} \hat{u}}{\sqrt{2}} L \cos(\omega_{LO} t) \cos(\omega_{IF} t) \,, \tag{6.43}$$

$$u_{RFb} = \frac{\hat{u}_{IF} \hat{u}}{\sqrt{2}} L \sin(\omega_{LO} t) \sin(\omega_{IF} t) \,. \tag{6.44}$$

Addition by means of the 0° coupler using the addition theorems results in

$$\cos(\alpha - \beta) = \cos(\alpha) \cos(\beta) + \sin(\alpha) \sin(\beta) \tag{6.45}$$

the transmit signal:

$$u_{RF} = \frac{\hat{u_{IF}}}{\sqrt{2}} \hat{u} L \cos(\omega_{LO} t - \omega_{IF} t) \,. \tag{6.46}$$

The amplitude and phase of the transmitted signal is directly proportional to the IF signal. The information that may be contained in a modulation of the IF amplitude and/or the IF phase is retained in the high-frequency signal of the lower sideband (image position). The upper sideband was canceled out by the two 90° phase rotations. In practice, the suppression of the upper sideband is typically 40 dBc.

A similar calculation can be performed for the single sideband converter for the receive case.

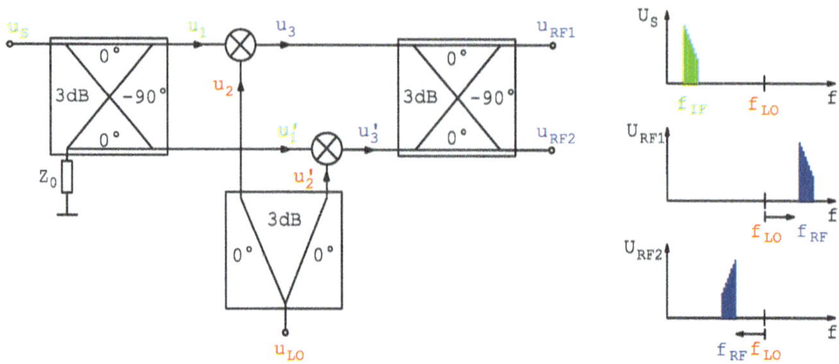

Fig. 6.31 Single sideband upward mixer ($u_S = u_{IF}$) with constant position output (RF2) and reverse position output (RF1)

An alternative setup with two RF outputs is shown in Fig. 6.31.

The signals of the single-sideband mixer with two RF outputs are analyzed as follows.

$$u_1 = \frac{u_S}{\sqrt{2}} = \frac{U_S}{\sqrt{2}} \cos(\omega_S t) \tag{6.47}$$

$$u_1{}' = \frac{u_S}{\sqrt{2}} \angle -\pi/2 = \frac{U_S}{\sqrt{2}} \cos(\omega_S t - \pi/2) = \frac{U_S}{\sqrt{2}} \sin(\omega_S t) \tag{6.48}$$

$$u_2 = u_2{}' = \frac{U_{LO}}{\sqrt{2}} \cos(\omega_{LO} t) \tag{6.49}$$

$$\begin{aligned} u_3 &= k\, u_1\, u_2 \\ &= \frac{k}{2} U_S U_{LO} \cos(\omega_S t)\, \cos(\omega_{LO} t) \\ &= \underbrace{\frac{k}{4} U_S U_{LO} \cos\left((\omega_{LO} + \omega_S)t\right)}_{\text{upper sideband}} + \underbrace{\frac{k}{4} U_S U_{LO} \cos\left((\omega_{LO} - \omega_S)t\right)}_{\text{lower sideband}} \end{aligned} \tag{6.50}$$

$$\begin{aligned} u_3{}' &= k\, u_1{}'\, u_2{}' \\ &= \frac{k}{2} U_S U_{LO} \sin(\omega_s t)\, \cos(\omega_{LO} t) \\ &= \underbrace{\frac{k}{4} U_S U_{LO} \sin\left((\omega_{LO} + \omega_S)t\right)}_{\text{upper sideband}} - \underbrace{\frac{k}{4} U_S U_{LO} \sin\left((\omega_{LO} - \omega_S)t\right)}_{\text{lower sideband}} \end{aligned} \tag{6.51}$$

$$u_{RF1} = \frac{1}{\sqrt{2}} \left(u_3 + u_3{}' \angle -\pi/2\right) \tag{6.52}$$

$$= \frac{1}{\sqrt{2}} \frac{k}{4} U_S U_{LO} \Big[\cos\left((\omega_{LO} + \omega_S)t\right) \cos\left((\omega_{LO} - \omega_S)t\right) \tag{6.53}$$

$$+\sin\left((\omega_{LO}+\omega_S)t-\pi/2\right)-\sin\left((\omega_{LO}-\omega_S)t-\pi/2\right)\Big] \tag{6.54}$$

$$=\underbrace{\frac{1}{\sqrt{2}}\frac{k}{2}U_S U_{LO}\cos\left((\omega_{LO}-\omega_S)t\right)}_{\text{lower sideband}}$$

$$u_{RF2}=\frac{1}{\sqrt{2}}\left(u_3\angle-\pi/2+u_3'\right) \tag{6.55}$$

$$=\frac{1}{\sqrt{2}}\frac{k}{4}U_S U_{LO}\Big[\cos\left((\omega_{LO}+\omega_S)t-\pi/2\right)+\cos\left((\omega_{LO}-\omega_S)t-\pi/2\right)$$

$$+\sin\left((\omega_{LO}+\omega_S)t\right)-\sin\left((\omega_{LO}-\omega_S)t\right)\Big] \tag{6.56}$$

$$=\underbrace{\sqrt{2}kU_S U_{LO}\sin\left((\omega_{LO}+\omega_S)t\right)}_{\text{upper sideband}} \tag{6.57}$$

The only difference between an IQ modulator and a single-sideband converter is the lack of a phase shifter in the IF branch, Fig. 6.32. However, the fact that the appearance of this IQ modulator is clearly different is partly due to the fact that IQ modulators are manufactured entirely using semiconductor technology. This is why there are no large distributed reciprocal couplers in these circuits and why the IQ transmitter modulator must be distinguished from the IQ receiver when handling them. Meanwhile, the transmitter and receiver units, as well as the oscillator technology described in the next section, are integrated on a single so-called transceiver IC (consisting of transmitter and receiver).

The mathematical description of the IQ modulator is similar to that of the single sideband transmitter. Modern data transmission and communication systems such as WLAN and LTE work with simultaneous amplitude ($A_{(t)}$) and phase modulation ($\phi_{(t)}$), which, however, can only assume discrete values. The modulated and highly mixed signal should be transmitted in the same position as the transmitted signal ($s_{(t)}$):

$$s_{(t)} = A_{(t)}\cos\left(\omega_{LO}t+\phi_{(t)}\right). \tag{6.58}$$

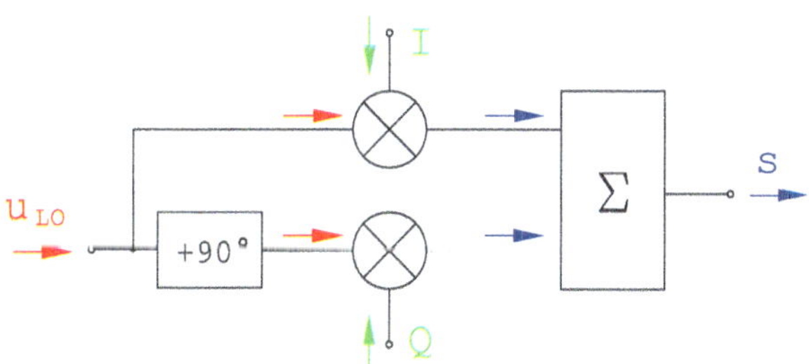

Fig. 6.32 Set-up of an IQ modulator for the transmission case in the equal position

If we apply the addition theorems to Eq. (6.58) again, we can divide the transmitted signal into a component $I_{(t)}$ that is in phase with the LO signal and a component $Q_{(t)}$ that is rotated by exactly 90° in phase with the LO signal:

$$s_{(t)} = I_{(t)} \cos(\omega_{LO} t) - Q_{(t)} \sin(\omega_{LO} t) \qquad \text{with} \tag{6.59}$$

$$I_{(t)} = A_{(t)} \cos\left(\phi_{(t)}\right) \qquad \text{and} \qquad Q_{(t)} = -A_{(t)} \sin\left(\phi_{(t)}\right) . \tag{6.60}$$

This generates the missing phase shifter in the IF branch in the computer. The fact that the Q signal is shifted by 90° to the I signal results in a conversion to the DC position.

Chapter 7
Phase-Locked Loops and Synthesis Generators

The connection of a high-frequency signal to a quartz oscillator with high frequency stability and low noise brought a huge leap in innovation for radio systems in the 1970s. The associated circuit is known as a phase-locked loop or phase-locked loop. However, the basic concept had already been developed 15 years earlier, [54].

This connection is standard for almost every oscillator or generator used. The (free-running) oscillators shown above are only a sub-component of the stabilized oscillator component.

PLL has become established for the standard abbreviation of the phase locked loop, which is also frequently used here. This chapter provides an introduction to the basics of PLL technology and presents the current state of circuit technology for the high-frequency range.

A PLL has two main tasks:

a) Stabilization of an RF signal in the frequency.
b) Reduction of the phase noise of the RF signal.

Both of these applications are achieved by controlling the output frequency or phase of the included controllable oscillator. Stabilization is necessary as the transmitter and receiver must operate in the same frequency band. The low phase noise is necessary to ensure the transmission dynamics.

7.1 Basics of Phase-Locked Loops

The task of a phase-locked loop is to connect the output signal of a frequency-adjustable oscillator (VCO, voltage controlled oscillator) with the frequency f_v to a reference oscillator with the stabilized frequency f_r and phase. In the following, the stabilization of the phase with an associated control. Later, we will discuss how this control can be extended to also control the frequency as well.

The block diagram of a pure phase control loop is shown in Fig. 7.1.

The output signal of the RF VCO with the frequency f_v (or the angular frequency ω_v) is divided to a lower frequency using an RF divider (1/N) and is then applied to the phase discriminator (PD). The signal with of the angular frequency ω_r of the quartz reference oscillator (XCO), which is very low-phase noise and temperature-stable, is applied to a second input of the phase discriminator (PD). This PD provides the output signal $u_d(t)$ if the two input signals are not in phase. This output signal

H. Heuermann, *Microwave Technology*,
https://doi.org/10.1007/978-3-658-45686-3_7

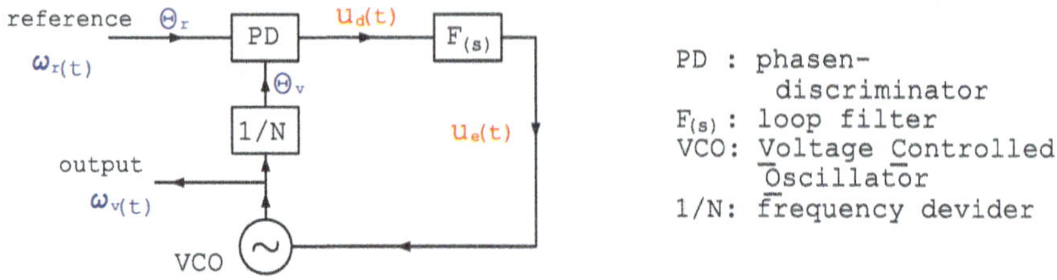

Fig. 7.1 Structure of a pure phase-locked loop (PLL)

$u_d(t)$ is smoothed using the filter F(s) and fed to the control input of the RF VCO as the error voltage $u_e(t)$.

In practice, the divider has digitally switchable divider values. It is also shown below that a frequency comparison is also carried out.

The following parameters are partly shown in Fig. 7.1 and partly introduced below:

Θ_v	: Phase of the tunable oscillator
Θ_r	: Phase of the reference signal
$\Theta_e = \Theta_r - \Theta_v$	: Error phase
ω_N	: Regulation frequency
$\Delta\omega$	: Frequency error
$u_d(t)$	: Output voltage of the PD
$u_e(t)$	: Error voltage
N	: Division factor
K_v	: Slope of the VCO in $\frac{\text{rad}}{V}$
K_d	: Slope of the discriminator in $\frac{V}{\text{rad}}$
$G(s)$	: Gain of the open control loop
$H(s)$	: Transfer function of the control loop
s	: Complex frequency ($s = \sigma + j\omega$)

The further derivation is based on the Laplace transformation:

$$Y(s) = \int_0^\infty y(t) \cdot e^{-st} \cdot dt \quad \Rightarrow \quad y(t) = 0 \quad \text{for } t < 0\,. \tag{7.1}$$

The original function $y_{(t)}$ is usually a time function. As the complex variable s contains the frequency ω, the image function $Y_{(s)}$ is often also referred to as a frequency function.

For the following derivation of the phase control, the frequency control of the PLL should be carried out first. This means that the frequencies at both inputs of the PD match.

If the frequency is "locked", the following applies: $f_v - N \cdot f_r = 0$.

The block diagram 7.1 and the variables listed above provide the following relationships for the time and frequency ranges:

$$\begin{array}{rcl} \text{Time domain} & \bullet\!\!-\!\!\circ & \text{Frequency domain} \\ u_d(t) = K_d \cdot (\Theta_r(t) - \Theta_v(t)) & \bullet\!\!-\!\!\circ & U_d(s) = K_d\,(\Theta_r(s) - \Theta_v(s)) \quad (7.2) \\ u_e(t) = f(t) * u_d(t) & \bullet\!\!-\!\!\circ & U_e(s) = F_{(s)} \cdot U_d(s) \quad (7.3) \\ \Delta\omega(t) = \dfrac{d\Theta_v(t)}{dt} = \dfrac{K_v \cdot u_e(t)}{N} & \bullet\!\!-\!\!\circ & s \cdot \Theta_v(s) = \dfrac{K_v \cdot U_e(s)}{N} \quad (7.4) \end{array}$$

With $\Delta\omega(t)$, an additional variable was introduced at the output of the VCO for the frequency error that occurs briefly due to a changing error voltage $u_e(t)$.

Derivation of the Transfer Function $H(s)$

Eq. (7.2) can be used in (7.3) to calculate the error voltage in the frequency range:

$$U_e(s) = F(s) \cdot K_d \cdot (\Theta_r(s) - \Theta_v(s))\,. \tag{7.5}$$

If you insert this expression of the error voltage (7.5) into Eq. (7.4), you get:

$$\Theta_v(s) = \underbrace{\frac{F(s) \cdot K_d \cdot K_v}{s \cdot N}}_{=G(s)} \cdot (\Theta_r(s) - \Theta_v(s)) \tag{7.6}$$

with the abbreviation $G(s)$ for the gain of the open control loop. The factor $1/s$ appears in this transfer function because it is not the phase of the VCO that is adjusted directly but the frequency. This additional s makes stable operation of the phase-locked loops more difficult because 90° phase-locked loops are already used up.

If the equation (7.6) is rearranged, the transfer function of the control loop is obtained:

$$H(s) = \frac{\Theta_v(s)}{\Theta_r(s)} = \frac{G(s)}{1 + G(s)}\,. \tag{7.7}$$

7.2 Control Behavior

The development of the control filter or at least the determination of the filter coef ficients is a fundamental problem when implementing a PLL in a system.

For a more detailed analytical description, a PI controller is used as a simple 1st order filter, Fig. 7.2.

The corresponding transfer function of this filter is:

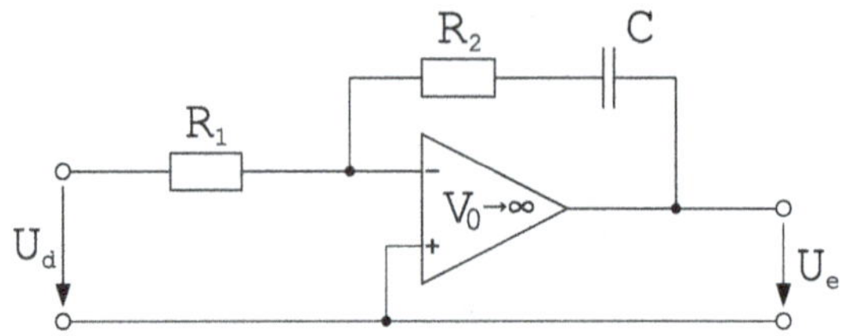

Fig. 7.2 PI controller as active 1st order control filter

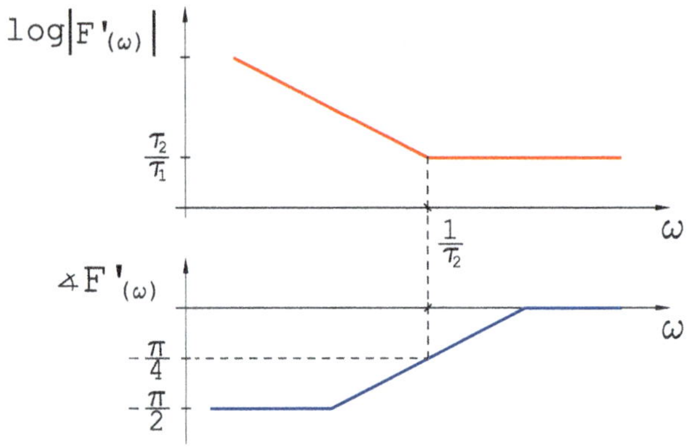

Fig. 7.3 Bode diagram of the 1st order filter as a phase-reducing element

$$F'(s) = \frac{U_e(s)}{U_d(s)} = -\frac{s\,\tau_2 + 1}{s\,\tau_1} \tag{7.8}$$

with $\tau_1 = R_1 \cdot C$ and $\tau_2 = R_2 \cdot C$.

The corresponding frequency behavior of $F'(s)$ is shown in Fig. 7.3.

The minus sign in $F'(s)$ can be eliminated by swapping the inputs on the phase discriminator. The next step is to continue with $F(s) = -F'(s)$.

Based on the general transfer function of the PLL

$$H(s) = \frac{G(s)}{1 + G(s)} \quad \text{with} \quad G(s) = \frac{K_d \cdot K_v}{s \cdot N} \cdot F(s) \tag{7.9}$$

results in the special gain of the open loop for the 1st order filter used

$$G(s) = \frac{2 \cdot \zeta \cdot \omega_N \cdot s + \omega_N^2}{s^2} \quad \text{resp.} \tag{7.10}$$

the special transfer function of the control loop

$$H(s) = \frac{2 \cdot \zeta \cdot \omega_N \cdot s + \omega_N^2}{s^2 + 2 \cdot \zeta \cdot \omega_N \cdot s + \omega_N^2} \tag{7.11}$$

with of the attenuation constant[1]

$$\zeta = \frac{\tau_2}{2}\sqrt{\frac{K_V \cdot K_d}{N \cdot \tau_1}} \quad (\text{typ. } \frac{1}{\sqrt{2}} \cdots 1) \tag{7.12}$$

and the control frequency

$$\omega_N = \sqrt{\frac{K_V \cdot K_d}{N \cdot \tau_1}} \quad ! \sim \frac{1}{\sqrt{N}}\,, \tag{7.13}$$

the design of which should be as high as possible in practice. Within the frequency range of the control frequency, the phase noise of the free-running oscillator is reduced immensely by the control. It is reduced to the value of the phase noise of the XCO plus the divider value in dB ($N = 100$ corresponds to 20 dB).

Numerical Example

The XCO frequency of 10 MHz and the divider value N of 100 result in an RF frequency of 1 GHz. The XCO phase noise of −150 dBm/Hz at 100 kHz spacing results in an RF oscillator noise of −130 dBm/Hz at 100 kHz spacing for this example. The division factor of 100 causes the degradation by this 20 dB.
It is important to examine the stability of $H(s)$.

Investigation: Frequency sweep with a PLL

An important prerequisite for the PD is that the following must apply $\left|\Theta_{e(t)}\right| < \pi$. Details on this follow in the PD realization.

A frequency sweep is initiated by changing the divider value. This results in the error phase from Eq. (7.6) in

$$\Theta_{e(s)} = \Theta_{r(s)} - \Theta_{v(s)} = \frac{-1}{G(s)} \cdot \Theta_{v(s)}\,. \tag{7.14}$$

For the frequency sweep via continuous change of the divider factor applies at the divider output (e.g. [88]):

$$\Theta_{v(s)} = \frac{\Delta\omega}{s^3}\,. \tag{7.15}$$

The double derivative is included here. The following already applies for a frequency jump:
$\Theta_{v(s)} = \frac{\Delta\Theta}{s} = \frac{\Delta\omega}{s^2}$. For a sweep, the product $\frac{1}{s}$ is added.

It generally applies to the Laplace transformation:

$$\lim_{t\to\infty} y_{(t)} = \lim_{s\to 0} s \cdot Y(s)\,. \tag{7.16}$$

[1] The Greek symbol ζ stands for zeta

Applied to our control problem, this results in the following:

$$\lim_{t\to\infty} \Theta_{e(t)} = \lim_{s\to 0} \left\{ \frac{-s}{G(s)} \Theta_v(s) \right\}, \tag{7.17}$$

$$= \lim_{s\to 0} \left\{ \frac{s^3}{2 \cdot \zeta \cdot \omega_N^2 \cdot s + \omega_N^2} \cdot \frac{\Delta\omega}{s^3} \right\}, \tag{7.18}$$

$$= \frac{\Delta\omega}{\omega_N^2} \overset{!}{<} \pi . \tag{7.19}$$

The Bode criterion for a control loop states that a transfer function must have an attenuation (magnitude less than 1) if a phase rotation of 180° is achieved. The error phase Θ_e describes the phase rotation of the transfer function $H(s)$. This means that the value π for Θ_e must not be reached. Consequently, the maximum change frequency for a sweep may only be the value $\Delta\omega_{max} = \omega_N^2 \cdot \pi$. The maximum tuning rate is therefore $\sim \omega_N^2$ i. h. $\sim \frac{1}{N}$.

7.3 Frequency Divider

The frequency dividers (dividers for short) of a PLL are the high-speed digital circuits and, with regard to the upper frequency that can be used, the module that limits the PLL.

Fixed dividers have a non-variable division ratio. Synchronous counters made from the master-slave JK flip-flops are often used here.

Switchable dividers are divided into two classes:

1. Dividers for frequency range changes,
2. Fractional divider.

The former contain multi-stage fixed dividers whose outputs are switched depending on the frequency range selection. Fractional dividers switch between a divider ratio N and at least one further divider ratio (e.g. N + 1). This leads to synthesizers with small frequency steps, as explained below.

It is very important to understand that the size of the division factor has a very large influence on the phase noise of a PLL. At best, the VCO can have a phase noise as low as that of the reference oscillator. However, due to the frequency division, changes only reach the phase discriminator *later*. This attenuation is reflected in a lower phase noise.

A frequency division of N = 10 corresponds to a deterioration of the phase noise by 10 dB. N = 100 corresponds to 20 dB less phase noise compensation.

This is why modern phase-locked loops rely on low division factors.

7.4 Discriminators

Es there are two types of discriminators in PLLs:

1. Phase discriminator (PD) also Phase detector,
2. Phase-frequency discriminator (PFD).

These are available in analog and digital circuit technology.

7.4.1 Phase Discriminators

The use of pure phase discriminators is rarer than that of phase-frequency discriminators. However, phase discriminators are at the heart of the PLL and are particularly important for understanding it.

Mixer

A Mixer is a special form. It replaces a fixed divider and has the PD function. Advantageous is that, in contrast to the fixed divider, no additional phase noise is added to the XCO. The disadvantage of the mixer as a PD is that the signals must already be in phase with $\pm 90^\circ$, otherwise otherwise mixing will occur at the wrong sideband. More details can be found in [88].

Exclusive-OR Gate

A simple exclusive-OR gate can be used as a phase discriminator. In contrast to the mixer, it has a desired linear characteristic curve. However, the two input signals must have a duty cycle of 50 %, which often cannot be guaranteed. Furthermore, as with the mixer, the error phase is symmetrical. This means that an exclusive-OR gate has the same restricted $\pm 90^\circ$ range as a mixer.

RS resp. JK Flip-Flop

In contrast to the two previously mentioned discriminators, an edge-controlled RS flip-flop (Fig. 7.4) has a uniqueness range over 360°.

With the edge-controlled RS flip-flop, the output signal Q is generated with the rising or falling edge according to the logic

	Q
S from $0 \rightarrow 1$	1
V from $0 \rightarrow 1$	0

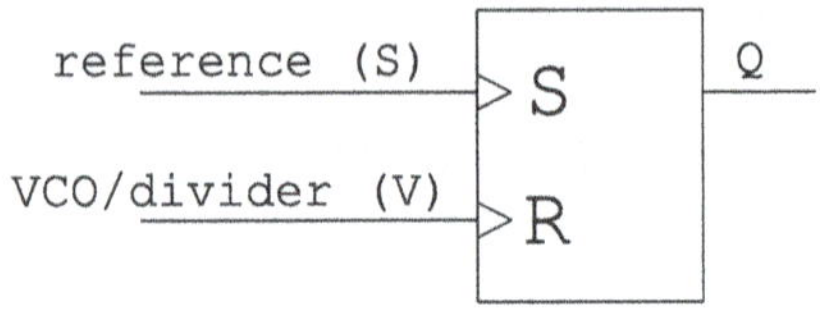

Fig. 7.4 RS flip-Flop as PD

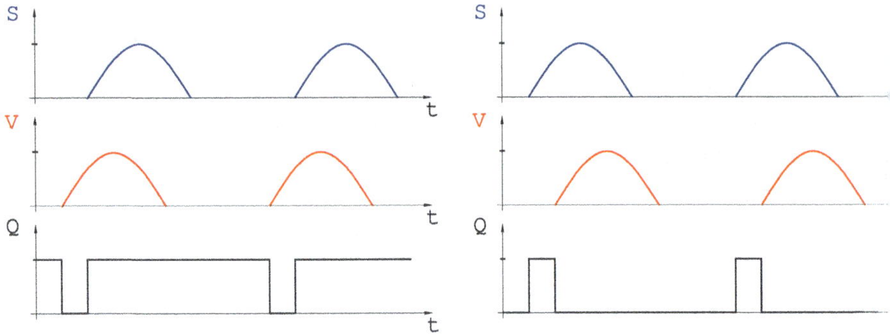

Fig. 7.5 Examples of the use of the RS or JK flip-flop for different phase positions

is controlled. The JK flip-flop is set with J = 1 and reset with K = 1 is reset:

J	K	Q^{n+1}	
0	0	Q^n	Storage
0	1	0	Reset
1	0	1	Set
1	1	$\overline{Q}^n$	Shift

The two examples illustrated in Fig. 7.5 show the positive signal components of the two input signals S and V. The edge control starts when the baseline is crossed. Simultaneous setting and resetting is therefore unlikely. When using a JK master-slave flip-flop, this unauthorized state does not exist at all.

The low-pass filtered signal of Q is proportional to the error phase Θ_e. From this signal, you can easily derive the error phase shown in Fig. 7.5 can be derived from this signal.

The smoothed output voltage $\overline{U}_q$ with the peak value U_u over the error phase Θ_e is shown in the following Fig. 7.6.

You can see the uniqueness range over 360°. Due to this asymmetrical characteristic curve, a PLL with mixer can safely lock onto a sideband.

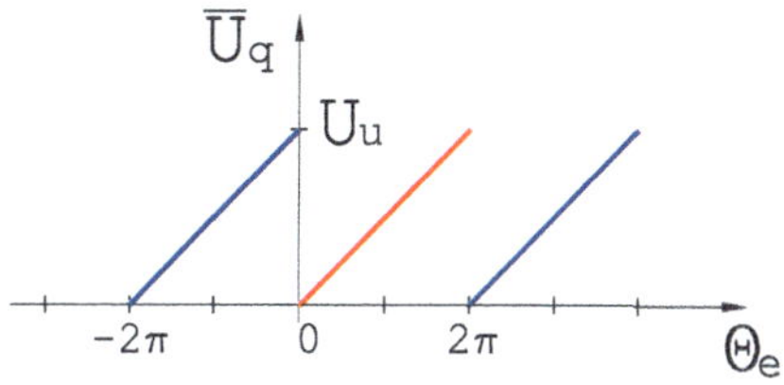

Fig. 7.6 Mean value of U_q over the error phase

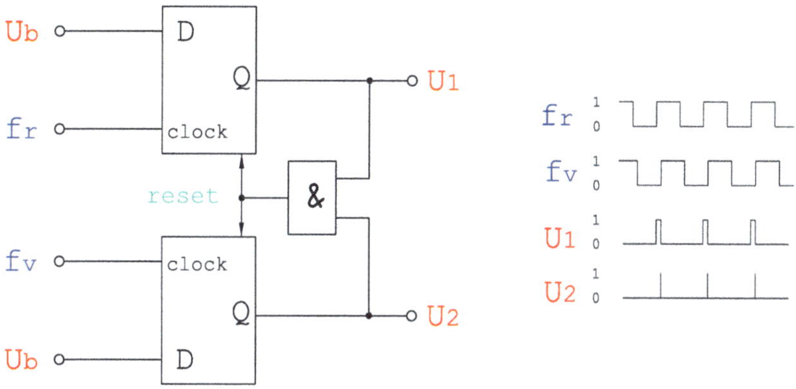

Fig. 7.7 Phase frequency discriminator with D flip-flops (positive edge control) and logic output signals

7.4.2 Phase-Frequency-Discriminators

In the last decade, complex digital circuits have been used as phase frequency discriminators (PFDs for short). have been used. The most important representative of these comparators is explained in [84] or [88]. This IC from Motorola (type MC12044) in ECL-Logic has a frequency capture range of up to 800 MHz.

Currently, a different architecture, shown in Fig. 7.7, is increasingly being used.

Both D flip-flops are edge-controlled and have a reset input.

Originally, this circuit was used with low-pass filters and the evaluation of the differential signal at the output, Fig. 7.8.

These circuits have the disadvantage that they do not have a constant gain or slope (K_D). This disadvantage no longer occurs if switched current sources with a charge capacitor C_p are used, Fig. 7.9.

The so-called *PFD dead zone* must be mentioned as a minor disadvantage of charge pumps. If the PLL is almost engaged, there are only very small output signals that no longer generate an output voltage due to the finite switching times of the switches and low-pass effects of the current sources.

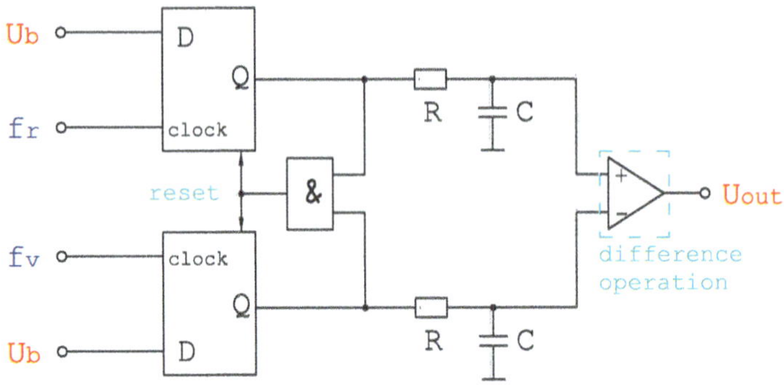

Fig. 7.8 PFD with low-pass filters

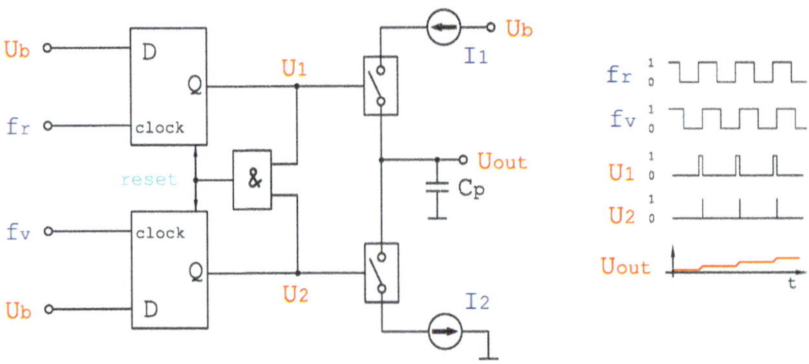

Fig. 7.9 PFD with charge pump (CP)

7.4.3 Design of Loop Filters

The output of a modern CP-PLL generates DC pulses. The time difference between the pulse intervals is proportional to the phase error. This pulsed signal cannot be used directly to control the VCO. It must first be smoothed by a low-pass filter. The design of the filter is decisive for the PLL characteristics. The filter has an influence on

stability, frequency reserve, control bandwidth, phase noise,
lock-in time and harmonic generation.

Active filters have the disadvantage that they generate additional noise. Active filters are no longer necessary, especially for CP-PLLs.

The general transfer function for a passive 4th order filter is as follows:

$$F_{(s)} = \frac{1 + s \cdot \tau_2}{A_0 \cdot s \cdot (1 + s \cdot \tau_1) \cdot (1 + s \cdot \tau_3) \cdot (1 + s \cdot \tau_4)} \,. \qquad (7.20)$$

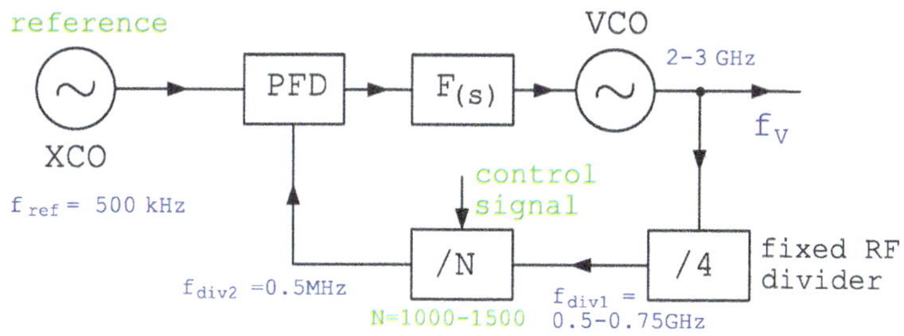

Fig. 7.10 Block diagram of the single-loop control circuit

The values for three different filter orders with the corresponding time constants are shown in Tab. 7.1.

In practice, the higher order filters have no negative effects on stability or other variables. In particular, these filters help to suppress unwanted signals above the loop bandwidth.

7.5 Single-Loop Control Circuit

A very simple synthesizer is shown in Fig. 7.10.

The output signal of the RF VCO is divided to a lower frequency using a fixed RF divider (/4) and a switchable divider (/N) and is then applied to the phase frequency discriminator. The signal of the quartz (crystal) reference oscillator (XCO), which is very low-phase noise and temperature-stable, is applied to a second input of the phase frequency discriminator. This PFD provides an output signal if the both input signals are not in phase or the frequency is not locked. This output signal is smoothed using the filter F(s) and fed to the control input of the RF VCO.

This control loop shows almost no improvement in phase noise due to the high division factors. It is primarily used for precise frequency linking. The stabilized output frequency of the VCO is:

$$f_v = \underbrace{N_{fix} \cdot N}_{=Ntot} \cdot f_{ref} \qquad \text{(here with} \quad N_{fix} = 4) \,. \tag{7.21}$$

The smallest step width

$$\Delta f = f_{ref} N_{fix} \qquad (\text{here} = 500\,\text{kHz}\, N_{fix}) \tag{7.22}$$

is determined by the output frequency of the reference signal and the fixed divider.

Alternatively, the smallest step width can be calculated using the bandwidth and the number of switchable divider values:

Tab. 7.1 Table for the design of passive loop filters

Parameter	Passive loop filters and their time constants		
	Filter 2. order	Filter 3. order	Filter 4. order
Filter	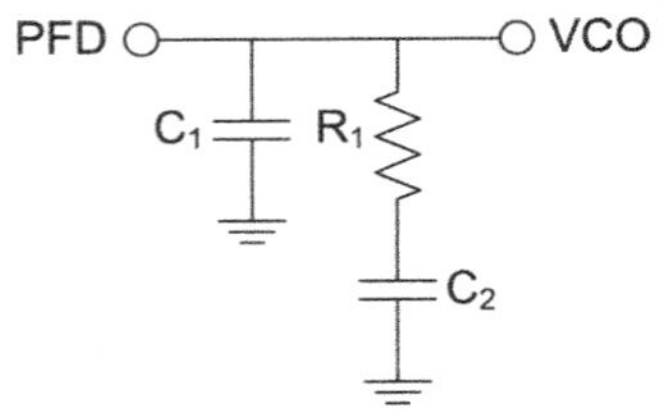	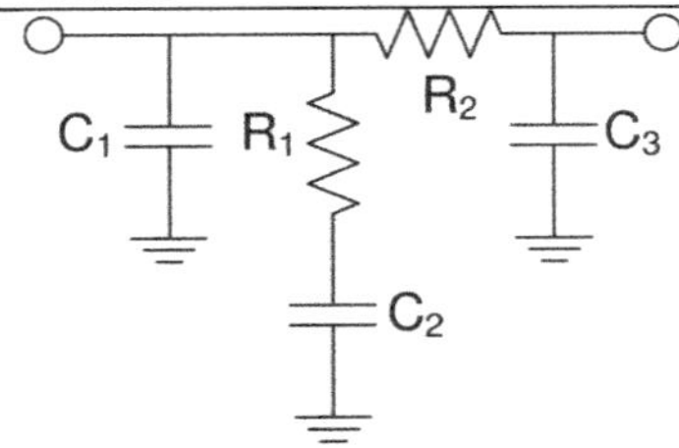	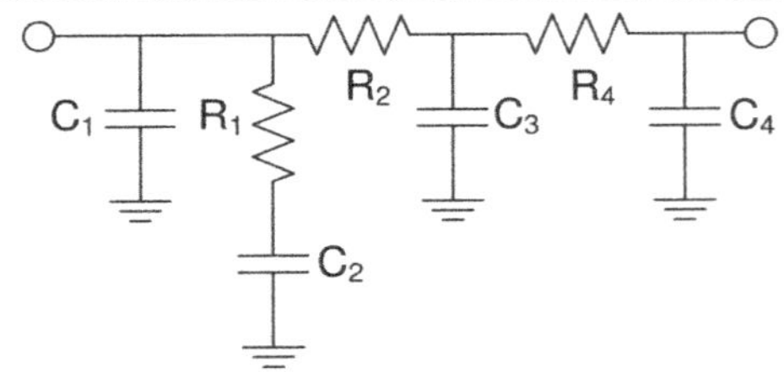
τ_1	$R_1 \cdot C_2 \cdot C_1 / A_o$	$R_1 \cdot C_2 \cdot C_1 / A_o$	$R_1 \cdot C_2 \cdot C_1 / A_o$
τ_2	$R_1 \cdot C_2$	$R_1 \cdot C_2$	$R_1 \cdot C_2$
τ_3	0	$R_2 \cdot C_3$	$R_2 \cdot C_3$
τ_4	0	0	$R_4 \cdot C_4$
A_o	$C_1 + C_2$	$C_1 + C_2 + C_3$	$C_1 + C_2 + C_3 + C_4$

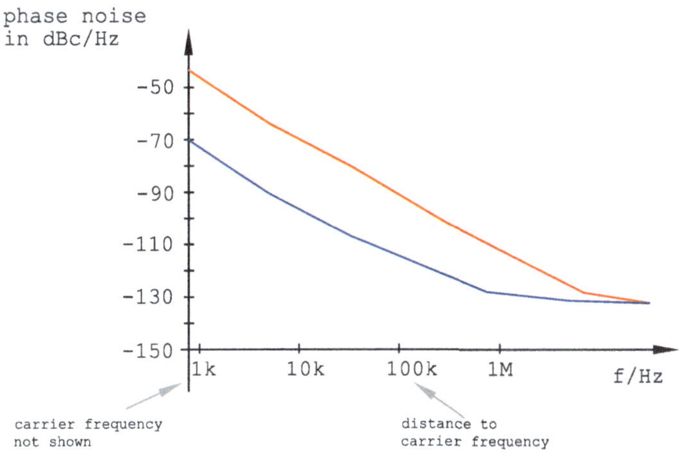

Fig. 7.11 Phase noise for low and high reference frequency (blue)

$$\Delta f = \frac{f_{max} - f_{min}}{N_{max} - N_{min}} \qquad \left(\text{here} = \frac{1\,\text{GHz}}{500}\right) . \tag{7.23}$$

However, only with a high control frequency $\omega_N \sim \frac{1}{\sqrt{N}}$ can the spectral purity of the VCO be brought close to that of the XCO and thus achieve low phase noise. Fig. 7.11 illustrates the differences in phase noise at low and high reference frequencies.

The noise power increases for large division factors $\sim N^2$.

$\Rightarrow$ Contradiction: small step size & good phase noise

Remedies: 1st Control loop with mixer
2nd Multi-loop control loops
3rd Control loops with fractional dividers

7.6 Control Loop with Mixer

Constant step widths with a smaller division factor can be obtained by using an additional fixed oscillator (LO) and a mixer, as shown in the Fig. 7.12.

The output frequency for mixers in equal position is

$$f_v = f_{LO} + N \cdot f_{ref} . \tag{7.24}$$

The step size is now f_{ref}.

In terms of phase noise, this circuit has very good characteristics. However, the total circuit complexity including the stabilized fixed oscillator is very high and therefore such mixers are only used for PLLs with high output frequencies. Furthermore, the control loop with mixer is only suitable for narrow-band applications.

7.7 Multi-Loop Control Loops

For synthesis frequencies in the lower GHz range, the nested control loops are the better solution.

The basic idea is that the XCO with a fixed output frequency is replaced by a reference oscillator with a variable output frequency. Its signal can be generated by direct signal generation via a D/A converter or another classic PLL.

If, for example, the XCO is replaced by a second PLL with the output frequency $f_v' = f_{ref} = 10 - 20\,\text{MHz}$, dividers are required over the range N = 100 – 200 with the associated low noise[2], Fig. 7.13. A switchable divider with the variable division factor M is used for the new reference PLL. With $f_v' = M f_{ref}'$ with the range M = 200 – 400 and only $f_{ref}' = 50\,\text{kHz}$ follows

$$f_v = N \cdot M \cdot f_{ref}' \,, \tag{7.25}$$

$$\Delta f_{min} = f_{ref}' \,, \tag{7.26}$$

$$\Delta f_{typ} = 10 f_{ref}' \,. \tag{7.27}$$

You can see that the step size has been reduced to a tenth. Furthermore, both loops have a significantly smaller divider factor. As a result, both loops control faster, which means that the phase noise of the individual loops is significantly reduced. For the multi-loop control circuit, the bandwidth in which the noise is reduced is improved.

However, the output signal close to the carrier only has a possibly improved phase noise due to the better VCO in the MHz loop. In practice, the HF VCO has by far the worst phase noise.

In terms of circuitry, the choice of a DDS (direct digital synthesizer) is a simple setup as an alternative to the first control loop with a low reference frequency shown, Fig. 7.14.

However, the phase noise of the DDS is so poor that low-noise synthesizers cannot be produced in this way. However, frequency changes can be made in the μHz range.

[2] The larger divider values with the larger range ensure that the minimum frequency step f_{ref}' is achieved in all combinations

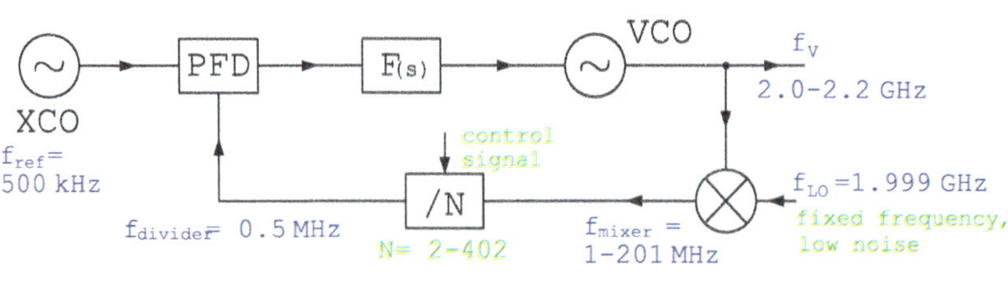

Fig. 7.12 Block diagram of a control loop with single-sideband mixer in equal position

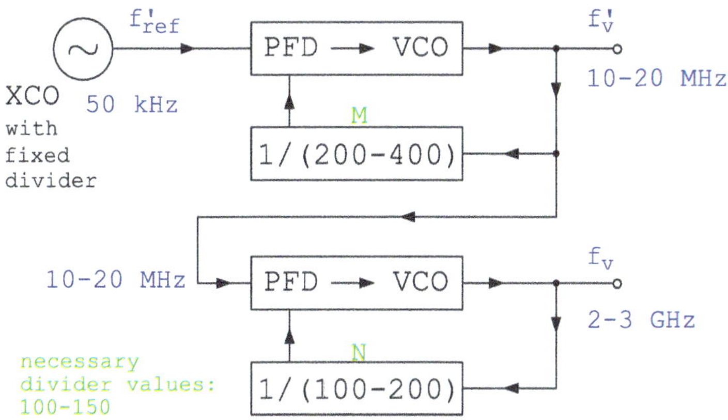

Fig. 7.13 Block diagram of a multi-loop control circuit

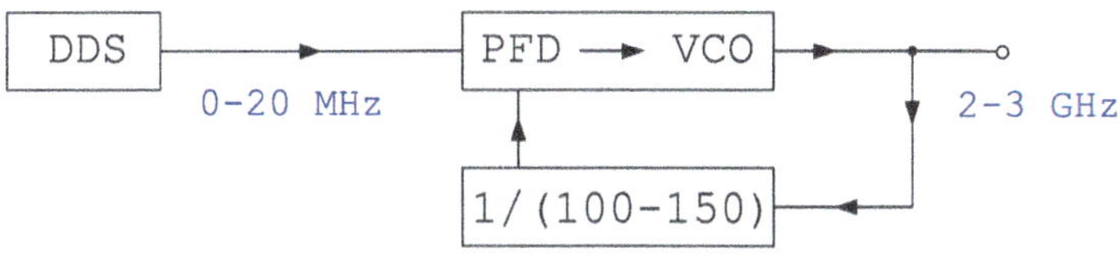

Fig. 7.14 Block diagram of a multi-loop control circuit with DDS

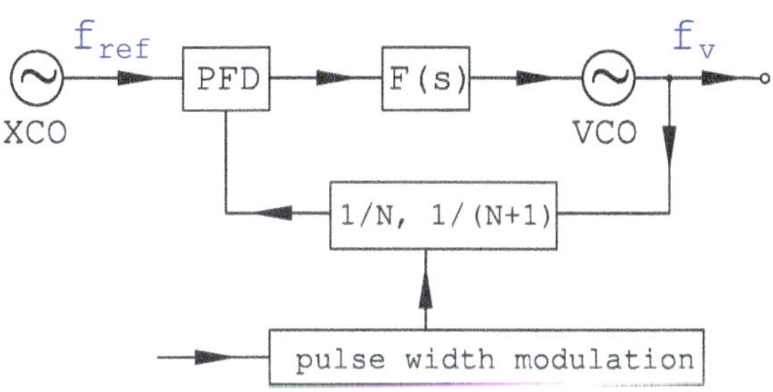

Fig. 7.15 Block diagram of a fractional-N synthesizer

7.8 Control Loops with Fractional Dividers

With a divider that is not limited to integer divider ratios, but also allows fractional values, the step size of the synthesizer can be made smaller than the comparison frequency [25, 84]. The calculation of f_{VCO} according to Eq. (7.21) remains valid. However, the effective division ratio is then no longer restricted to integer values, while the value N is still fixed. The only slightly changed PLL shows the Fig. 7.15.

A fractional divider ratio is achieved by switching between two or more integer values (a pulse width modulation of the divider ratio, so to speak). The typical progression is shown in Fig. 7.16.

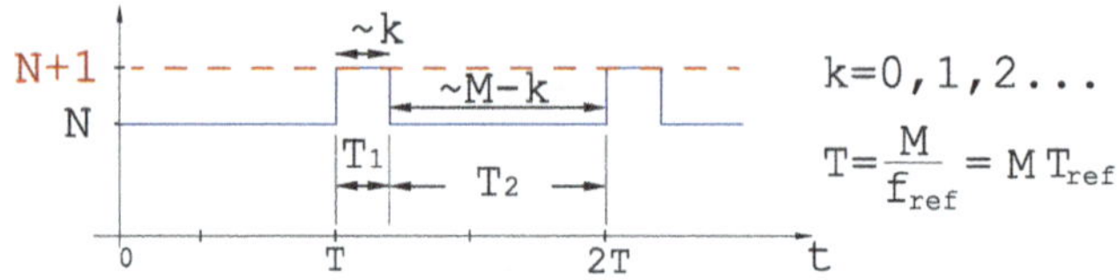

Fig. 7.16 Switching sequence for fractional divider

Choice: k = 1: The divider is switched to N + 1 for the shortest possible period. This period corresponds to T_{ref}, the period duration of the reference frequency f_{ref}.

$$T_{k=1} = \frac{1}{f_{ref}} = T_{ref} \tag{7.28}$$

The quantity M is to be introduced as the divider factor rate and indicates the fraction of a period T with which the divider can be controlled.

When switching between two divider values N_1 and N_2 with the pulse durations

$$T_1 = \frac{k}{f_{ref}} \quad \text{and} \quad T_2 = \frac{M-k}{f_{ref}} \tag{7.29}$$

the mean value of the divider ratio and the smallest increment is then

$$\overline{N} = \frac{N(M-k)+(N+1)\cdot k}{M} = N + \frac{k}{M}, \tag{7.30}$$

$$\Rightarrow f_V = \overline{N}\cdot f_{ref} = \left(N+\frac{k}{M}\right)\cdot f_{ref}, \tag{7.31}$$

$$\Delta f = \frac{1}{M}\cdot f_{ref} \quad \text{resp.} \quad \Delta f = \frac{1}{T}. \tag{7.32}$$

The frequency of the VCO should then be at the reference frequency multiplied by the average value of the divider ratio. Switching the divider ratio causes interference phase modulation at the phase discriminator. The smallest time unit in which T_1 and T_2 can vary is $T_{pulse} \sim 2k$. The minimum frequency of the interference phase modulation is therefore dependent on f_{pulse} and the required smallest adjustment step size ΔN of

$$f_{stör} = f_{Puls}\cdot \Delta N. \tag{7.33}$$

The minimum frequency of the phase disturbance caused by the divider switching therefore corresponds to the smallest step size Δf of the frequency synthesizer. So far, the fractional-N method differs only slightly from the classic synthesizer with a fixed divider in terms of interference.

If an attempt is made to suppress the interference using a narrow loop filter the same compromises have to be made as in the previous section. The only advantage would then be the multiplication factor of the phase comparator and reference

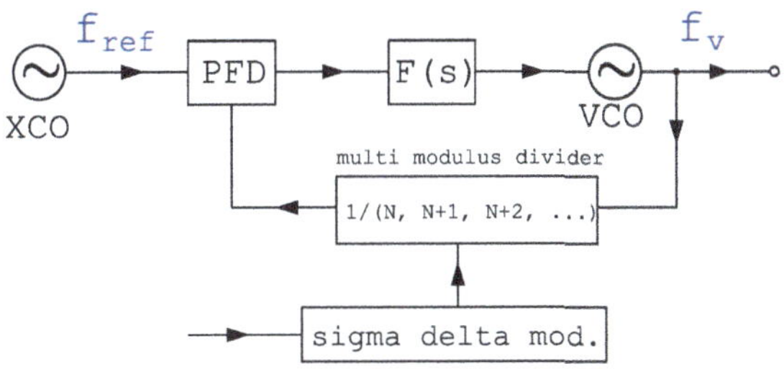

Fig. 7.17 Block diagram of a ΣΔ-Fractional-N synthesizer

noise. Since the course of the interference can be predicted is predictable, it can be compensated by adding an antiphase voltage. The loop filter bandwidth can then be greater than the smallest frequency step size f_s can be selected. This means that digital emulation of the modulation signal, direct digital angle modulation is possible. The small step size with a short settling time allows, for example, even predictable Doppler shifts to be precisely to be readjusted. The compensation of the phase disturbance by the phase disturbance through divider switching requires, in addition to the calculation logic, a good D/A converter to generate the compensation voltage. A further development now often used in the RF sector is is the ΣΔ method, which is also used in CD audio technology for digital-to-analog and analog-to-digital conversion has prevailed. Due to its simple integrability, it already plays an important role in RF technology as a digital frequency.
In shortened form:
Disadvantage: Periodic switching generates interference lines
Remedy: Pseudo-random switching of the sigma-delta modulation

7.8.1 ΣΔ-Fractional-N Synthesizer Generator

As shown in above, the filtering away of the disturbances brings interference brings few advantages, as the loop filter bandwidth must be made smaller than the frequency step size, [81]. The ΣΔ-Fractional-N method causes, like the conventional Fractional-N method, secondary lines at the frequency step width. However, instead of the usual pulse width modulator, a ΣΔ modulator is used, whose spectral energy maximum is at high frequencies (Fig 7.17).

The changes in the divider value of the PLL are controlled by the ΣΔ modulator. This provides a quasi-random data stream whose arithmetic mean value can be set digitally. Because the energy is not evenly distributed spectrally, but is largely in the higher frequency range, so little interference energy is obtained near the carrier that the analog noise of the phase discriminator and the reference predominates. The loop filter can therefore be selected wider than the frequency step width without compensation measures. This typical behavior is shown in Fig. 7.18.

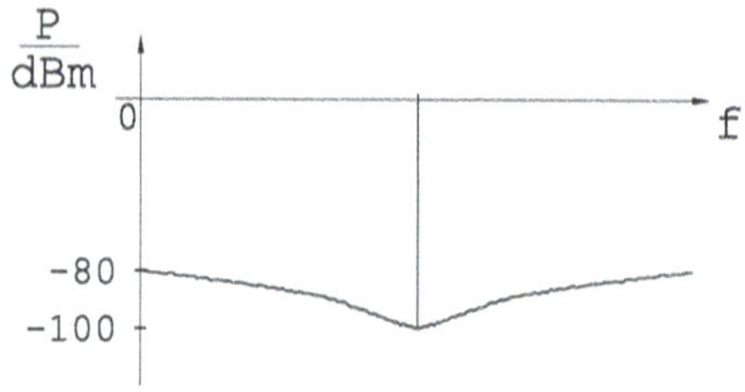

Fig. 7.18 Fixed signal and typical phase noise of a ΣΔ-Fractional-N synthesizer

ΣΔ-Modulators can be realized in different structures. One interesting variant is the MASH modulator, as described in [71], because it does not require a feedback network and therefore therefore also has no stability problems.

7.8.2 Direct Digital Modulation

The small step size of a ΣΔ-Fractional-N synthesizer enables direct digital frequency and phase modulation [25]. This is an exact D/F converter. The advantages of a direct digital modulator are obvious:

- No D/A conversion of digital modulation data is necessary.
- The modulation type and deviation can be changed via software.
- Large modulation ranges without distortion can be realized (non-linearities of the VCO characteristic curve have no influence).
- There are no longer any analog setting elements for defining the stroke.
- The modulation range is independent of component tolerances.

The ΣΔ synthesizer has been used for a long time for GMSK modulation in GSM handsets on a large industrial scale.

The ΣΔ-Transmitter

If you have two ΣΔ-Fractional-N synthesizers each for direct digital modulation, you can read as in [38] page 349, you can read how to use them to produce a vector signal generator. This ΣΔ-transmitter again performs the phase and amplitude modulation and thus modulation methods such as 64 QAM perform.

It is very advantageous that the power amplifiers can be driven much more into the compression than with the usual operating modes, which are usually direct or close to the inefficient class A operation. Further literature can be found in [50], as LINC technology in [2] and many other publications.

Chapter 8
Technically Generated GHz-Plasma and Applications

This chapter first explains the basics of plasma theory and technology, then briefly describes typical technical applications such as lighting technology, scalpel and spark plugs as the state of the art, and, as a microwave topic, deals in particular with the novel micro and microwave plasmas and their advantages in the known technical applications. A brand new feature is the illustration of how to generate decoupled plasmas. Among other things, these make it possible to manufacture industrial furnaces whose efficiency is significantly higher than that of furnaces operated using hydrogen burners.

8.1 Basics of Technically Generated Plasmas

In everyday life, we are very familiar with the three states of matter: solid, liquid and gas. In the universe, however, most matter (atoms, molecules) is in the fourth state of matter, the plasma. More than 99 % of visible matter in the universe is in the plasma state.

Starting from a solid, matter is brought into the next state of aggregation by adding energy. The most common form of energy is temperature. Very strong heating of gases leads to thermal ionization. Extremely high pressure (as in the interior of the sun) leads to pressure ionization. Strong photon irradiation causes photoionization. However, any electromagnetic energy across the entire frequency spectrum from direct current to radioactivity is also very well suited to raising the energy level of matter. This is what we will focus on below.

Plasmas are thus created in gases to which energy is added. The resulting collision processes in the matter between the electrons and the atoms or molecules generate new charge carriers. The remaining atoms or molecules are now positively charged and are referred to as ions. Gases that contain a large number of ions are called plasma.

Necessary Ionization Energy

In the following, the energy is expressed in the unit eV.

The following applies: $1\,\mathrm{eV} = 0.16\,\mathrm{aJ} = 0.16 \cdot 10^{-18}\,\mathrm{J} = 0.16\,\mathrm{aWs}$.

H. Heuermann, *Microwave Technology*,
https://doi.org/10.1007/978-3-658-45686-3_8

The energy can also be converted into a temperature. If an atom has a low energy of only 1 eV, its temperature is around 11600 K.

In order to ionize an atom, a free electron must initiate the ionization. If this free electron has an energy of several eV (e.g. 15.8 eV for argon), it is able to initiate this ionization. After this successful ionization, two free electrons and one ion are available. Further ionization energies of some elements are shown in Tab. 8.1.

If no further energy is added, an electron will recombine again after a short time. Plasmas can only exist with a continuous supply of energy. In a plasma, the following relationship applies to the densities of the n_0 neutral parts, the n_e electrons, and the n_i ions:

$$n_0 > n_e = n_i \,. \tag{8.1}$$

In the following, we will always assume that the continuous supply of energy is always carried out by a technical, electromagnetic excitation.

Temperature in a Plasma

It is well known from semiconductor technology that the mobility of the electrons is much greater than that of the holes (atoms in the crystal lattice that are missing an electron). The same applies to the relationship between electrons and ions.

If a gas is heated further and further from the vaporization temperature, the molecules start to move more and more. They absorb more and more kinetic energy and move in all directions. When the temperature becomes so high that the ionization energy is reached, the gas turns into plasma.

In a technical plasma, heating takes place through electromagnetic excitation. It goes without saying that in this case the moving electrons heat up faster and more easily than the ions. Furthermore, in a technical plasma (for pressures above 100 mbar), there is always a boundary region. In this edge region, the ions experience a directed movement towards the wall due to the ambipolar field. This arises when the plasma is switched on, as the electrons reach the wall much faster than the inert ions and therefore charge the wall negatively. The boundary layer of a plasma is virtually collision-free due to this directional movement.

We can define the following four different temperatures in a plasma:

Electron, ion, neutral gas and wall temperature.

Masses in a Plasma

Electrons, in particular, are accelerated in technical microwave driven plasmas by electromagnetic fields. The field strengths are so high that the relativistic increase in mass must be taken into account. This results in the following for a body with the mass m_0 at rest has the greater mass m at the speed v over

Tab. 8.1 Atomic masses of gases and other important elements for technical plasmas

Element	Element symbol	Element mass mass in g	Ionization energies in eV
Helium	He	$6.64 \cdot 10^{-24}$	24.6
Neon	Ne	$3.35 \cdot 10^{-23}$	21.6
Argon	Ar	$6.62 \cdot 10^{-23}$	15.8
Krypton	Kr	$1.39 \cdot 10^{-22}$	14.0
Xenon	Xe	$2.18 \cdot 10^{-22}$	12.1
Nitrogen	N	$2.32 \cdot 10^{-23}$	14.5
Oxygen	O	$2.66 \cdot 10^{-23}$	13.6
Phosphorus	P	$5.11 \cdot 10^{-23}$	10.5
Sulphur	S	$5.32 \cdot 10^{-23}$	10.4
Hydrogen	H	$1.67 \cdot 10^{-24}$	13.6
Mercury	Hg	$3.33 \cdot 10^{-22}$	10.4

Tab. 8.2 Electron mass as a function of velocity

Continuous voltage in V	Electron velocity in m/s	Electron mass in g
10	$18.8 \cdot 10^{5}$	$9.11 \cdot 10^{-28}$
10^{3}	$18.7 \cdot 10^{6}$	$9.12 \cdot 10^{-28}$
10^{5}	$16.5 \cdot 10^{7}$	$10.9 \cdot 10^{-28}$
10^{6}	$28.3 \cdot 10^{8}$	$28.8 \cdot 10^{-28}$
$3.1 \cdot 10^{6}$	$29.7 \cdot 10^{9}$	$64.3 \cdot 10^{-28}$
∞	Speed of light	∞

$$m = \frac{m_0}{\sqrt{1 - v^2/c_0^2}} . \tag{8.2}$$

Some values of an electron are printed in Tab. 8.2.

For the most important gases and other elements that occur in technical plasmas, the masses and ionization energies are given in Tab. 8.1.

Air is a very important process gas. The main components of air are nitrogen (in the form of N_2) and oxygen (in the form of O_2) with a total of 99.03 % of the total volume.

Pressure Dependence of Plasmas

The ionization capacity of a gas is extremely pressure-dependent. The particle density correlating with the pressure determines the properties of a plasma. Therefore the classic technical applications are also divided into two classes: `low pressure plasmas` and `high pressure plasmas` are distinguished. Typical particle densities for low-pressure plasmas are between 10^9 and $10^{17}\,\mathrm{cm}^{-3}$ [21] and for high pressure plasmas between 10^{20} and $10^{22}\,\mathrm{cm}^{-3}$.

A typical example is shown in Fig. 8.1, where the proportion of neutral particles is many times greater in real terms.

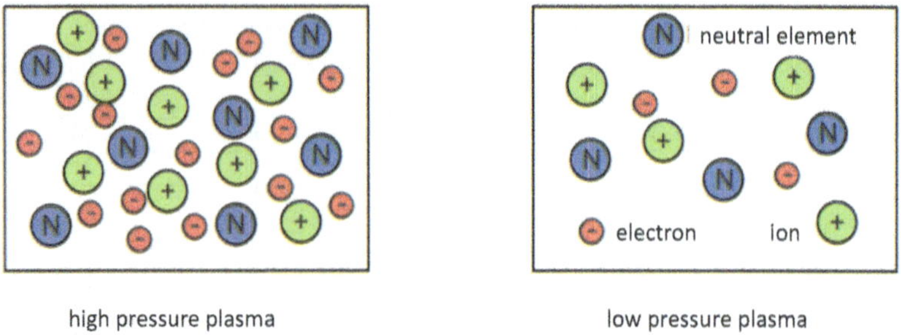

Fig. 8.1 Simple models of a high- and low-pressure plasma

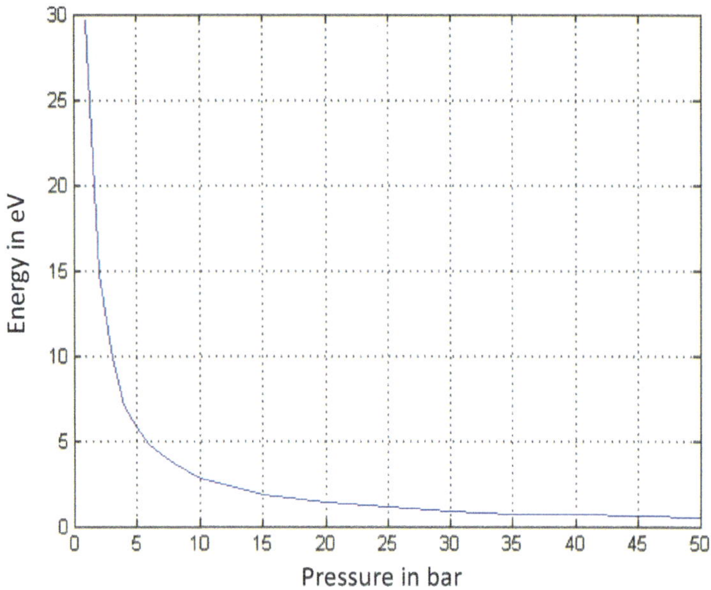

Fig. 8.2 Typical energy of a free electron in the HF field as a function of pressure at 2.45 GHz and $E_0 = 10\,\text{MVm}^{-1}$

Fig. 8.2 shows how much the energy of an electron decreases during the transition from the normal pressure range (1 bar $\simeq 10^5$ Pa) to the high pressure range despite constant electrical excitation.

If the pressure (the particle density) becomes too low (vacuum and slightly above), then an accelerated particle no longer hits another particle and no plasma can be generated or maintained. This effect can be calculated very well using the Debye length.

Debye Length

An important parameter for describing a plasma is the Debye length. This describes that potential disturbances within a plasma decay exponentially, with the Debye length λ_D as the decay constant. After a length of about $7 \cdot \lambda_D$ the potential disturbance has decayed.

It can be calculated with the Boltzmann constant k_B, the electric field constant ε_0 and the elementary charge e as a function of the temperature T and the electron density n_e as

$$\lambda_D = \sqrt{\frac{\varepsilon_0 k_B T}{n_e e^2}} \,. \tag{8.3}$$

The prerequisite for these equations is that the electron density is equal to the ion density. Since a plasma, whose dimensions are considerably greater than the Debye length, behaves in an electrically neutral manner towards the outside, this condition is met. This quasineutrality can only be violated within a sphere with the radius λ_D, [21]. The plasma therefore shields itself against external interference fields.

Plasma Frequency

Another important parameter in plasma physics is the plasma frequency ω_P. The application of an electric field leads to a fluctuation of the charge carriers. If the frequency of this electric field is high enough, the ions, due to their relatively high inertia, can no longer follow the field and are assumed to be at rest. The displacement of the electrons now leads to charge separation, which is compensated by the Coulomb force, and the electrons are accelerated in the opposite direction. Due to the inertia of the electrons, they are accelerated beyond their equilibrium position and in turn and generate strong electric fields. The result is an oscillation (resonance effect) with very little loss if the applied alternating voltage signal exceeds the angular frequency

$$\omega_P = \sqrt{\frac{n_e e^2}{\varepsilon_0 m_e}} \,. \tag{8.4}$$

Since Eq. (8.4) contains almost only constants, this expression can be reduced to

$$\omega_P = 5.64 \cdot 10^4 \cdot \sqrt{n_e} \;\; \text{resp.} \;\; f_p = 8976 \cdot \sqrt{n_e} \,, \tag{8.5}$$

where the electron density is given in cm^{-3} and the plasma frequency has the unit Hz (Fig. 8.3).

If the product of plasma frequency and Debye length is formed, as shown in [21], the result is, that the charges in a plasma are only displaced in a range that is in the order of magnitude of the Debye length.

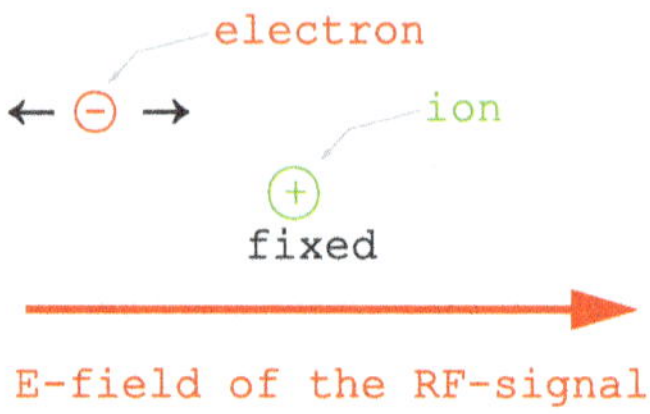

Fig. 8.3 $\omega = \omega_P$: The ions are stationary and the electrons are accelerated by the externally applied electric field and the Coulomb force

This plasma frequency is of great importance for the propagation of electromagnetic waves in a plasma. Only electromagnetic waves with angular frequencies $\omega > \omega_P$ can propagate in a plasma. Below this *cut-off* frequency, the waves are reflected by the plasma. This in turn means that an electromagnetic wave with $\omega > \omega_P$ transmits its energy directly into the interior of the plasma, while a wave with $\omega < \omega_P$ only transmits the energy to the surface of the plasma.

It should be noted that only low-loss, low-pressure plasmas such as the ionisphere allow the transmission of electromagnetic radiation.

To get a feeling for the order of magnitude of the plasma frequency, a few values are listed in Tab. 8.3.

The following also applies:

- For $\omega < \omega_P$: There is an electron current and an ion current.
- For $\omega > \omega_P$: There is a pure electron current (ions are stationary).

Mean Free Path Length

The mean free path length, also referred to as MFP for "Mean Free Path" is characterized by

$$\lambda_{MFP} = \frac{1}{n\,\sigma_{Streu}} \quad \text{with} \quad n = n_0 + n_i \; . \tag{8.6}$$

The MFP indicates the path length traveled by an electron before it collides with a particle on a statistical average. The atomic cross-section σ_{Streu} is actually weakly temperature-dependent, but is assumed to be constant here for the sake of simplicity and is calculated with the radii of the colliding particles:

$$\sigma_{Streu} = \pi \cdot (r_1^2 + r_2^2) \; , \tag{8.7}$$

where r_1 and r_2 are the radii of the colliding species (e.g. argon atom: $r_{Ar} = 158\,pm$ and electron: $r_{e^-} = 2.8\,fm$).

To calculate the mean free path, the particle density n is calculated using the equation of the state of ideal gases

Tab. 8.3 Plasma frequency f_P for different particle densities

n_e in cm^{-3}	f_P in GHz
$1 \cdot 10^9$	0.28
$1 \cdot 10^{10}$	0.89
$1 \cdot 10^{12}$	8.97
$1 \cdot 10^{14}$	89.76

$$n = \frac{p}{k_B T} \,. \tag{8.8}$$

An ideal gas at room temperature is assumed for this. If the Eqs. (8.8) and (8.7) are used in the Eq. (8.6), the following relationship can be established:

$$\lambda_{MFP} = \frac{k_B T}{\sigma_{Streu} \cdot p} \,. \tag{8.9}$$

This equation can be used to calculate the mean path length as a function of pressure, shown here using an electron in pure argon gas.

Eq. (8.9) and Tab. 8.4 clearly show that the mean free path is inversely proportional to the pressure. This change in MFP leads to the problem that if the pressure is too high, not enough energy can be absorbed by the $\vec{E}$ field before the electrons collide with neutral particles again on a statistical average. Conversely, the pressure must not be too low either; otherwise, there will be too few collisions to ensure an effective energy exchange.

In high-pressure plasmas, however, the temperature is considerably higher than in low-pressure plasmas. This results in a larger undirected drift movement of the particles, which in turn ensures a higher energy of individual particles, as the undirected drift overlaps constructively, but also destructively, with the directed drift movement of the $\vec{E}$ field.

Plasma Beta

The plasma beta is a highly interesting quantity if you want to plasma in as small a space as possible, which is of interest for industrially used ion sources through fusion technology.

Plasma beta β: The ratio between the thermal and magnetic pressure in a plasma.

The thermal pressure tries to expand the plasma, and the magnetic pressure tries to compress it.

The calculation is made using

$$\beta = \frac{n \, k_B \, T}{B^2/(2\,\mu_0)} \,. \tag{8.10}$$

For a plasma beta enclosed by the magnetic field, the following must apply: $\beta < 1$
.

Tab. 8.4 Average free path length of an electron in argon and argon density for various pressures at room temperature and 2.5 GHz

n in cm^{-3}	p in pa	p in bar	λ_{MFP} in m
$2.41 \cdot 10^{16}$	$1 \cdot 10^{2}$	0.001	$2.6 \cdot 10^{-3}$
$2.41 \cdot 10^{19}$	$1 \cdot 10^{5}$	1	$2.6 \cdot 10^{-6}$
$1.21 \cdot 10^{21}$	$5 \cdot 10^{6}$	50	$5.2 \cdot 10^{-8}$
$4.83 \cdot 10^{21}$	$2 \cdot 10^{7}$	200	$1.3 \cdot 10^{-8}$

ECR-Plasma

An ECR plasma (Electron Cyclotron Resonance Plasma) is a gas discharge plasma that is generated in a magnetic field, which allows the electrons to spin around the magnetic field lines. electron gyrations around the magnetic field lines. The plasma is excited by microwave power generators (typically at 2.45 GHz) to amplify the electron gyrations. Due to the high electron energy level and high electron density, the ECR plasma is an effective tool for generating ionized matter for various applications such as surface coating, material synthesis, plasma diagnostics and fundamental plasma research. ECR plasmas are usually very hot and have temperatures of several thousand Kelvin, giving them high reactivity and chemical activity. These plasmas are generated in low pressure or vacuum chambers in order to have control over the pressure and chemical environment of the plasma. The properties of the ECR plasma depend on many factors, such as the magnetic field strength, the gas composition, the microwave power level and the pressure environment. The ECR plasmas are also used in astrophysics as a model for the study of magnetic fields in cosmic objects such as stars and planets. The microwave-based ECR plasmas are an important field of research in plasma physics (e.g. Cern) and have a variety of applications in materials science, the semiconductor industry, nanotechnology and biomedicine.

At Cern, superconducting magnets are used to confine the plasma. In [44] an approach (DCR: Dipole Channel Resonance) is described how energy can be confined by an electromagnetic GHz field with a linear plasma current in the center without additional effort by magnets. This plasma is confined purely by the high magnetic flux density of the microwave energy (Fig. 8.4).

The application of the basics will be demonstrated by the design of a potential fusion reactor in the following example.

Exercise Example: Design of a DCR Fusion Reactor

Opposite: There are many semiconductor-based power sources whose output signal is freely adjustable in magnitude and phase, and which can also be arranged in several layers. The system can be designed according to [44] that a plasma in line form with an extremely thin diameter is created when the transmitters are controlled

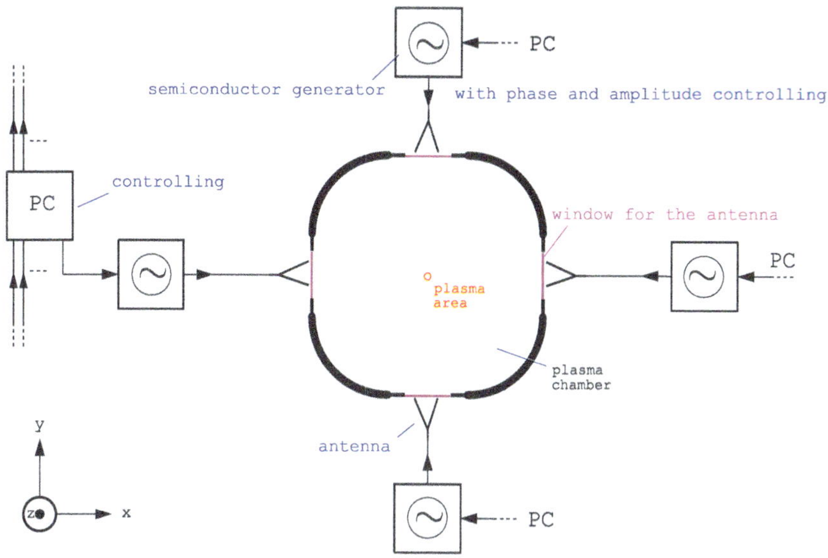

Fig. 8.4 Schematic structure of a DCR system in a section at half height (line plasma spreads in the z-direction)

correctly. The fusion technique requires densities of $n = 10\,10^{22}$ electrons per m^2 and energies of 10 keV (100 million degree) are required.

Quested: 1) Which boundary conditions must be met to ensure compliance with the plasma gas? 2) Under which boundary conditions can the target values required for fusion technology be achieved?

1. The line current corresponds to the current of a dipole antenna and has the initial values I = 31 A, and f = 2.45 GHz, whereby the dipole length is $\ell = \lambda/2$. The selected distance of the B-field to the center is 0.1 mm. The magnetic flux density is calculated from the basic equation of the Hertz dipole:

$$B = j\,\mu_0\,I\,\frac{\ell}{2\,\lambda_0\,r}\,e^{-j\,k_0\,r}\left(1 + \frac{1}{j\,k_0\,r}\right) . \tag{8.11}$$

   This results in a flux density of almost 19 Tesla. If the Eq. (8.10) is converted according to the temperature and this flux density value is used and $\beta = 1$, the temperature of 104 mill. degree required for confinement is obtained, which corresponds to the requirement of fusion technology.
   This calculation states: If the "effective magnetic flux density" corresponds to the flux density of a Hertzian dipole at a distance of of 0.1 mm, then the outer antennas only need to generate a current of 31 A in the dipole-shaped plasma to fulfill the plasma beta. The flux density of the Hertzian dipole is significantly greater at short distances and decreases at longer distances.

2. For the further rough estimates, it is assumed that the current is constant over the entire conductor (which is not true). Furthermore, the entire coupled energy is assigned to the moving electrons, which is a good approximation.
   Under these simplified conditions for an estimated diameter of the dipole of $d = 2\,\mu\text{m}$, the volume of the dipole Vd can be calculated using $Vd = \pi d^2/4\ell$.
   The number of charge carriers is $x = nVd$ and the base resistance of the antenna should be $R = 150\,\Omega$.
   The work per half period (one acceleration phase) results from $Wg = RI^2\,1/f/2$ and the work per particle from $Wx = Wg/x$.
   In the end, the total work in eV is calculated via $WxeV = Wx/0.16\,10^{-18}$ and is within the desired range at $9.57\,\text{keV}$.
   Ultimately, the absorbed power can be calculated based on the footpoint resistance and the antenna current and is 144 kW. In practice, the power amplifiers have efficiencies of only 70 % and therefore the required grid power is 205 kW, which is very low for a fusion system.
   That such a realization is within the realm of possibility is shown at the end of the chapter with the scientific results of the decoupled plasma. In addition, a first free-standing plasma had already been generated at the time this book was completed.

8.2 Low-Pressure Plasmas

Low-pressure plasmas are plasmas that are generated in negative pressure conditions by means of DC electric fields or alternating fields up to the medium GHz range.

Typical of these low-pressure plasmas are

- Low particle density of the charge carriers
- Large mean free path length
- Collision processes are rare
- The energy of electrons and ions is different
- Low average temperature (typically 300 - 2000 K)

In low-pressure plasmas, the particle density of the charge carriers is low because, on the one hand, the neutral particle density is also low and, on the other hand, the probability of collision processes is low.

This means that the degree of ionization of a gas with a total of n_n particles is also low:

$$n_e/n_n \approx 10^{-6} - 10^{-2} \quad (\text{therefore applies } n_n \approx n_0)\ .$$

Typical pressures in low-pressure systems are below 1 mbar, which are achieved with a vacuum pump. Fig. 8.5 shows a modern system for the application purposes.

The following applications of low-pressure plasma in industry are already state of the art:

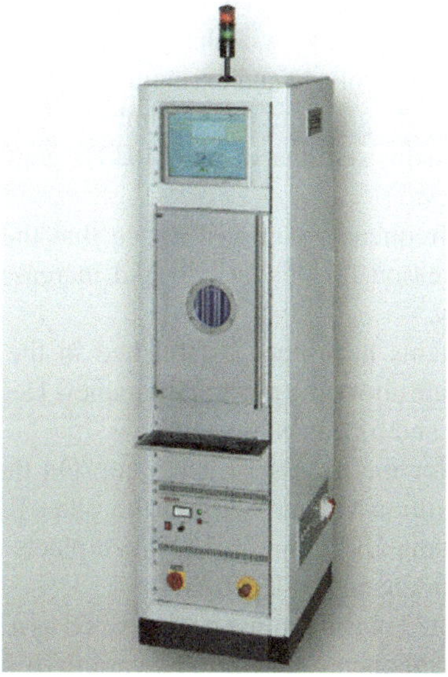

Fig. 8.5 Low-pressure plasma system from Diener electronic

SURFACE CLEANING: The surface of the components is cleaned physically by the ion bombardment and, depending on the type of gas, also by chemical reactions. The contamination is converted into the gas phase and extracted.

PLASTIC ACTIVATION: The plastic surfaces of the components are plasma-treated with oxygen or air, for example. Radical points are formed to which paint or adhesive systems adhere well.

SURFACE ETCHING: The surface of the components is etched with a reactive process gas. Material is specifically removed, converted into the gas phase and extracted. The surface is enlarged and is very easy to wet. Etching is used for pre-treatment before printing, bonding and painting as well as for roughening the material.

SURFACE COATING: A gas (e.g. hexamethyldisiloxane) is introduced into the plasma chamber. Plasma polymerization[1] are used to generate material sheets on surfaces.

Various continuously adjustable (often via pulse operation) generators with different outputs are used in such systems according to Fig. 8.5. Typical technical specifications are discribed in the following.
The three operating frequencies each have advantages and disadvantages for the various applications.

As with the microwave oven, the **2.45 GHz** technology involves an antenna that transmits into the enclosed space. The GHz energy is also generated by a magnetron.

[1] Polymerization is a chemical reaction in which monomers, usually unsaturated organic compounds, react under the influence of catalysts and by breaking the multiple bonds to form polymers (molecules with long chains consisting of monomers linked together).

The other two frequency ranges belong to the class of capacitively coupled plasmas. At low frequencies below **10 kHz**, a plasma is made to glow by the high voltage in the range of the maximum voltage and extinguishes in the range of the zero crossing of the voltage. This extinguishing (by recombination) no longer takes place at **40 kHz**. The plasma remains stationary. With 40 kHz generators, you work far below the plasma frequency.

If you increase the frequency, you will notice that the ignition voltage drops. Certainly, at least one reason for this is the rapid increase in voltage of the high-frequency energy supply.

For this reason, systems have been established in the free frequency band at **13.56 MHz**. These operate above the plasma frequency. Depending on the two states

a) Gas is not ionized and b) Gas is ionized

there are two tasks for the necessary generator. On the one hand, it must initially provide a high ignition voltage and, on the other hand, the energy must be transported into the plasma after ignition has taken place. However, both are to be carried out at the fixed frequency of 13.56 MHz.

Here, the load (activated plasma) must be described as a lossy capacitance whose capacitive and ohmic components depend on a variety of factors such as gas selection and pressure setting. In order to fulfill this task, the reflection values (S_{11}) of the plasma are often measured and an adjustable transformation network is optimally set. Adjustable capacitors are used here via stepper motors. Figure 8.6 shows the associated network.

Another very large application of low-pressure plasmas is low-pressure lamps.

8.2.1 Low-Pressure Lamps

A large number of conventional lamps work under low-pressure conditions with a plasma discharge. These include the

- neon tubes and the related compact fluorescent lamps (energy-saving lamps),
- Vapor discharge lamps, sodium vapor lamps,
- quartz lamps or mercury vapor lamps, gas discharge tubes and many more.

The compact fluorescent lamp is discussed in more detail below.

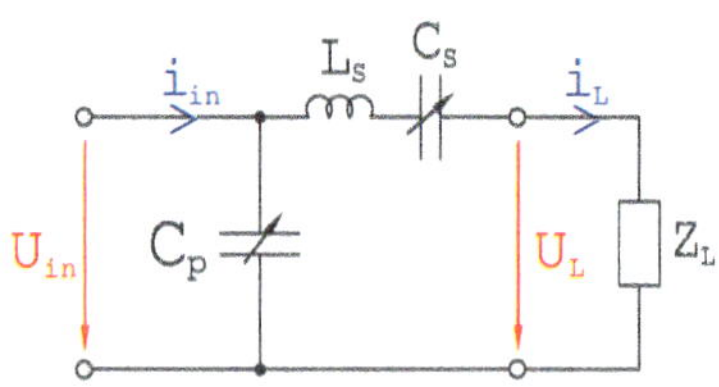

Fig. 8.6 Γ-transformer for low-pressure plasma systems at 13.56 MHz

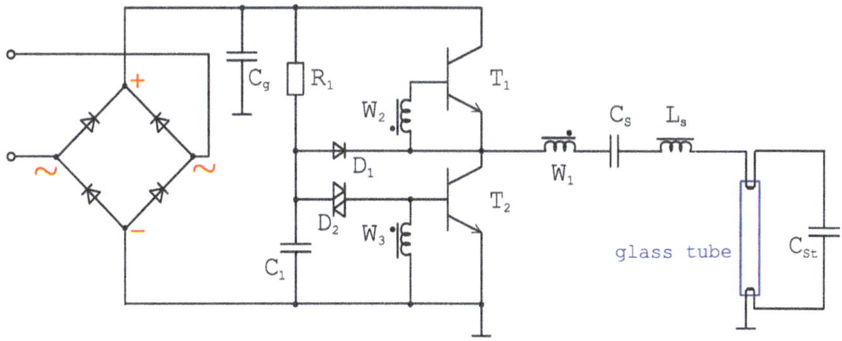

Fig. 8.7 Circuit diagram of an electronic ballast (ECG) with a switching frequency of switching frequency of 45 kHz (all W coils are magnetically coupled)

As fluorescent lamps, compact fluorescent lamps are mercury vapor lamps. To reduce the dimensions, the gas discharge tube is not straight, but bent (several times) into a U-shape or designed as a filament. A further reduction in size and higher luminance is achieved by increasing the internal pressure.

At approx. 60 lm/W (lumens/watt), compact fluorescent lamps have a luminous efficacy that is around four to five times higher than normal incandescent lamps with 12 to 15 lm/W. They therefore require 75 % to 80 % less electrical power for the same luminous flux. However, their luminous efficacy decreases over the course of their service life.

To operate a compact fluorescent lamp, as well as other gas discharge lamps, an electronic ballast (EB) is usually used nowadays. Some electronic ballasts initially heat the cathodes when the lamp is started by connecting them in series to a PTC resistor in the circuit. Once this has heated up due to current flow, it becomes highly resistive and releases the discharge path for the ballast and the lamp ignites. The pressure build-up, i.e. the evaporation of the mercury, takes place when the lamp is switched on by preheating the cathodes or by heating filaments (directly heated cathodes) and subsequent self-heating. For this reason, compact fluorescent lamps do not reach their full luminosity immediately.

The gas discharge path itself works on a resonant converter, i.e. the AC mains voltage is first rectified and then converted back into an AC voltage of higher frequency (approx. 45 kHz). The alternating direction is achieved with two switching transistors, the alternating voltage is then fed to the lamp circuit via a ferrite core choke.

Due to the higher operating frequency, the choke is very small, has lower losses and saves material compared to the 50 Hz chokes of older conventional ballasts (CCG). In addition, the higher operating frequency leads to higher lamp efficiency than with fluorescent lamps with CCGs, as the gas discharge itself works more effectively and the losses in the choke are lower. In addition, the human eye cannot perceive the frequency of $f_0 = 45$ kHz as flickering.

Tab. 8.5 Technical applications for high-pressure discharge plasmas

application	Typical pressure in bar	Typical electrical power in W
Spark plug	100	50-500
Beamer lamp	220	150
Xenon lamp	40	10—35
Welding technology	1	500–5000
Plasma jets/ -torches	1	20–15000
Furnace	1–10	500–200000

Fig. 8.7 shows a circuit diagram of such an electronic ballast[2] is shown. After switching on, the capacitors C_1, C_s and C_{St} are charged via R_1. As W_3 is initially high-impedance, T_2 conducts later for a short time. The series resonant circuit with the main elements W_1, L_s and C_s and the unignited plasma in the glass tube or C_{St} is closed.

A large voltage builds up across the glass tube via the driving inductance W_1 and the plasma is ignited. From then on, it can only be seen as an ohmic resistance in the first approximation. The base voltage for T_2 collapses at W_3 and T_2 blocks. W_3 releases the stored energy into C_1. At the same time, the base voltage could build up behind W_2 and T_1 is conductive. The resonant circuit receives new energy from the positive DC voltage, which is rectified directly from the mains voltage. The leading switching at 45 kHz comes from the resonant circuit of C_1 and W_3.

The very close relationship to the class D amplifier with two transistors can be seen. These circuits typically have an efficacy of more than 90 %. It is important in circuit optimization that T_1 and T_2 never open at the same time.

8.3 High-Pressure Plasmas

High-pressure plasmas are plasmas that are generated in normal and high-pressure conditions by means of DC electric fields or alternating fields up to the medium MHz range.

Typical for these high-pressure plasmas are

- High particle density of the charge carriers
- Small mean free path length
- Impact processes are frequent
- Energy of electrons and ions rather equal
- High mean temperature (typically 5000 K)

High-pressure plasmas are based on arc discharge in technical applications. Tab. 8.5 presents the most important technical applications and the associated pressures.

A typical arc discharge is shown in Fig. 8.8.

[2] This circuit is also known as a "resonant converter".

Fig. 8.8 Photo of a current spark plug with arc discharge plasma

In the following, the example of the spark plug (which is mainly used in cars) will be discussed in more detail.

8.3.1 Classic Spark Plug

The ignition in the petrol engine has the following tasks and effects: 1) The ignition of the petrol-air mixture by means of an ignition spark; 2) The ignition has a significant influence on the combustion process; 3) Effect on consumption, emission behavior and performance.

Before the ignition is triggered, a pressure of typically 10 bar has been built up by the compression of the piston. When the arc flashes over and the ignition spark is present, the pressure increases to up to 100 bar. Under this condition, the arc discharge should be maintained. The energy stored in the ignition coil allows an ignition spark duration (burning time) of typically 1.5 ms (Fig. 8.9).

8.4 Corona Plasmas

Corona discharges or plasmas occur at excitation frequencies in the lower and medium MHz range and at very high voltages. These can be generated with a Tesla transformer, for example.

Technically, these plasmas are used in industrial low-pressure and atmospheric-pressure systems for surface treatment.

Recently, the company Borg-Warner also pursued a corona spark plug. However, this still has some disadvantages, as can be clearly seen in the following picture.

This plasma is otherwise known as large decorative glass beads. Isolated conductive paths are formed, but these do not have to penetrate through to the mass. The

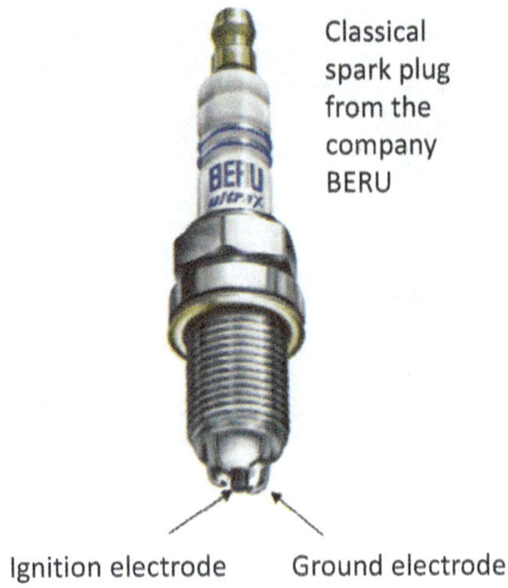

Fig. 8.9 Photo of a current spark plug

spark plug	characteristics
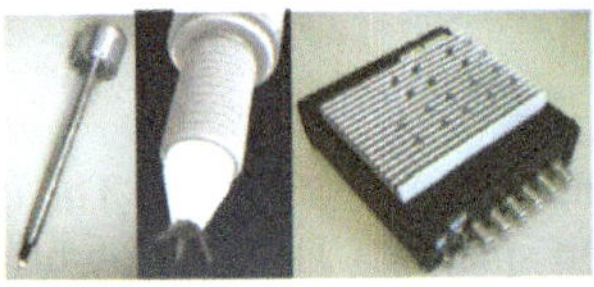 Figure 6: Igniter (left), igniter tips (center), controller (right) of Borg Warner's Corona Ignition System EcoFlash	☑ can ignite lean, combustible mixtures over a wide pressure range ☑ free in time of ignition, quick re-ignition ☑ a spatial pattern of sparks with a large number of spark channels - volume ignition effect ☑ minimized electrode erosion ☒ expensive high voltage electronic ☒ adaption of the piston to the ignition system for best results ☒ no plug and play technology ☒ very large controller

Fig. 8.10 Comparison of the BorgWarner spark plug with the conventional spark plug

performance parameters of the sparc plug with this corona plasma from BorgWarner is presented in Fig. 8.10.

8.5 Microwave Plasmas

Plasmas generated in low, normal and high-pressure conditions by means of alternating electromagnetic fields in the upper MHz and GHz range are referred to as microwave plasmas or microplasmas.

Typical for these microwave or GHz plasmas are

- Relatively high particle density of the charge carriers (after pressure)
- Relatively small mean free path

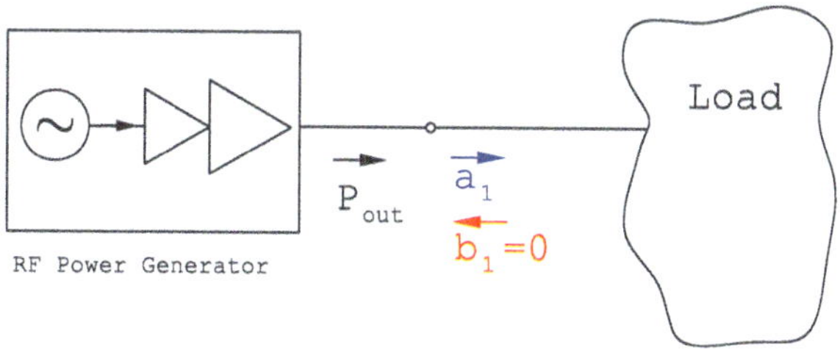

Fig. 8.11 The goal/desire of all GHz power transmissions

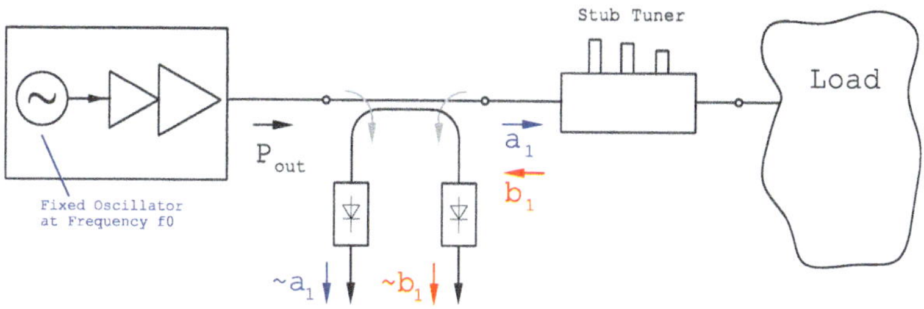

Fig. 8.12 The currently most commonly used method for a good adaptation of the GHz power source to the (unchanging) load

- Impact processes are frequent
- Energy of electrons and ions different
- Relatively low mean temperature

The basic problem in GHz plasma technology is the same as in GHz heating technology and other applications up to radar technology: As shown in Fig. 8.11.

The state of the art, particularly in GHz heating technology, uses magnetron sources as fixed oscillators and tuners for matching (Fig. 8.12), as already described in the power amplifier chapter.

The technology developed at the IMP for generating GHz plasmas has special voltage transformers that require control electronics with special regulation. These microwave plasmas are generated with very high efficiency for start and operation via so-called bi-static matching. The basic concept "multi-stage transformers" is illustrated in Fig. 8.13.

With this technology, which is easy to manufacture in practice, GHz plasmas are now competitive with arc discharge, low and high pressure plasmas in many applications, as will be shown in many examples below.

First, however, the theory for the transformers and the various solutions for the control electronics are presented.

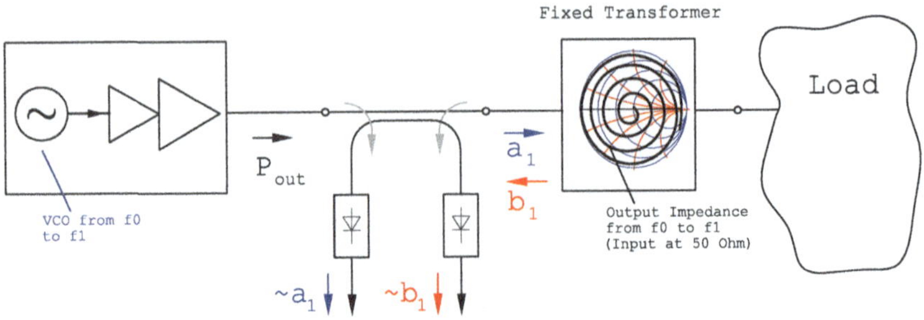

Fig. 8.13 The path used by the IMP for a good adaptation of the GHz power source to the load, which may now have two different impedances

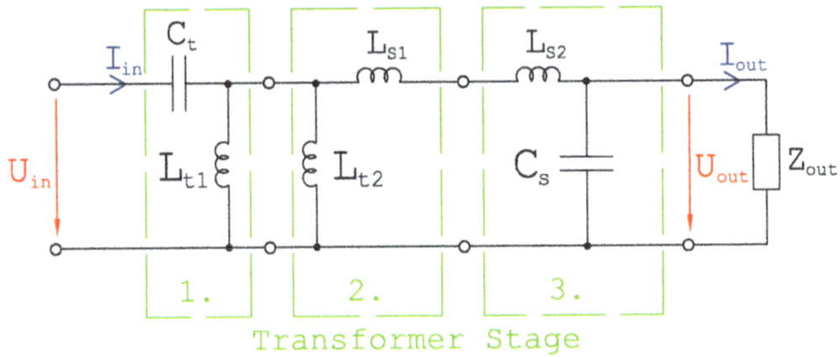

Fig. 8.14 Simplified equivalent circuit of a three-stage impedance transformation circuit for igniting and operating a microwave plasma

8.5.1 Theory of the Three-Stage Transformation

The concept of the now frequently used three-stage transformation (for the impedance and the voltage) is shown as a circuit diagram in Fig. 8.14.

Both the first and third transformation are realized by a so-called Γ-transformer, [38]. The second stage contains a simple autotransformer. The entire impedance transformation runs from the system impedance Z_0 of 50 Ω via the intermediate values to the output impedance Z_{out} of typically $0.5 \cdot 10^6$ Ω (for the ignition case) or 10–300 Ω (for the operating case).

The detailed derivation for calculating the component values of the Γ transformer is printed in [38]. The inductance L_{t1} of the first Γ transformer is calculated via

$$L_{t1} = R_1 R_2 C_t \,. \tag{8.12}$$

It should be noted that in the implementation the L_{t1} and L_{t2} (Fig. 8.15) are realized by a single bridge that is connected to the ground. R_1 corresponds to the system impedance Z_0 typically 50 Ω, R_2 to the output impedance of the first transformation stage and C_t to the capacitance between the 50 Ω input connector and the shunt inductance L_{t1}. The capacitor C_t is calculated as follows:

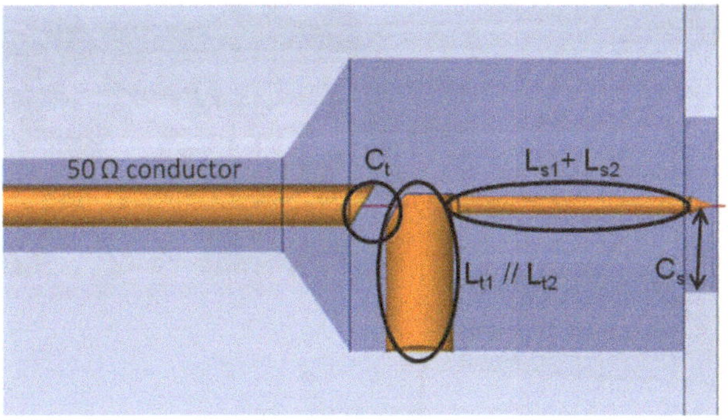

Fig. 8.15 Realization of the three-stage impedance transformer by means of low-loss coaxial components

$$C_t = \frac{1}{\omega_0}\sqrt{\frac{1/R_1}{R_2 - R_1}} \quad . \tag{8.13}$$

Here, ω_0 describes the resonant frequency.

The following equation provides the shunt capacitance C_s at the end of the third impedance transformation stage:

$$C_s = \frac{1}{\omega_0}\sqrt{(1-t)\cdot t} \cdot \frac{1}{R_3} \quad . \tag{8.14}$$

The variable t_3 represents the transformation ratio, where R_3 indicates the input resistance of the third stage and R_4 represents the output value (corresponds to Z_{out}).

$$t_3 = \frac{R_3}{R_4} \tag{8.15}$$

The series inductance L_{s2} is built up in the conversion (Fig. 8.15) together with L_{s1} and is calculated from

$$L_{s2} = \frac{1}{\omega_0}\sqrt{(1-t)/t} \cdot R_3 \quad . \tag{8.16}$$

Taking into account the two easy-to-calculate inductances L_{s1} and L_{t2} of the autotransformer, all components of this three-stage impedance transformation are now calculated in Eqs. (8.12)–(8.16).

Fig. 8.15 shows a coaxial distributed network designed based on the simple equivalent circuit and Eqs. (8.12)–(8.16).

The microwave power is fed to the left side via the 50 Ω line.

The first impedance transformation stage is created by an air gap for C_t and a ground stub L_{t1} for the same earth stub and the first section of the following high-impedance coaxial line form the autotransformer.

Fig. 8.16 200 W Generator using PC control and evaluation under Matlab

The third transformer stage is formed by the second part of the high-impedance coaxial line and the terminating capacitance to earth.

8.5.2 Processor-Based Control Electronics for Microwave Plasmas

Special microwave generators are required to control a microwave plasma with the impedance transformation networks presented above.

These microwave sources generate the necessary power in the frequency range of 2.40 GHz to 2.50 GHz. To support ignition and the operating mode (optimum power feed into the generated plasma), these generators must also be able to measure the reflection factors of the plasma source.

Figure 8.16 shows a generator that is controlled by PC-controlled Matlab software and can, among other things, display the sclar S_{11} values over time.

This generator contains a scalar network analyzer for so-called hot S parameter measurement. This means that the S-parameters of power signals are measured.

Based on these reflection parameter values, various search and optimization algorithms can be used in the frequency band to find the optimum matching point in this PC-supported solution.

More compact solutions with small control units were also developed at the IMP. These solutions included

- a microcontroller for sequence control and evaluation,
- a D/A converter to generate the control voltage,
- a VCO to generate the RF signal,
- an electronically switchable attenuator for power control,
- an amplifier chain and
- at least one coupler and one detector for measuring the reflected power.

A corresponding block diagram is shown in Fig. 8.17.

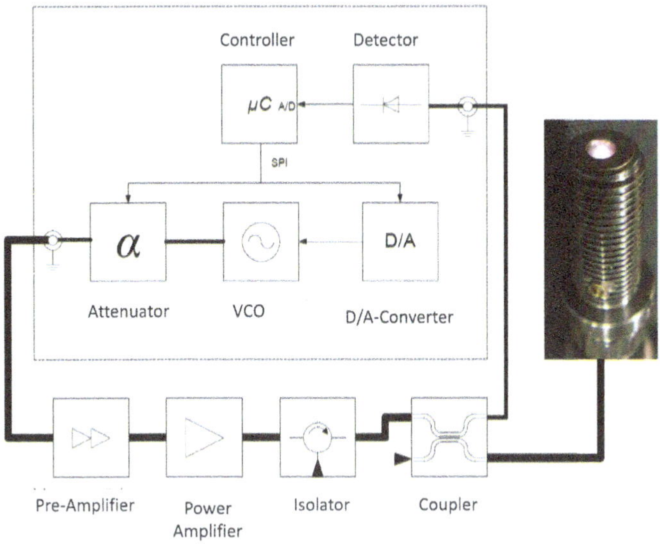

Fig. 8.17 Representation of the block diagram of the control unit using a microprocessor

As a further processor solution, work was also carried out on a chipset based on state-of-the-art technology, which is shown in Fig. 8.18.

However, processor developments were discontinued because control electronics based on an analog control loop, like the PLL, were a breakthrough in terms of the variables

- control speed,
- high integration level and
- costs

brought. In these three points, GHz plasma technology can now beat all other plasma technologies and laser technology anyway.

This patented amplitude-locked loop circuit was tested using SMD technology and implemented as an IC design, which is explained in detail below (Fig. 8.18).

8.5.3 Control Electronics Using an Amplitude-Locked Loop Circuit

In a PLL, as shown in chapter 7, the phase of a voltage-controlled oscillator signal is stabilized in a control loop with the phase of a reference oscillator.

In the ALL (Amplitude-Locked Loop) presented below, the frequency of a voltage-controlled oscillator is detuned in such a way that the optimum adjustment is achieved in a frequency band.

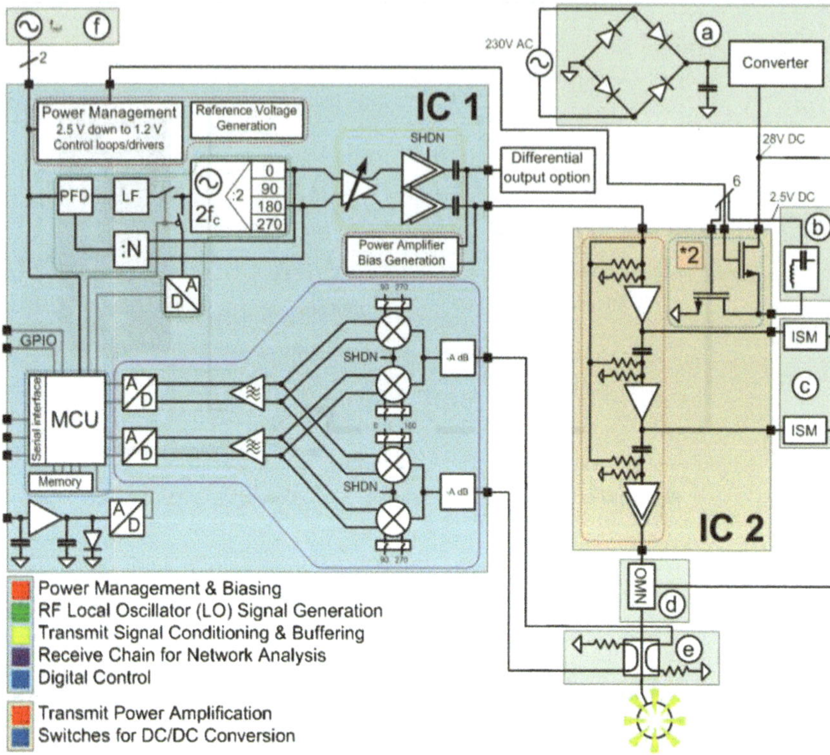

Fig. 8.18 Block diagram of a possible chip set for a highly integrated control unit with microprocessor control

The PLL technique presented in chapter 7 was developed as early as 1955 and published in [54]. It was not until 1969 that this new technology was used in a compact circuit.

It is shown below that this new ALL technology has already been developed and manufactured in a high-integrity CMOS IC, [46].

At the heart of a PLL is the phase discriminator. The amplitude discriminator is also at the heart of the ALL, Fig. 8.19.

The function blocks loop filter, voltage-controlled oscillator (VCO) and power amplifier (PA) have already been described in this book. book have already been described. The coupler can be found in many basic HF books including [38] can be found. The amplitude discriminator (AD or ADi) is completely new.

In the ALL, the output signal of the VCO is brought into the desired power range via the multi-stage amplifier and fed to the load (in this case the lamp). At the coupler, some energy from the wave running towards the load and the wave reflected by the load reflected from the load to the AD. In this centerpiece (the AD) of the ALL, an output signal is generated that is smoothed via the loop filter. The

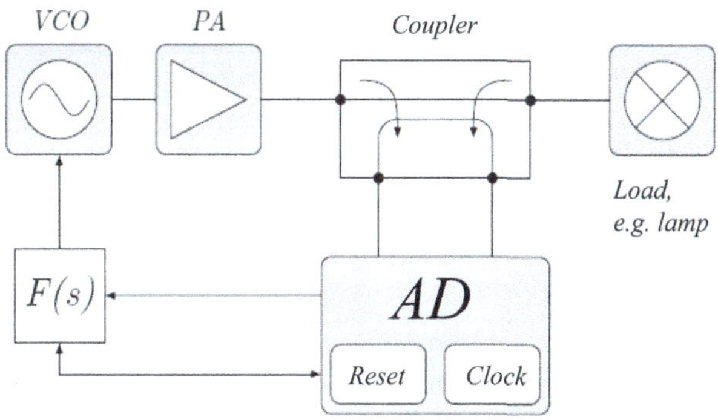

Fig. 8.19 Block diagram of the ALL control loop with the five function blocks and one lamp as load, [87]

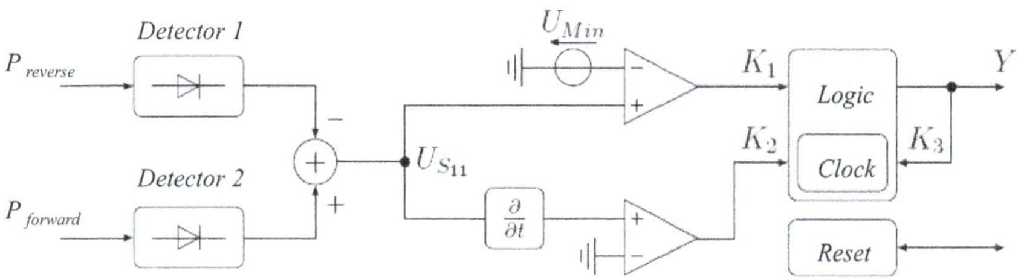

Fig. 8.20 Design of the amplitude discriminator, [87]

output signal is smoothed via the loop filter and adjusts the VCO so that the smallest reflection value occurring in the frequency band is found.

Structure of the Amplitude Discriminator

The input of the amplitude discriminator is formed by the two microwave detectors, which supply a DC voltage at the output, which is proportional to the microwave power at the input, Fig. 8.20. These two voltages are related in an operational amplifier circuit (symbol +) so that the output voltage U_{S11} corresponds to the input reflection factor of the load.

In the upper branch, which ends with the digital output information K_1, the voltage of U_{S11} is compared with a reference voltage. In the lower branch, the slope of U_{S11} is determined to check whether there has been a change and this is fed to the logic unit as digital information K_2. K_2 is 1 if the voltage U_{S11} droppes and the matching improves! $Y = K_3$ is the output signal of the logic unit.

The three digital signals K_1, K_2 and K_3 are input signals for the logic unit, which, in addition to a residual logic and a clock generator for operating this digital signal processing, has the logic core shown in Fig. 8.21.

The truth table associated with this rather simple logic is printed in Fig. 8.22.

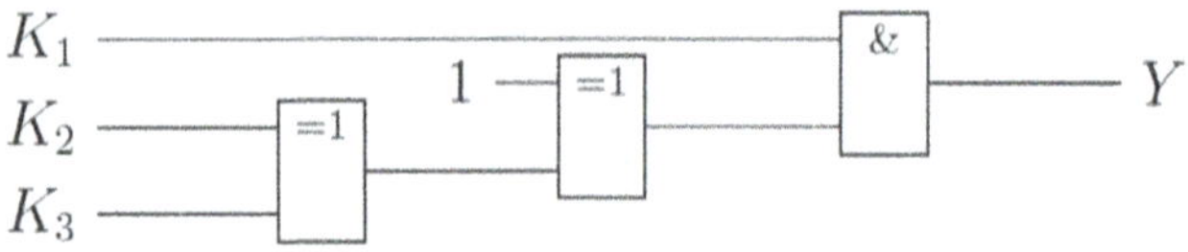

Fig. 8.21 Circuit diagram of the gate structure with only two EXOR gates and an AND gate for the logic of the amplitude discriminator, [87]

K_1	K_2	K_3	Y
0	0	0	0
0	0	1	0
0	1	0	0
0	1	1	0
1	0	0	1
1	0	1	0
1	1	0	0
1	1	1	1

Fig. 8.22 The truth table of the logic of the amplitude discriminator, [87]

Background/Function of the Logic

- If the output signal Y is 1 and therefore e.g. 2 V is present at the output, then this voltage pulse (smoothed via the filter) will increase the voltage at the VCO input and thus the frequency at the VCO output.
- For this, K_1 must have the value 1, which is given if U_{S11} has a sufficient adjustment (e.g. −2.5 dB), which is set with U_{Min}.
- Case 1: Furthermore, K_2 and K_3 are both at 0, which means there was no improvement in the fit and Y was previously at 0 (which corresponds to 0 V).
- Case 2: K_2 and K_3 are both at 1, which means that the full output voltage already resulted in an improvement in the matching in the previous work cycle.

In the blue blocks of Fig. 8.22, the VCO input voltage is only changed (reduced) in one direction, while in the red block the VCO input voltage is repeatedly reduced and increased. Here you end up in the "lock range".

If such a control is running in "lock mode", the control only ever jumps to the three frequency values of the very best fit and the fit of the two neighboring values, as shown in Fig. 8.23 using a real measurement with 10 MHz frequency steps.

With such a control loop, there are many other points to consider, such as for example the time behavior. The points are explained in the associated publication [46].

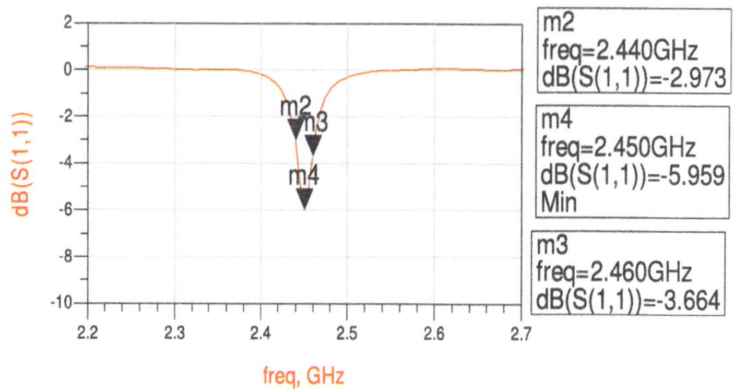

Fig. 8.23 Measurement of the matching S_{11} and the three frequency points of the "lock mode" of the ALL

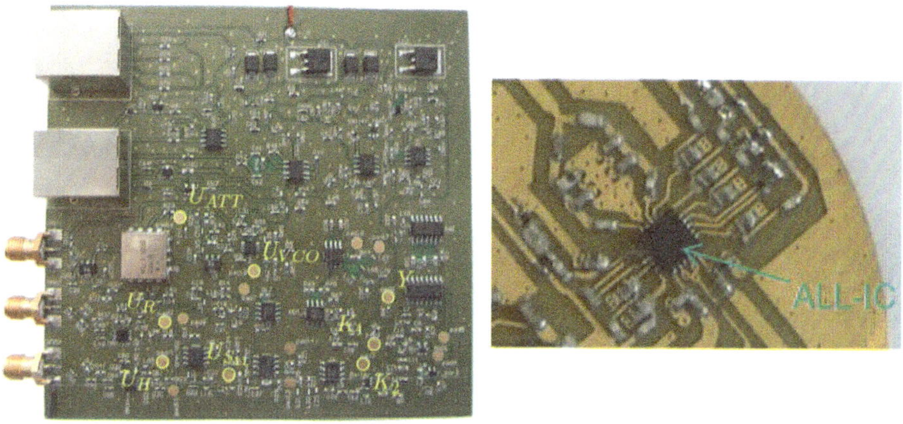

Fig. 8.24 Link: SMD solution of the ALL and right: CMOS implementation of the ALL

Hardware Realization and Measurement

This ALL was first set up and optimized at the IMP using individual commercially available electronic components. In the second step, this circuit was implemented in a CMOS IC on the Infineon C11N process in close cooperation with the IMST (Kamp-Lintfort). Figure 8.24 shows these two setups.

In 12.2016, this development was purchased by the US semiconductor company MACOM. In addition to the application in this GHz plasma technology, this technology can also be used in microwave ovens with semiconductor amplifiers.

The internal structure of the CMOS IC can be seen in Fig. 8.25.

The complete circuit for a 15 W application of the integrated ALL circuit is shown in Fig. 8.34.

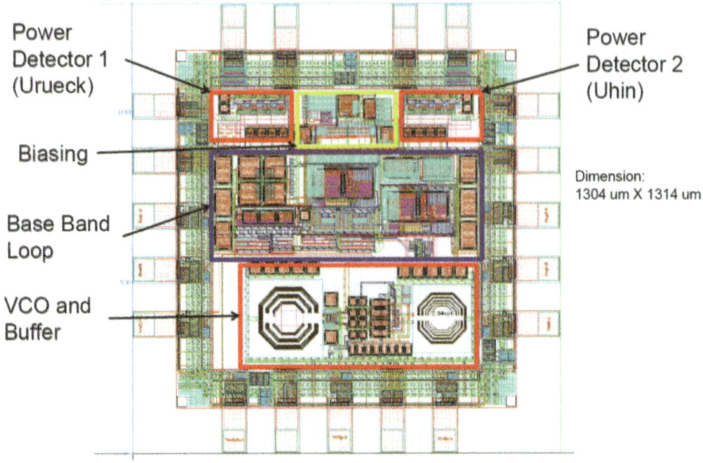

Fig. 8.25 Internal structure of the CMOS-ALL-IC

Fig. 8.26 Photo of the first HF spark plug with different drive powers

8.5.4 Microwave Spark Plug

The classic spark plug has already been briefly presented in section 8.3.1. With the corona plasma, the latest research work of the BorgWarner company was also briefly introduced in 8.4 in comparison to the conventional spark plug.

Since 2006, the IMP at FH Aachen has been researching the microwave spark plug (MW spark plug for short), which was introduced as an HF spark plug, and the associated control electronics in the frequency range around 2.45 GHz.

As early as 2006, quite remarkable results were achieved in a first diploma thesis with a new type of transformation network located in the spark plug, Fig. 8.26.

A spark plug with microwave plasma is expected to have many advantages:

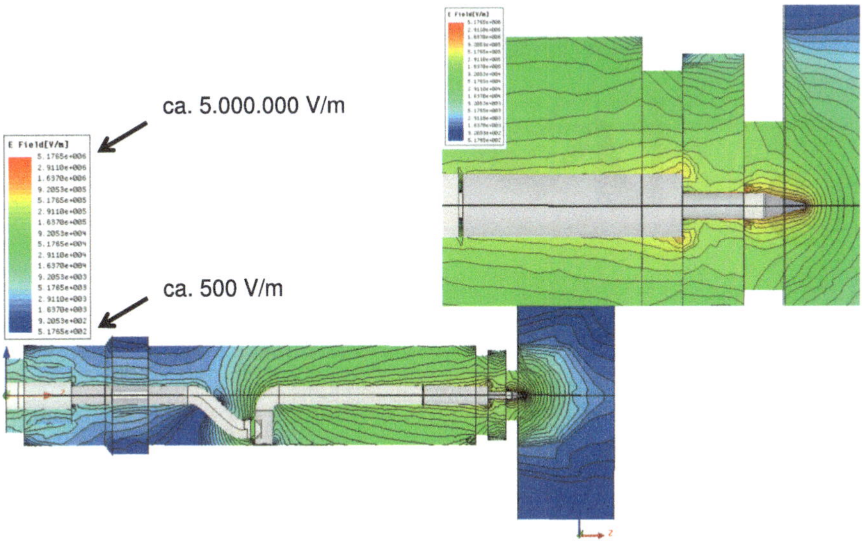

Fig. 8.27 Field simulation of the first HF spark plug

1. The injection jet can be guided to the electrode;
2. The ignition duration can be set variably;
3. It can be checked whether ignition has actually taken place;
4. The longest possible service life;
5. High efficiency of the entire ignition system.

This image also perfectly reflects the character of the GHz plasmas. While a corona plasma has discharges with individual "streamers in the high MHz and GHz range, a plasma ball is formed.

The hypothesis is that the number of streamers becomes more frequent and the length shorter at higher frequencies in electrodes for plasma generation that can be used in identical broadband applications. Above a certain frequency, plasma with a large number of streamers turns into spherical plasma.

The internal structure of the first one developed at the IMP and the increase in field strength can be seen very clearly in Fig. 8.27.

In the meantime, many technological steps have been taken to turn the first illustrative examples (see Fig. 8.26 without insulation) into spark plugs that can be used in the engine. These new MW spark plugs have not only proven their functionality in pressure tests, various special measurements and engine runs, but they also far exceed the entire state of the art.

For example, tests have been carried out up to 40 bar starting pressure under nitrogen at room temperature. The spark plug works perfectly even with input signals of 120 W pulses.

Generation 2b can be operated with 400 W pulses and 75 % duty cycle. This means that this GHz spark plug can withstand 300 W in continuous operation.

	spark plug	characteristics
	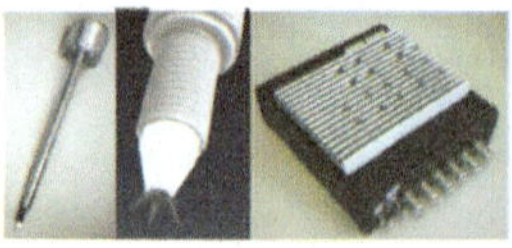 Figure 1: Igniter (left), igniter tips (center), controller (right) of Borg Warner`s Corona Ignition System EcoFlash	☑ can ignite lean, combustible mixtures over a wide pressure range ☑ free in time of ignition, quick re-ignition ☑ a spatial pattern of sparks with a large number of spark channels - volume ignition effect ☑ minimized electrode erosion ☒ requires internal high-pressure gas filling ☒ adaption of the piston to the ignition system for best results ☒ no plug and play technology ☒ very large controller and expensive high voltage electronic
status quo	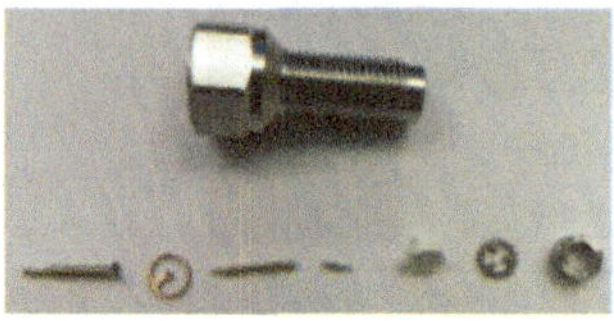 Figure 2: MW-spark from the IMP	☑ cheap and compact electronic and spark plug, easy to integrate ☑ supports pulses up to 400 W ☑ no need to adapt the piston ☑ high integrated CMOS control loop (ALL) is available (but not tested) ☑ no internal gas filling ☑ high voltage only in the spark plug ☑ no electrode material in the plasma ☑ free in time of ignition ☑ very quick re-ignition ☑ free in duration of ignition ☑ successful motor run ☑ volume ignition effect dependent on wind-blown dispersal

Fig. 8.28 Performance characteristics of the 1st and 2nd generation MW spark plug compared to the conventional spark plug

The amount of energy for an ignition spark over just 1 ms is 400 mJ and therefore far greater than that of state-of-the-art spark plugs. The efficiency is so good that the energy in the GHz plasma almost corresponds to the energy of the supplied microwave energy.

Conventional spark plugs have a maximum energy of 30 mJ in the plasma.

Fig. 8.28 shows the various properties of MW spark plugs.

Further details on the spark plugs can be found in numerous technical publications, such as [45].

A visualization of the spark plug electronics for large-scale production is shown in the Fig. 8.34.

8.5.5 Lamp Technology

The presented technology with the impedance transformers can also be used very advantageously for the development of new types of inexpensive plasma lamps.

Fig. 8.29 shows a single-sided controlled projector lamp in the 120 W power class, which can be dimmed, among other things.

Fig. 8.29 Innovative 120 W projector lamp with only one-sided control, operated via a 2.45 GHz generator

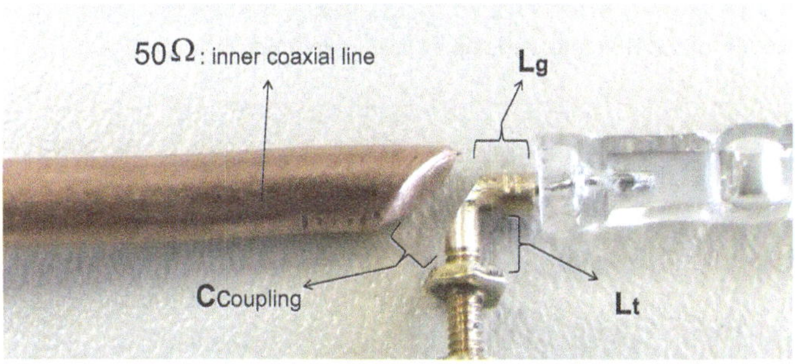

Fig. 8.30 Photo of the internal components of the first HF projector lamp manufactured in 2006

This projector lamp with the standard commercial filling (mainly argon and mercury) reaches an internal pressure of 220 bar during operation. Measurements of this new lamp show the state of the art:

- significantly faster starting behavior,
- best restart characteristics when hot,
- better luminance and higher proportion of light in the visible range.

The biggest advantage, however, is that only one electrode gets hot and there is only one single point with the highest radiation density.

With conventional projector lamps, which are based on an arc discharge there are two equivalent light points. For use in the projector, the efficiency is improved by a factor of almost 2. Further details are published in [28].

The internal structure of this new HF projector lamp is shown in the Fig. 8.30.

The first lamp developed in a diploma thesis was optimized in a research project lasting just one year. The results compared to a standard projector lamp are summarized in the Fig. 8.31.

In the standard, the brightest light points occur at the beginning and end of the arc discharge. The projector optics are focused on one of these two points. As a result, only 50% of the light output is used. The great advantage of this HF projector lamp

Properties	Reference lamp: Philips TOP 120 W / 132 W 1.0	RF-Plasma lamp: OSRAM PVIP 120 W / 132 W 1.0 modified
Luminous flux [lm]	7825	8521
Power of lamp [W]	132	149
Efficiency of lamp [lm/W]	60,2	57,2
Leuchtdensity [Gcd/m^2]	2,59	2,79
Color rending index Ra [%]	62,5	66,8
Starting time (cold) [s]	55	16
Re-ignition [s]	ca. 120	ca. 0 – 40

Fig. 8.31 Results of the HF projector lamp by year of development

Fig. 8.32 First demonstrator of an electrodeless low-pressure lamp operated at 2.45 GHz

is that it only has one very strong light spot and therefore the projector also has a noticeably improved efficiency.

After the projector lamp, research work on new types of low-pressure lamps, in particular energy-saving lamps, began at FH Aachen. The first innovative step was the construction of energy-saving lamps in which the metal electrode no longer protrudes into the plasma (Fig. 8.33). This lamp already stood out as a quick-start lamp (Fig. 8.32).

In the course of further research work, a first mercury-free energy-saving lamp was created on this basis, which is shown in the Fig. 8.33. The LIT at KIT, headed by Prof. R. Kling, developed the special lamp filling and the highly efficient AC-DC converter electronics, which are also integrated.

The associated inner GHz electronics are shown in Fig. 8.34. These electronics require a DC power supply with 28 V and a very small, simple line coupler just before the lamp to measure outgoing and reflected energy.

If a further power transistor is added, this compact and inexpensive electronic system can also be used as spark plug electronics in large quantities.

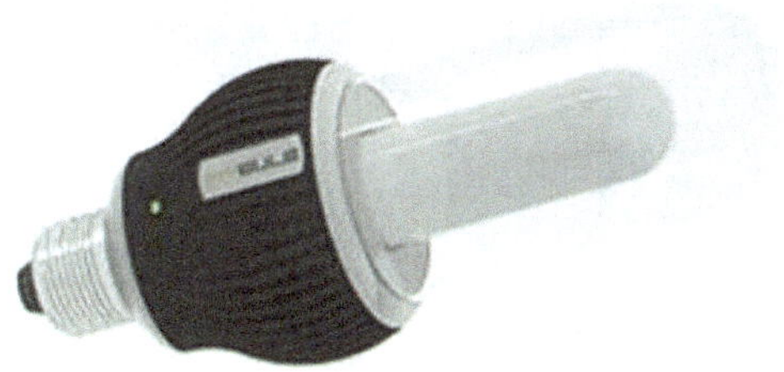

Fig. 8.33 First mercury-free energy-saving lamp from the IMP and the LIT

Fig. 8.34 Internal GHz electronics of the mercury-free energy-saving lamp

MW Lamps with Great Future Potential

The MW lamps with the greatest potential for the future are seen in very small high-pressure lamps in which the microwave energy is is coupled in purely capacitively. Such lamps have been scientifically investigated in detail at the IMP and published in [96], among others. These lamps could be dimmed from 6 W to 200 W and had an efficiency of 129 lum/W at 43 W.

The lamp volumes can be further reduced so that the power values of 6–15 W for household and shopping light applications can be realized with such high efficiencies and the best spectral light quality.

8.5.6 GHz Plasma Jets

There are only two companies that produce plasma jets or emitters that operate with a 2.45 GHz signal. The company Aurion was the first to launch an emitter with a microwave input power of around 10 W. Since 2010, the Aachen University of

Fig. 8.35 Versatile plasma jet (here process gas: air, 100 W and wide opening)

Applied Sciences spin-off (www.hhft.de) has been introducing various atmospheric jets that can be used up to 200 W for many years and now far beyond.

One of the first plasma jets developed at university FH Aachen shows Fig. 8.35 at an incident microwave power of 100 W.

The physical advantages of 2.45 GHz plasmas (microwave plasmas or microplasmas) for users have been demonstrated in many publications. These new jets now offer scientists as well as practical users the opportunity to utilize these advantages for a variety of possible applications. Properties of microwave plasma emitters at a glance:

- Atmospheric 2.45 GHz plasma with powers up to 500 W or 2500 W
- Very low energy requirement: around only 20 % of the energy requirement of a hydrogen burner!!!
- Suitable for almost all gases,
- Line and area emitters,
- Ideally suited for manual and robotic use (genuine, 100 % potential-free),
- Activation/cleaning of various materials,
- Maintenance and wear-free,
- Small and handy design.

The images of the HF spark plug already clearly showed the big difference between a plasma generated by an arc discharge and a plasma generated by a 2.45 GHz signal. The first-ever generation of cannula jets uses pure plasma generation at the tip of the electrode. The gas is fed through a cannula and 100 % of it is forced through the plasma. The jets with the most compact design are shown in Fig. 8.36.

The fact that a microplasma does not strike against the mass has many advantages, such as absolute freedom from potential and low waste heat losses. Furthermore, a microplasma is not created by electrons leaving and entering the electrodes, but only by capacitive coupling. This means that the electrodes are not stressed as much as with an arc discharge.

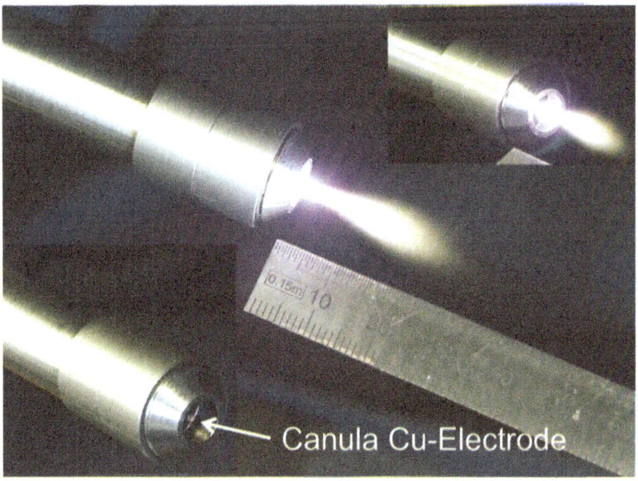

Fig. 8.36 HHF plasma jet of the PC series with cannula

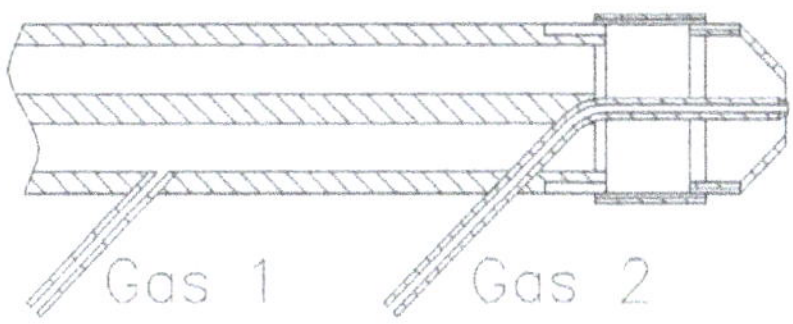

Fig. 8.37 Sectional view of the head section of the PCA series plasma jet with cannula

Especially for process gases that are difficult to ionize, such as air, the degree of ionization is much higher with a microplasma emitter than with an emitter that works with an arc discharge, as the arc only ionizes a small part of the gas. Scientific studies that have already been carried out on 10 W systems show the great effectiveness of microplasmas in technical applications. Spectroscopic investigations carried out by the IMP show that almost no material escapes into the plasma at the electrodes. Due to the lower internal heating, all HHF plasma emitters are equipped with virtually maintenance-free copper electrodes.

In addition, the two gas connections (Fig. 8.37) allow the user many more options.

In Brief: Technological Applications for Surface Processes

Plasmas are increasingly being used in all areas of industry as well as in medicine. Important applications for low power (under 100 W):

- Activation of plastic surfaces for welding, melting, painting, printing, coating, bonding and other processing steps,

- Cleaning of various metal surfaces for soldering, painting, printing, coating, gluing, bonding and other processing steps,
- Disinfection and sterilization in medical technology from instruments to wounds.
- Ageing of components.

Additional areas of application for larger cables (>100 W) are:

- Soldering of metals with soft or hard solder,
- Melting of surfaces and drilling of holes,
- Labeling of various materials,
- Baking of lacquers,
- Cutting of various materials,
- Welding of various materials,
- Coating of various materials.

The plasma jets already have temperatures of over 3400 °C at a 200 W microwave power in the plasma jet.

SF-Jet with Great Future Potential

The first cannula jets have a geometry that is based purely on the design of the standard jets. The design is optimized for transferring microwave power through the jet with the least amount of reflection possible.

The new SF-jets (SF: Straight Forward) are based on the fact that the cannula runs straight through the jet. Therefore, in addition to the process gas, you can also

- rod-shaped solid materials such as wires, glass fiber, PAN, etc. as well as
- threads,
- powder and
- liquids

into the plasma and push them through lengthwise, which significantly increases the dwell time and saves process time and therefore energy.

When working with powder, yields of up to 100 % can be achieved!

The jet shown in Fig. 8.38 is operated at 200 W and now serves as the basis for developing 3D printers for metal and ceramics in the international PlasmaPrint research project. For metal printing, only wires should be used as far as possible and for ceramic printing, only suspensions should be used as far as possible in order to avoid the expensive powder feed.

Of course, a 500 W or better 1 kW microwave generator is also orders of magnitude cheaper than a comparable laser.

MW-Jet for Rapid Heating

Many heating processes in large-scale industry run via hot air, which does not have a good thermal coupling to the heating surface.

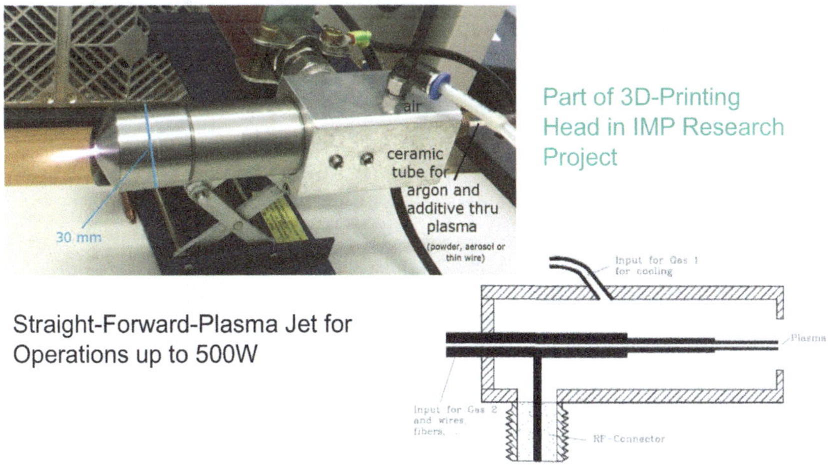

Fig. 8.38 Foto and sectional view of the SF jet, which can be operated with 500 W at 2.45 GHz can be operated

In contrast, the greatest heat is generated in the atmospheric plasma where most recombination takes place. If you shine a jet onto a surface, it is precisely this surface that is heated very strongly.

For these applications, the so-called MagJet, which can work in continuous operation with a microwave power of 2500 W, was launched on the market in 2019. This jet, shown in Fig. 8.39, outperforms all arc and spark discharge jets in terms of power compatibility by a factor of 4.

This jet can save a lot of energy in industry. Of course, hot processes using gas burners can also be replaced by this jet. In contrast to combustion, no soot is produced and the energy input and energy costs are significantly lower.

The generators for these MagJets are the same as those used in large-scale industry for microwave heating of ovens. The magnetron generators are available in power ranges from 3 to 25 kW (and more).

At these high power levels there is no longer any frequency switching. Details of this jet are [86] including the special technology for ignition.

kW-Jets for Welding

A special jet that is also currently being researched is a jet, that is also being researched and is to be used for welding.

The first functional welding head, which can be operated with up to 6 kW at the input, is shown in Fig. 8.40 after just 1.5 years of development.

The corresponding welding results are shown in Fig. 8.41. These already very good weld seams were produced without an optimization process and without the use of filler materials (welding rods).

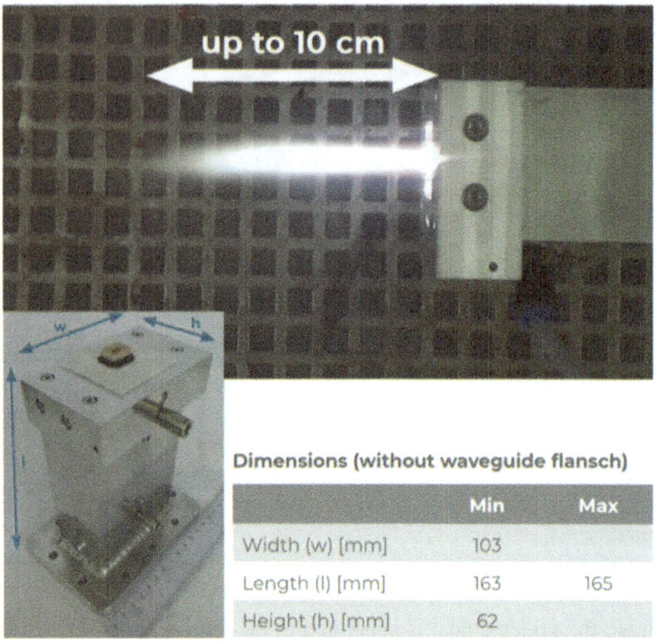

Fig. 8.39 Photo of the plasma and a 3D view of the MagJet, which operates at up to 2500 W at 2.45 GHz can be operated continuously

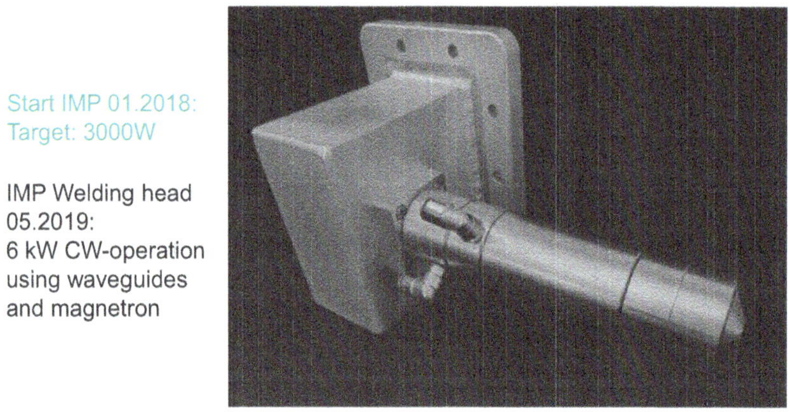

Fig. 8.40 Photo of the welding head being developed as part of a ZIM project

8.5.7 MW Surgery

A state-of-the-art HF scalpel is often used in surgery. In this HF surgery, voltages of up to 10,000 V and electrical power of typically 400 W are used via a high-frequency signal (therefore HF) of typically 1 MHz.

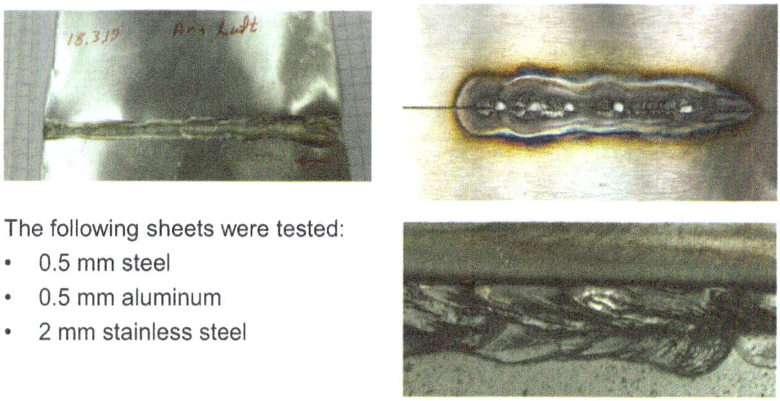

Fig. 8.41 Welding results for different metals

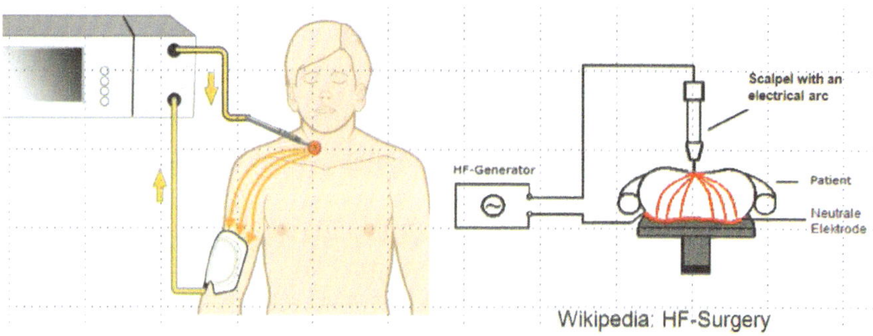

Fig. 8.42 Actual HF scalpel with arc discharge and 400 W at typ. 1 MHz

The person is placed on an insulated table. An arc discharge occurs at the HF scalpel, which has the advantage that bleeding is stopped at the same time as the incision by closing the affected vessels. The current required for this then flows through the body and back to the generator via a second large-area neutral electrode, Fig. 8.42.

With the development of modern transistor technology in the 1970s, electrosurgery devices could be made small and compact and designed for numerous new applications. In order to keep pace with the rapid developments in medicine, microprocessors were used in the 80s, which were then used in the 90s to develop automatically controlled electrosurgical devices. Automatically controlled HF surgical devices became established in the could prevail, [61].

The use of this HF scalpel technique is dangerous due to the current flow in the body and all normal medical technology limits are exceeded. Nevertheless, the HF scalpel is now predominantly used in clinics, as the healing process is orders of magnitude shorter.

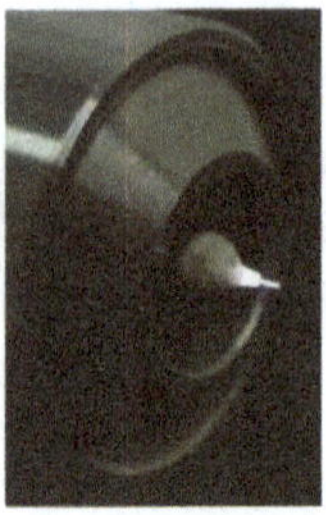

$I = P_{in} / A = 8 \cdot 10^4\ \mathrm{W/cm^2}$ **Energy density**

P_{op} = 50 W CW	Power of the MW Plasma jet V3
$S^{dB}{}_{11}$ = -3 dB	Reflexion value
P_{in} = 25 W CW	Power in the plasma
$A = \pi \cdot d_i^2/4$	Diameter of the beam d_i = 0.2 mm

Fig. 8.43 MW scalpel version 3 for pure CW operation with 50 W at typ. 2.45 GHz

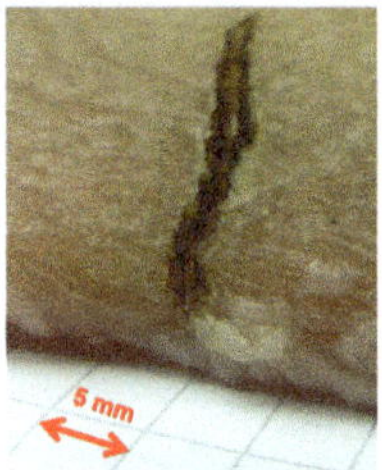

Fig. 8.44 Cutting result with the MW scalpel version 3 for pure CW operation with 50 W at typ. 2.45 GHz

At the IMP, a scientific study has successfully investigated whether microwave plasma has the potential to replace the HF scalpel, [99].

In simplified terms, the MW scalpel is a cannula jet with an extremely small cannula. Basic results were achieved that showed in the first step that a plasma can be realized that comes close to the energy densities of laser, Fig. 8.43. This basic result naturally reduces the R&D risk for further applications that require a very high power density.

The cut achieved (tested on pork) with CW operation only shows too much blackening. It is expected that the blackening from Fig. 8.44 will no longer occur due to the pulse mode, which is used with the HF scalpel.

8.5.8 GHz Plasma Technology as Basic Technology

The application examples presented in the previous subsections show that GHz plasma technology is a basic technology that enables many applications. The laser is a similar basic technology.

The following overview presents the applications carried out at the author's institute and those still open/planned (Fig. 8.45).

The basic approach for the flow technology can be found in the following chapter and also in the patent specification of FH Aachen.

In the field of 3D printing of metal and ceramics with this GHz plasma technology, a doctorate and several scientific publications such as [93] and [59] have already been published.

8.5.9 Decoupled GHz Plasma Technology for Furnaces and other Applications

The latest major innovative step in GHz plasma technology is the demonstration that the plasma can be shaped by the specification in a 3D field simulation. To do this, the microwave properties of the complex dielectric behavior of the desired plasma must be known. The modeling of a microwave plasma takes place in three steps:

- Step 1: A plasma is generated through the use of tuners and external ignition sources.
- Step 2: This plasma is measured using a hot S-parameter measuring station.
- Step 3: In a field simulation that describes the generated plasma, the complex (sometimes negative in the real part) dielectric parameters are optimized until this simulation matches the measured complex scattering parameters.

In the following, a basic development is presented, which is a direct basis for many possible applications.

Ovens that can convert electrical energy into heat with an efficiency of around 87 % using a 915 MHz magnetron source and are therefore significantly more efficient than hydrogen burners are very obvious.

The Fig. 8.46 shows this realized demonstrator from the outside.

Applications GHz Plasma	Status	Era	Eng.-years	Comments	IMP project partner
Spark plugs	High Power Spark Plugs – Patent and realization by the Company Weissgerber, Dortmund, Germany	2006 -	5	IMP support your special variants	Weissgerber
Lamps	Different kinds of lamps were successfully developed, i.e. with the LED-efficiency and a perfect light	2007 -	6	These high-end lamps have a market	pinkRF, NL and more open
Plasma jets u. coupler for plasma chambers	Plasma jets from 10 to 3000W, variants with cannula electrode form sputtering and more	2006 -	8	R&D by IMP spin-off www.HHFT.de	HHF, Ionics, B., Fricke u. Mallah, Peine
Microwave scalpel	Demonstrator shows good first results	2013 -	4	Is „on-going"	BOWA, Gomeringen
Welding equipment	Successfully welding tests, without ground electrode	2018 – 2020	2	Incudes benefits of WIG- and laser welding	Aixcon, Stolberg; Germany
3D-printer	Printer for metal and ceramic with 500W	2019 – 2021	3	Low cost in comparison to the laser printer	Ionics and other, open
Broad band ultra-sonic generator	Ultrasonic generator for measurement technique and more	open	0	More than one application	open
Ion sources and fusion technology	Fusions technology will have 12m (diameter) and needs only 150kW of input power	open	0	The potential of this technology is gigantic	open

Fig. 8.45 Research projects at the Institute for Microwave and Plasma Technology (IMP) at FH Aachen

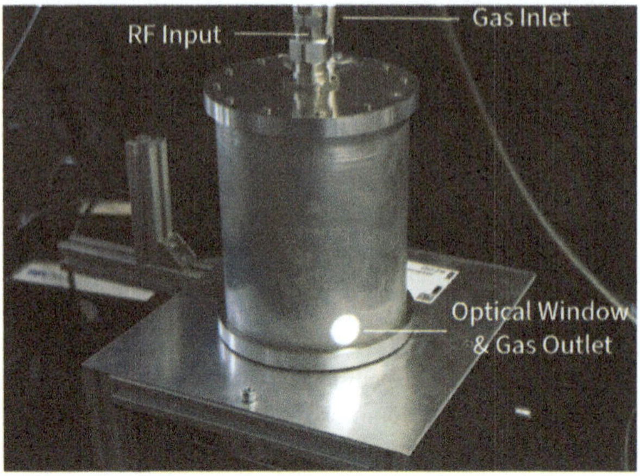

Fig. 8.46 Photograph capturing the operating decoupled plasma generating device at a frequency of 2.41 GHz and an input power of 100 W

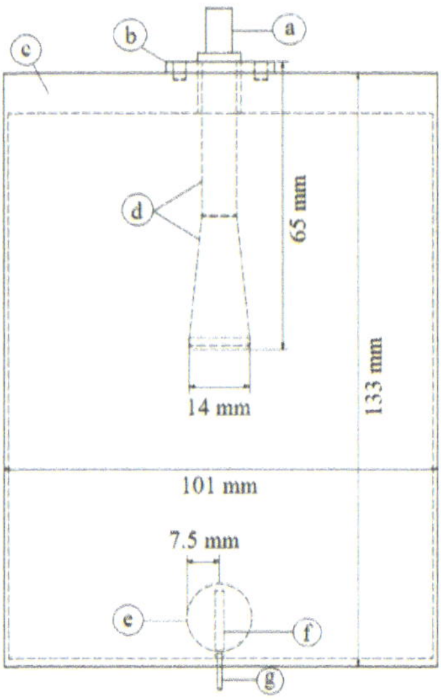

Fig. 8.47 Schematic illustrating the decoupled plasma generating device. The device operates by transmitting a microwave signal through (a) a coaxial transmission line, connected to (b) an N-connector leading to (c) the cavity. The schematic includes specifications of (d) the mode-converting monopole antenna, showcasing its dimensions, as well as (e) a 15 mm diameter hole for a camera. The illustration highlights (f) the electromagnetic plasma model and (g) the ignition pin, which is removed after plasma ignition.

Fig. 8.47 shows the internal structure of the demonstrator. This research work pioneers the introduction of the decoupled microwave plasma principle combined with plasma shaping. The approach involves the use of two types of antennas within the circular cavity resonator: a low-loss coupling monopole antenna and a lossy

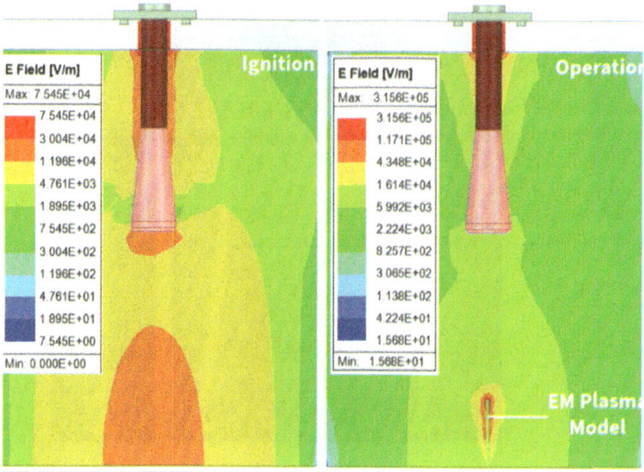

Fig. 8.48 Scalar electric field distribution screenshots in a microwave cavity resonator for the ignition phase (left images) and operation phase (right images), elevation view near the plasma region. The ignition phase illustrates the scalar electric field distribution without a plasma model, whereas the operation phase showcase the scalar electric field distribution with a plasma model.

antenna with the electrical properties of a line plasma. However, a challenge arises in determining the plasma properties without preliminary measurements, a situation that will be elucidated in subsequent chapters. It is crucial to emphasize that the cavity has been designed specifically for the operating case, featuring the insertion of a thin wire for ignition through a small central opening at the bottom plate of the cavity.

A plasma is now specified here as a line plasma. For ignition, a small hole is drilled in the bottom center of the chamber through, which an ignition wire is inserted. Once ignition is complete, this wire is pulled out. A plasma model is used for operation, which is explained briefly and in more detail above in [108] and [44]. The associated field simulation is shown in Fig. 8.48. The plasma created inside corresponds pretty much exactly to the specification. The associated publication cited also shows that the complex scattering parameters for both operating modes match very well. Fig. 8.49 shows the resulting plasma. The researcher attributes the observed beam decay in argon plasma to gas circulation, a phenomenon that is prominent in argon plasma but absent in air plasma (due to the lack of flow), where no decay was observed. To investigate this, they conducted experiments involving the temporary cessation of argon gas flow, resulting in a perpendicular argon plasma configuration. To ensure the filling of the entire resonator with argon, we maintained a constant inflow of argon, creating gas circulation that slightly deformed the argon plasma beam.

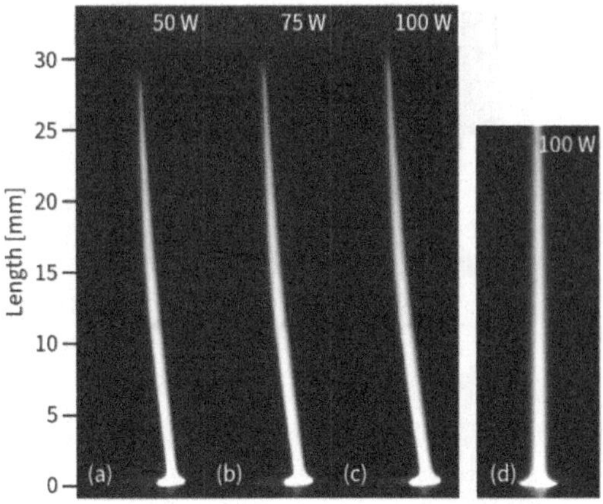

Fig. 8.49 Three images vividly depict the line shape of decoupled plasma at input power values of (a) 50 W, (b) 75 W, and (c) 100 W. It is noteworthy that only 20 W, 25 W, and 29 W of the input power are absorbed by the plasma. The areas corresponding to active species are measured as 15 mm^2, 16.7 mm^2, and 18 mm^2, respectively, and are highlighted with red lines. Argon flow is maintained in (a)-(c) to sustain an argon atmosphere, resulting in a subtle beam decay. The fourth image (d) captures the plasma at input power value 100 W with the argon flow momentarily turned off, showcasing the effect of gas flow on the plasma beam decay.

Chapter 9
Basic Concepts of Antenna Theory

In this chapter on the basic concepts of antenna theory starts with an overall system in the form of radar systems. The system basics are easily transferable to other systems such as radio relay, satellite radio and also mobile radio. mobile radio[1].

In addition, important basics such as the various elementary steels, the far field, the equivalence principle and especially aperture antennas are discussed. Finally, the characteristics of the antennas are explained.

While the assemblies of the transmit and receive branches are similar for various applications, the application of the antenna of a radar device to a specific area of use often poses new problems.

The exact calculation of the radiation of electromagnetic waves must be based on the solution of Maxwell's equations, taking into account the boundary conditions. If a solution fulfills Maxwell's equations, this solution is the only one due to the uniqueness theorem. This procedure is only possible in special cases such as the elementary emitter and the open rectangular waveguide.

For this reason, numerical methods are generally used even for simple arrangements.

In many cases, only the current distribution on the antenna is calculated numerically, as the radiated field can be calculated in a closed form for a given current distribution.

For some so-called planar antennas, such as the horn radiator and the parabolic antenna, the current distribution can be specified as a good approximation and their radiation can therefore be calculated without numerical aids.

The following chapters provide a very basic and detailed description of the most important aspects of modern antenna technology. However, if you would like to learn more about the multitude of possible antenna implementations, [9] is an up-to-date work.

---

[1] Chapters 9 and 11 are based on parts of the manuscript "Microwave Technology" and "Radar Technology" by Prof. H. Chaloupka and have been slightly modified in terms of content and in particular in a new layout and with new pictures. Reprinted here with the kind permission of the author.

H. Heuermann, *Microwave Technology*,
https://doi.org/10.1007/978-3-658-45686-3_9

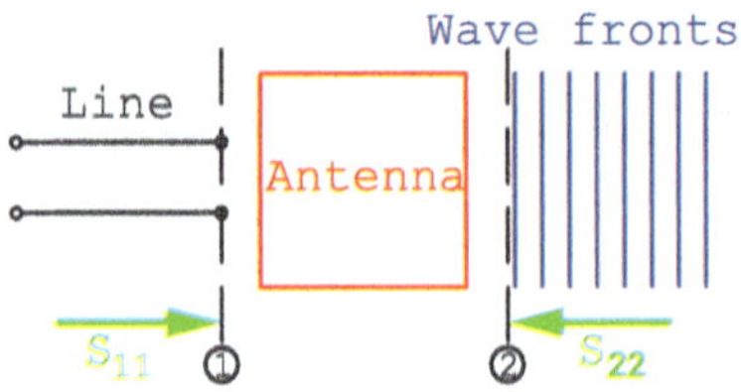

Fig. 9.1 Antennas as a reciprocal ($S_{21} = S_{12}$) mode converter between the transmissions lines of the microwave electronic circuits and the free space

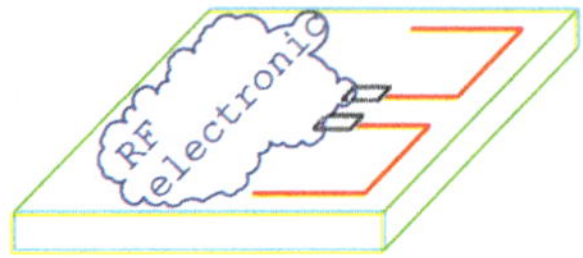

Fig. 9.2 Substrate antennas are inexpensive and support compact solutions

9.1 Introduction in the Antenna Technology

Apart from high-speed digital circuits and converters for fiber optic technology, the majority of GHz circuits have antennas[2].

The antenna is the converter from the conducted mode to the free-space mode, Fig. 9.1. We can discribe this process with the shown S-parameters as well as the M-parameters, which indicates the modes[3]. However, in the practice it is not easy to measure the transfer function of this mode converter. Calibration procedures that support such measurements are not usually programmed into network analyzers. The names of the procudures are "Unknown Thru" and "Without Thru" (chapter 3).

Fig. 9.1 also illustrates the metrological challenge of precisely measuring the antenna, which is very important information, especially with regard to its in-band dispersion (group delay). Modern numerical field simulation (chapter 2) is very helpful here, as it simply delivers very good results in this respect.

The fact that the antenna is often very closely connected to the electronics in practice is shown in Fig. 9.2. This will be the trend in the future to come to higher integration levels.

In [38], chapter 10 describes how such a symmetrical antenna is measured in practice and how it is precisely precisely adapted by the line. However, the illustration on the circuit board ends there. In this chapter and the subsequent chapters will show what needs to be considered in the free space.

[2] The author would like to apologize right at the beginning of the chapter for the fact that some indices are still derived from the German language. However, the mathematical depth of the description of antenna theory exceeds many other book descriptions and a translation of the indices would result in many errors.

[3] A detailed description of the M-parameters can be fund in the RF book of the author, [38].

Fig. 9.3 Wire antennas, arranged vertically above the ground or a conductive surface (car roof)

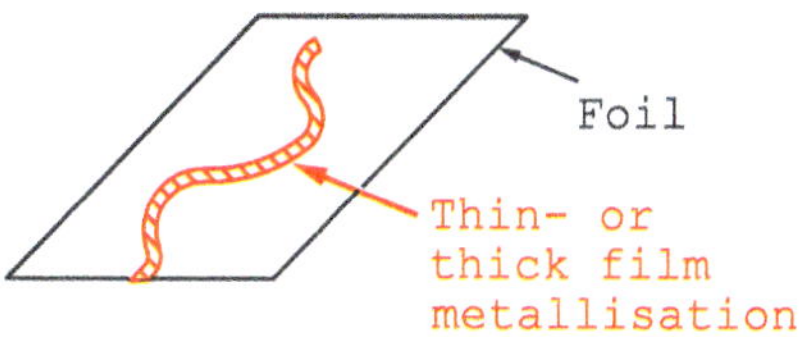

Fig. 9.4 Foil antennas used in very compact devices

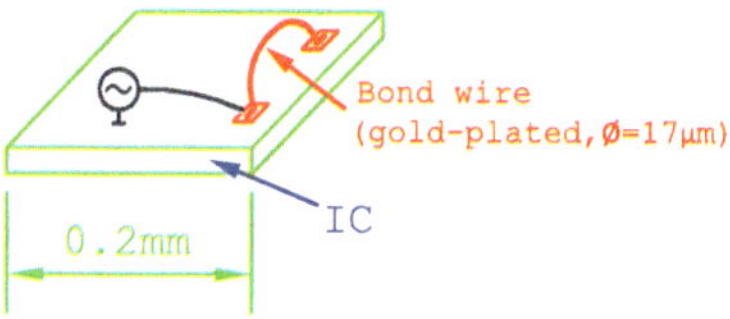

Fig. 9.5 Miniature antennas on semiconductors (ICs) for the two- and three-digit GHz range

Traditionally, radio technology started with wire antennas, as shown in Fig. 9.3. These wire antennas take up a lot of installation space, but have low ohmic losses.

Modern radio systems (e.g. smartphones), on the other hand, have foil antennas. These are light and compact. However, this antenna must always be calculated with the mass of the environment, which must not change. Furthermore, the ohmic losses are often very high, which limits the range of the radio system (Fig. 9.4). In practice, it is possible, but not easy to measure the antenna losses. As described in the section above, a network analyzer must be calibrated using procedures that are not programmed. Furthermore, the provision of the calibration standard for the free space causes additional problems.

The trend of the future for the two-digit GHz range will again be wire antennas that are mounted directly from the bonding wires on the semiconductors (Fig. 9.5).

Common for all these antennas, the same basic principles of antenna technology apply.

Before we go into these basics in detail, a brief introduction to a complete radar system will illustrate the great relevance of antennas.

9.2 Short Introduction to Radar Technology

Under the artificial term RADAR (RAdio Detection And Ranging) is used to summarize the methods for detecting objects and determining their spatial position and their state of motion with the aid of electromagnetic waves.

Despite their relatively low resolution, radar methods are used very frequently as they are able to provide distance information over long ranges regardless of weather conditions.

In 1904, the first radar method for locating ships was registered as a patent and realized. After 1940, there was rapid system and technological development in military technology. Radar technology then established itself in the aerospace sector, which was also less cost-sensitive.

Technological developments in recent years have enabled the cost-effective realization of radar devices for a wide range of industrial applications, e.g. for level measurement and for the transport sector with applications for non-contact distance and distance measurement. While impulse radar is primarily used for aircraft navigation due to the need to be able to monitor long ranges, systems based on the so-called FMCW principle are becoming increasingly popular in industrial measurement technology, (short form: **f**requency **m**odulated **c**ontinuous **w**ave).

FMCW methods can be used to create precision rangefinders with accuracy down to the submillimeter range. A detailed description of the FMCW-functionality is given in [38].

9.2.1 Radar Equation

After as before, large areas of application for radar devices are airspace surveillance and the detection of ships. A key performance feature of such radar systems is their maximum range. The relationship between the maximum detectable range R_{max} and the parameters of the radar system, such as the transmission power and the reflection properties of the radar target, is given by the **radar equation**.

To derive this radar equation, the Fig. 9.6 should be considered first. The radar device emits a short pulse of electromagnetic energy via the antenna. This antenna has the gain G. The gain is the ratio of the maximum radiation density of the antenna to the radiation density of an isotropic radiator (omnidirectional radiator).

The radar target is practically always in the far field of the antenna and has dimensions that are small compared to the beam width of the antenna. This means that a homogeneous plane wave with radiation density S_a strikes the radar target. The intensity of the radiation incident on the target increases with increasing transmission power P_t and improved antenna gain G. As the distance R increases, the radiation or power density S_a decreases in inverse proportion to the increasing spherical surface $4\pi R^2$ of an isotropic radiator.

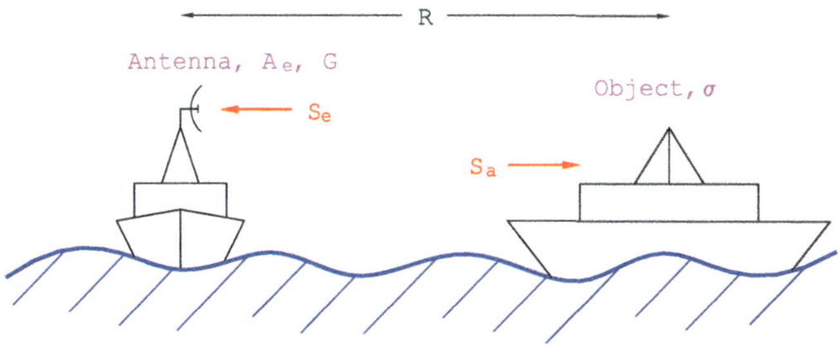

Fig. 9.6 Application of a radar device in shipping

$$S_a = P_t \, G \, \frac{1}{4\pi R^2} \tag{9.1}$$

The generally direction-dependent reflection behavior of the target is described by the so-called radar cross-section σ. Taking into account the distance R from the radar target to the receiving antenna, the received radiation density S_e is obtained:

$$S_e = \frac{P_t \, G}{4\pi R^2} \, \sigma \, \frac{1}{4\pi R^2} \,. \tag{9.2}$$

Using the antenna effective area A_w, which is directly linked to the gain G of the Antenna,

$$A_w = \frac{\lambda^2 \, G}{4\pi} \qquad (\lambda : \text{electrical wavelength}), \tag{9.3}$$

the received power is obtained $P_e = S_e \cdot A_w$:

$$P_e = \frac{P_t \, G \, A_w \, \sigma}{(4\pi)^2 \, R^4} \,. \tag{9.4}$$

Crosstalk, system noise and the filter bandwidth used reduce the receiving system to a minimum detectable receiving power $P_{e,min}$. The so-called radar equation specifies the maximum detectable distance R_{max}:

$$R_{max} = \left(\frac{P_t \, G \, A_w \, \sigma}{(4\pi)^2 \, P_{e,min}} \right)^{1/4} \,. \tag{9.5}$$

This maximum detectable distance depends on the system parameters P_t, G, A_w and $P_{e,min}$ as well as on the reflectivity of the measurement object with the radar cross-section σ .

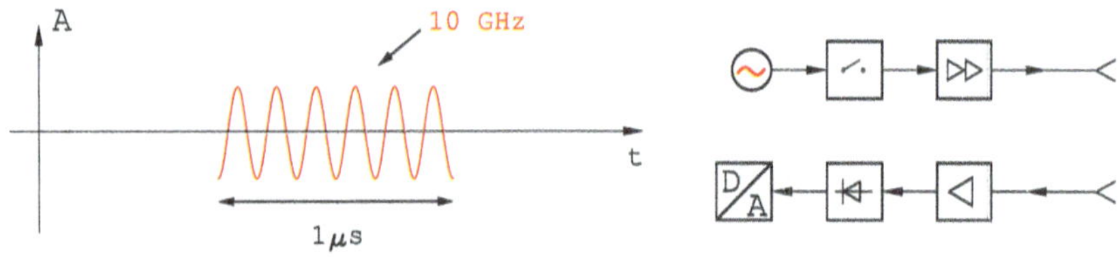

Fig. 9.7 Simplified illustration of a radar pulse and the pulse radar system

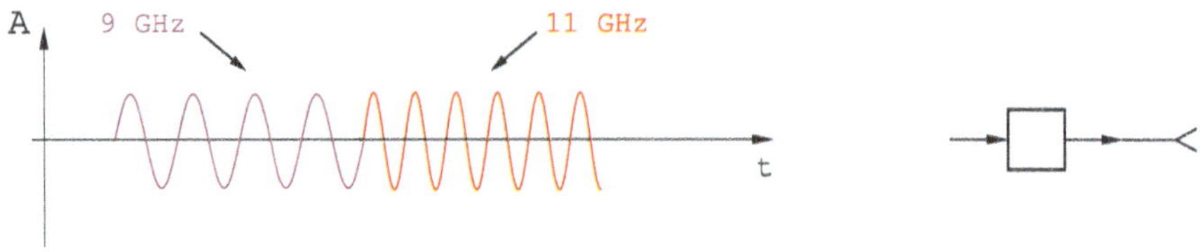

Fig. 9.8 Dispersive line delays 9 GHz signal by 0.5 μs

9.2.2 Basic Principle of Radar Technology

In classical pulse radar the distance to the radar target is determined from the pulse propagation time and its direction from the angular position of a well-focused antenna. The distance R results from the pulse propagation time τ_R to the target and back, under the assumption that the electromagnetic wave propagates at the speed of light c_0 (Fig. 9.7 and Fig. 9.8).

$$R = \frac{c_0 \, \tau_R}{2} \tag{9.6}$$

A electromagnetic wave requires just 1 ns for a distance of 30 cm or 1 μs for 300 m. After a single pulse has been sent, a certain amount of time should pass until the echoes from the most distant radar targets have also been received before another pulse is sent so that distances **unique** can be determined. The maximum unambiguous distance range R_{uniq} depends on the pulse repetition frequency f_p or the pulse repetition time $T_p = 1/f_p$:

$$R_{uniq} = \frac{c_0 \, T_p}{2} \, . \tag{9.7}$$

The basic principle of a typical pulse radar device for wide-range applications can be explained using the block diagram 9.9.

A clock controls the timing of the signal generation and the display on the display unit. This clock triggers the pulse modulator, which in turn amplitude modulates a high-power transmitter via the supply voltage. A wide-range radar device typically has a peak pulse power of several megawatts and pulse widths of several microseconds.

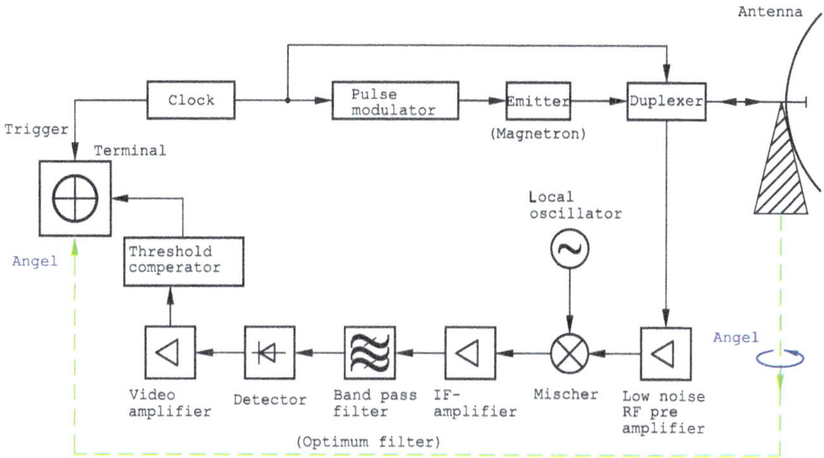

Fig. 9.9 Block diagram of a pulse radar

The transmission pulse is emitted via a rotating antenna, which in turn is also used for reception. A transmit–receive switch (duplexer) switches the transmitter through to the antenna for the duration of the transmit pulse and separates the receiver from the antenna, thus preventing the receiver from being destroyed by the transmit energy. When not transmitting, the duplexer feeds the signals received by the antenna to the low-noise receiver stage and decouples the transmitter from the receiver.

In order to keep the system noise as low as possible, the received signals must be as low-noise as possible and amplified as far as the largest possible received pulses allow.

With the help of a fixed or local oscillator and a mixer, the pulse, which in the simplest case consists of a short high-frequency sequence of sinusoidal oscillations, is converted into an intermediate frequency range (IF range). For further signal processing, the IF signal is first amplified and then filtered. The state of the art is to use a suitably shaped bandpass filter that optimizes the ratio of the signal energy passed through to the noise level remaining at the output. If white noise is present, the optimum filter (matched filter) must have an impulse response that corresponds to the mirror image of the transmitted pulse.

In the next step, the envelope of the filtered IF signal pulse is obtained by rectification with the detector, amplified and compared with a threshold value in a comparator circuit. All signals that exceed the threshold value are displayed as a function of the azimuth angle and the distance in this type of omnidirectional radar (Fig. 9.10).

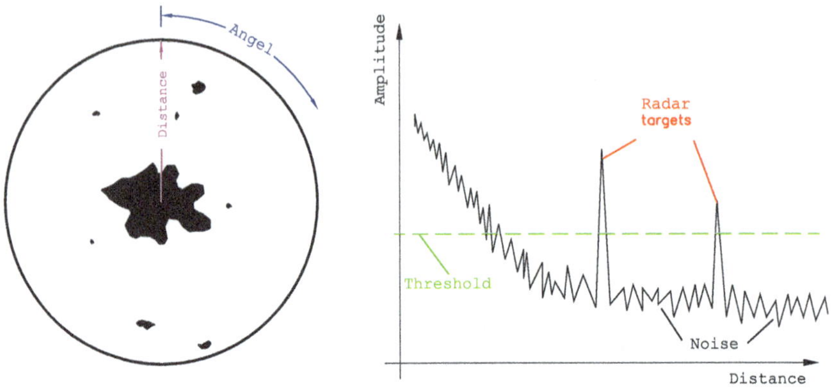

Fig. 9.10 Pictorial representation of the radar information: left: in polar coordinates with distance and angle, right: with fixed angle with amplitude as a function of distance

9.2.3 Radar Frequencies

The choice of frequency band for a radar device is initially determined by the requirements for the angular resolution. For example, a rectangular area antenna with the width L results in a so-called half-width $\Delta\varphi$ in the angular resolution of

$$\Delta\varphi = 0.88\,\frac{\lambda_0}{L} \qquad (\lambda_0 : \text{Wavelength in free space, } \Delta\varphi \text{ in radian})\,. \tag{9.8}$$

Thus, the specification of an angular resolution $\Delta\varphi$ and a maximum antenna size L enforces the smallest operating frequency f.

$$f = 0.88\,\frac{c_0}{\Delta\varphi\, L} \qquad (c_0 : \text{speed of light in free space}) \tag{9.9}$$

Another and often more important criterion for the choice of operating frequency is the legal regulations. For distance meters for industrial applications in particular, only three bands are approved by the European authorities (table in Fig. 9.12).

However, if you want to work at other frequencies, the device must either not emit any radiation into the environment when in use, such as in a steel tank, or must operate with a transmission power of less than approx. −60 dBm.

Furthermore, the most important free frequency bands for industrial applications in the USA and internationally agreed frequency bands for special applications in the shipping and aviation sectors are specified in Fig. 9.12.

In addition to these artificial restrictions, the influence of the atmosphere must also be taken into account (Fig. 9.11 and 9.12).

The attenuation characteristic of our atmosphere, which is caused by molecular resonances of the gases in the atmosphere, may be overlaid by the attenuating in-

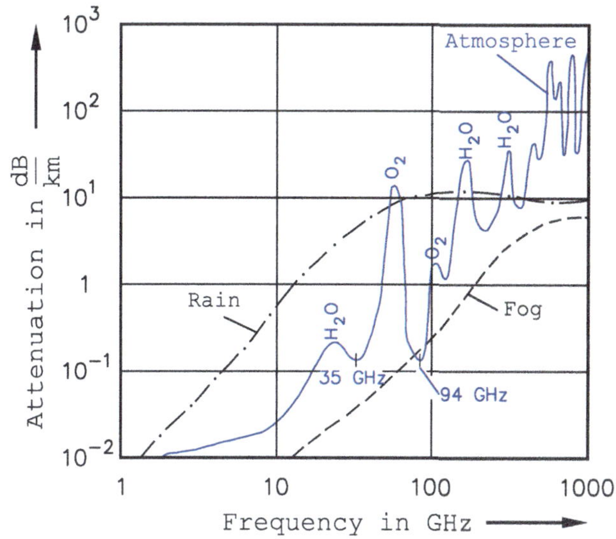

Fig. 9.11 General free space attenuation in the atmosphere

fluence of fog or rain. This attenuation depends heavily on the density of the fog or rain.

The oxygen resonance limits most applications to frequencies up to 36 GHz. However, high attenuation is desirable for special applications, such as distance warning radar. Since the fog in particular considerably attenuates electromagnetic waves at higher frequencies beyond the infrared range and far beyond the visible range, the window at 94 GHz is of great importance for many future applications.

9.3 The Field of an Elementary Emitter

In furthermore, the electric and magnetic elementary radiators are described in detail.

9.3.1 *Electric Elementary Emitter*

The static dipole is known from electrostatics. It is obtained from two point charges of charge q and $-q$ by allowing their distance ℓ to approach zero and increasing q so that the product, namely the dipole moment p, remains finite.

$$p = q\,\ell \tag{9.10}$$

Frequency band		Free bands for industry applications		International Frequency bands for radar applications	
Band name	Frequency area in GHz	in Europe in GHz	in USA in GHz	Frequency band in GHz	Typical radar application
L	1 … 2		1.722 …	1.215 … 1.40	Sekundar radar; midrange radar for air space monitoring
S	2 … 4	2.400 … 2.500	2.200 3.358 … 3.600	2.30 … 2.50 2.50 … 3.70	Airport radar; monitoring and longrange tracking; ship radar
C	4 … 8	5.725 … 5.875	5.460 … 7.250	5.25 … 5.925	Precition ship radars
X	8 … 12		8.500 … 9.000 9,500 … 10.60	8.50 … 10.68	Prëcision–landing–radar; Wether radar in air planes Ship radar
Ku	12 … 18		12.70 … 13.25 13.40 … 14.47	13,40 … 14.00 15.70 … 17.70	Doppler navigation
K	18 … 27	24.00 … 24.25	24.00 …		
Ka	27 … 40		31.20 31.80 … 36.50	33.40 … 36.00	Airports; Controlling of air and landing field
mm	40 … 300				

Fig. 9.12 List of frequency bands of the most important RF applications

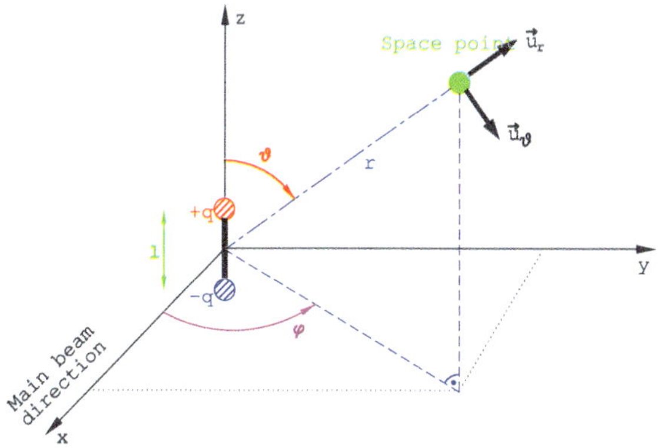

Fig. 9.13 Explanation of the static dipole

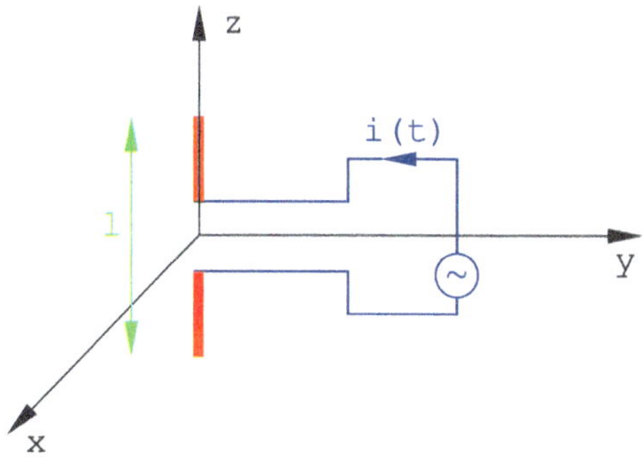

Fig. 9.14 Realization of the electric elementary emitter

The field excited by this dipole in free space is:

$$\vec{\mathbf{E}} = \frac{p}{4\pi\varepsilon_0 r^3}\left[\sin\vartheta\,\vec{\mathbf{u}}_\vartheta + 2\cos\vartheta\,\vec{\mathbf{u}}_r\right] . \tag{9.11}$$

Here are $\vec{\mathbf{u}}_\vartheta$ und $\vec{\mathbf{u}}_r$ unit vectors, r is the distance to the space piont, ϑ is the `elevation angle` and φ the `azimuth angle` (Fig. 9.13).

If between a time-dependent current $i(t)$ (positive counting direction $+z$) flows between the two point charges, the dipole moment p is also time-dependent and the following applies:

$$\frac{dp}{dt} = \dot{p} = \ell\,\frac{dq}{dt} = \ell\, i(t) . \tag{9.12}$$

An electric elementary radiator, which is also referred to as a Hertzian dipole, has such a time-dependent dipole moment and is realized as shown in Fig. 9.14.

The electromagnetic field of this dipole must fulfill the corresponding Maxwell's equations for $r \neq 0$ in source-free space.

$$\operatorname{rot}\vec{\mathbf{H}} = \varepsilon_0 \frac{\partial \vec{\mathbf{E}}}{\partial t} \qquad \text{and} \qquad \operatorname{rot}\vec{\mathbf{E}} = -\mu_0 \frac{\partial \vec{\mathbf{H}}}{\partial t} \tag{9.13}$$

This results in the solution

$$\vec{\mathbf{E}}(r,\vartheta,t) = \frac{1}{4\pi\varepsilon_0}\left\{\left[\frac{p(t-\frac{r}{c})}{r^3} + \frac{\dot{p}(t-\frac{r}{c})}{cr^2} + \frac{\ddot{p}(t-\frac{r}{c})}{c^2 r}\right]\sin\vartheta\,\vec{\mathbf{u}}_\vartheta \right.$$
$$\left. +2\left[\frac{p(t-\frac{r}{c})}{r^3} + \frac{\dot{p}(t-\frac{r}{c})}{cr^2}\right]\cos\vartheta\,\vec{\mathbf{u}}_r\right\}, \tag{9.14}$$

$$\vec{\mathbf{H}}(r,\vartheta,t) = \frac{1}{4\pi}\left[\frac{\dot{p}(t-\frac{r}{c})}{r^2} + \frac{\ddot{p}(t-\frac{r}{c})}{cr}\right]\sin\vartheta\,\vec{\mathbf{u}}_\varphi\,, \tag{9.15}$$

where $c = 1/\sqrt{\varepsilon_0\mu_0}$ is the speed of light. The validity of the solution with the equations (9.14) and (9.15) can be confirmed by showing that

- Maxwell's equations are fulfilled and
- in the limiting case $\dot{p} = \ddot{p} = 0$ and $p(t-\frac{r}{c}) = p$ the static solution results.

The solution that fulfills these two conditions is also the only solution to the radiation problem due to the uniqueness theorem.

From Eqs. (9.14) and (9.15) it can be seen that the time course of the electric and magnetic field is determined by the time course of the dipole moment $p(t)$ and its derivatives $\dot{p}$ and $\ddot{p}$. However, a time delay (retardation) of $\tau = r/c$ occurs due to the finite propagation speed of the electromagnetic field.

The retarded time τ thus indicates the propagation time that an electromagnetic wave requires from the source point to the incident point r when propagating at the speed of light c.

For the case of a harmonic time dependence (angular frequency ω, freespace-wavelength $\lambda_0 = 2\pi c/\omega$), the following is obtained after introducing the usual phasor notation

$$p(t) = \operatorname{Re}\left\{P\,e^{j\omega t}\right\}, \qquad i(t) = \operatorname{Re}\left\{I\,e^{j\omega t}\right\} \qquad \text{etc.} \tag{9.16}$$

and taking into account $I\ell = j\omega P$ (see Eq. (9.12)) from Eq. (9.14) and (9.15) for the phasors of the electric and magnetic field:

$$\vec{\mathbf{E}}(r,\vartheta) = jZ_0\,\frac{I\,\ell}{2\,\lambda_0\,r}\,e^{-jk_0 r}\left\{\left[1+\frac{1}{jk_0 r}+\frac{1}{(jk_0 r)^2}\right]\sin\vartheta\,\vec{\mathbf{u}}_\vartheta + 2\left[\frac{1}{jk_0 r}+\frac{1}{(jk_0 r)^2}\right]\cos\vartheta\,\vec{\mathbf{u}}_r\right\} \tag{9.17}$$

and

$$\vec{\mathbf{H}}(r,\vartheta) = j\,\frac{I\,\ell}{2\,\lambda_0\,r}\,e^{-jk_0 r}\left[1+\frac{1}{jk_0 r}\right]\sin\vartheta\,\vec{\mathbf{u}}_\varphi\,, \tag{9.18}$$

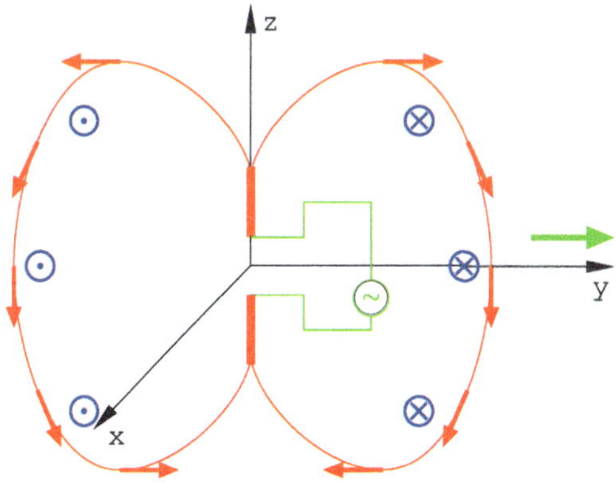

Fig. 9.15 Representation of the E-fields (red) and H-fields (blue) of the dipole antenna in the far field

with the wave number $k_0 = \omega\sqrt{\varepsilon_0\mu_0} = 2\pi/\lambda_0$ and the propagation impedance or field wave impedance in free space $Z_0 = \sqrt{\mu_0/\varepsilon_0} \approx 377\,\Omega$.

The near field is characterized by the fact that $k_0 r \ll 1$ applies. If this approximation is used, the result for the E-field is an expression that corresponds to the static field and for the H-field a field that corresponds to the stationary field of a current-carrying conductor (near field $\hat{=}$ static fields). In the far field, $k_0 r \gg 1$ and therefore $1 \gg 1/k_0 r \gg 1/(k_0 r)^2$. From Eqs. (9.17) and (9.18), the electromagnetic field for the Hertzian dipole at a distance of several wavelengths (far field) is a good approximation to (Fig. 9.15).

$$\vec{\mathbf{E}}(r,\vartheta) \approx jZ_0 \frac{I\,\ell}{2\,\lambda_0\,r}\, e^{-jk_0 r} \sin\vartheta\, \vec{\mathbf{u}}_\vartheta \tag{9.19}$$

and

$$\vec{\mathbf{H}}(r,\vartheta) \approx j \frac{I\,\ell}{2\,\lambda_0\,r}\, e^{-jk_0 r} \sin\vartheta\, \vec{\mathbf{u}}_\varphi\,. \tag{9.20}$$

Finally the Eqs. (9.19) and (9.20) can be formulated in coordinate-free vector notation, if the orientation of the dipole is described by the vector $\vec{\ell}$ with the magnitude $\ell = |\vec{\ell}\,|$ and the direction to the considered receptor point by $\vec{\mathbf{u}}_r$, which is shown graphically in Fig. 9.16. The results are

$$\vec{\mathbf{H}}(\vec{\mathbf{r}}) \approx j \frac{I\,\vec{\ell}\times\vec{\mathbf{u}}_r}{2\,\lambda_0\,r}\, e^{-jk_0 r} \tag{9.21}$$

and

$$\vec{\mathbf{E}}(\vec{\mathbf{r}}) \approx Z_0 \left(\vec{\mathbf{H}}\times\vec{\mathbf{u}}_r\right)\,. \tag{9.22}$$

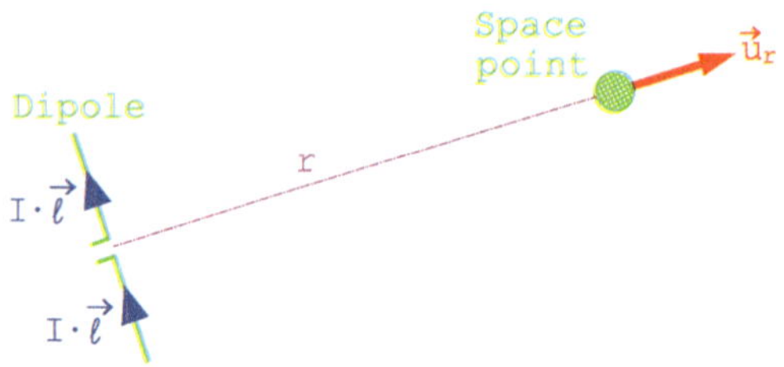

Fig. 9.16 Description of the dipole as a vector element

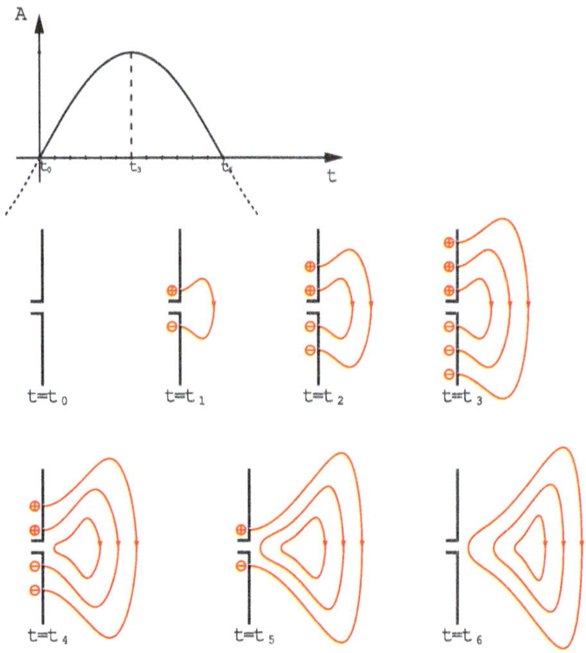

Fig. 9.17 Illustration of the field separation at the dipole over the period of a half period

The separation of the electromagnetic wave over time is shown in Fig. 9.17 in the form of the electric field at short range.

The radiant power density emitted through the space is obtained from the real part of the complex Poynting vector

$$\vec{\mathbf{S}} = \frac{1}{2}\left(\vec{\mathbf{E}} \times \vec{\mathbf{H}}^{*}\right) , \tag{9.23}$$

which has only one component in the r-direction for the elementary emitter:

$$S_r \approx \frac{Z_0}{2}\left(\frac{I\,\ell}{2\,\lambda_0\,r}\right)^2 \sin^2\vartheta\ . \tag{9.24}$$

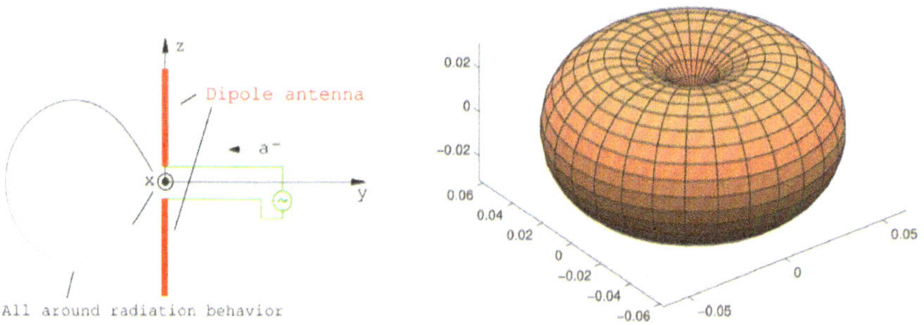

Fig. 9.18 Representation of a (symmetrical) dipole antenna along the z-axis and the associated radiation pattern (shape of a toroid) of the main lobe driven via a symmetrical source by means of of a push-pull signal (left: 2D, right: 3D)

From Eqs. (9.19), (9.20) and (9.24), the following properties of the field in the far zone of an electric elementary radiator can be read off (Fig. 9.18):

1. The surfaces of constant phase (phase surfaces) are formed by the spherical surfaces around the location of the dipole.
2. $\vec{\mathbf{E}}$ and $\vec{\mathbf{H}}$ are tangential to the spherical surface.
3. $\vec{\mathbf{H}}$ is perpendicular to the orientation of the dipole.
4. $\vec{\mathbf{E}}$ and $\vec{\mathbf{H}}$ are perpendicular to each other and the following applies as for plane homogeneous waves $|\vec{\mathbf{E}}| = Z_0 \cdot |\vec{\mathbf{H}}|$.
5. $|\vec{\mathbf{E}}|$ and $|\vec{\mathbf{H}}|$ decrease with increasing radius corresponding to $1/r$.
6. Pure active power transport takes place in the direction of propagation.
7. The field is zero on the axis of the dipole ($\vartheta = 0, \pi$) and is maximum in the plane through the dipole on which the dipole is perpendicular. The fields are independent of the azimuth angle φ.

9.3.2 *Magnetic Elementary Emitter*

An current-carrying conductor loop with small dimensions ($\ll \lambda_0$) generates a field that is dual to the field of the elementary electric radiator described above, i.e. the fields diverge except for constants by interchanging $\vec{\mathbf{E}}$ and $\vec{\mathbf{H}}$ (Fig. 9.19).

Formally introduce the magnetic current M,

$$M\,\vec{\ell} = j\,Z_0\,k_0\,A\,I\,\frac{\vec{\ell}}{\ell}\,, \tag{9.25}$$

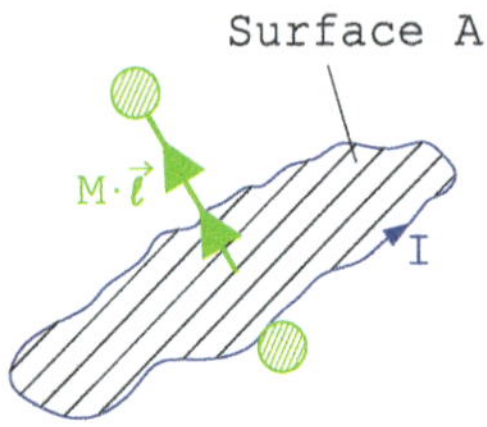

Fig. 9.19 Magnetic elementary emitter

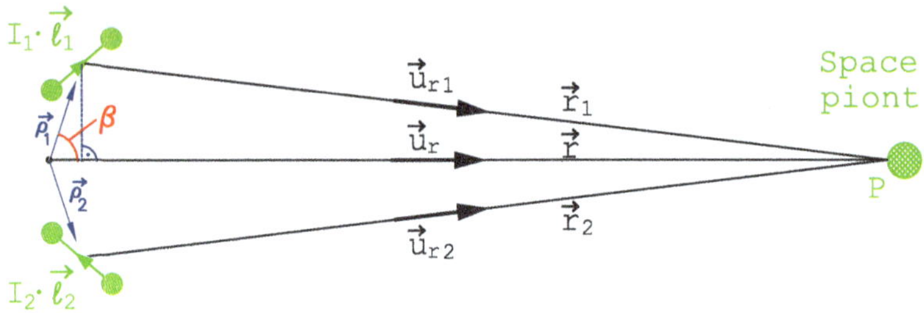

Fig. 9.20 Calculation of the far field of two elementary emitters

where A is to indicate the area of the conductor loop, then for the far zone, i.e. for $k_0 r \gg 1$, of this magnetic elementary radiator, analogous to Eqs. (9.21) and (9.22)

$$\vec{\mathbf{E}}(\vec{\mathbf{r}}) \approx -j\,\frac{M\;\vec{\ell}\times\vec{\mathbf{u}}_r}{2\,\lambda_0\,r}\,e^{-jk_0 r} \tag{9.26}$$

and

$$\vec{\mathbf{H}}(\vec{\mathbf{r}}) \approx \frac{1}{Z_0}\left(\vec{\mathbf{u}}_r\times\vec{\mathbf{E}}\right)\,. \tag{9.27}$$

9.4 The Far Field of an Arbitrary Current Distribution in Free Space

In this section considers the total field resulting from the superposition of the fields of several elementary radiators. It is assumed that a the distance r of the receptor point P to the elementary radiators (source points) fulfills the far-field condition $k_0 r \gg 1$.

The basic idea will first be explained using the special case of two electric elementary radiators. The two elementary radiators may be located in the points described by the position vectors $\vec{\rho}_1$ and $\vec{\rho}_2$ according to Fig. 9.20. The following applies to the amounts r, ρ_1 und ρ_2 as well as for the space point vector $\vec{\mathbf{u}}_r$:

$$r = \left|\vec{\mathbf{r}}\right| \; ; \qquad \rho_i = \left|\vec{\rho}_i\right| \; ; \qquad \vec{\mathbf{u}}_r = \frac{\vec{\mathbf{r}}}{r} \; .$$

The electric field at the space point P is obtained by superimposing the individual contributions with the space point vectors $\vec{\mathbf{u}}_{r_1}$ and $\vec{\mathbf{u}}_{r_2}$ with Eqs. (9.22) and (9.21):

$$\begin{aligned}\vec{\mathbf{E}}(\vec{\mathbf{r}}) = \frac{jZ_0}{2\lambda_0} \Bigg\{ &\left[\left(I_1\,\vec{\ell}_1 \times \vec{\mathbf{u}}_{r_1}\right) \times \vec{\mathbf{u}}_{r_1}\right] \frac{1}{r_1}\, e^{-jk_0 r_1} \\ &+ \left[\left(I_2\,\vec{\ell}_2 \times \vec{\mathbf{u}}_{r_2}\right) \times \vec{\mathbf{u}}_{r_2}\right] \frac{1}{r_2}\, e^{-jk_0 r_2} \Bigg\} \; .\end{aligned} \tag{9.28}$$

The quantities r_1, r_2, $\vec{\mathbf{u}}_{r_1}$ and $\vec{\mathbf{u}}_{r_2}$ are to be replaced by the quantities r, $\vec{\mathbf{u}}_r$ and the position vectors of the source points ρ_1 and ρ_2. Applying the Pythagorean theorem yields the following for the right-angled triangle with the hypotenuse r_1

$$r_1^2 = (r - \rho_1 \cos\beta)^2 \; + \; (\rho_1 \sin\beta)^2 \; . \tag{9.29}$$

If we multiply Eq. (9.29) and replace the scalar product $r\rho_1 \cos\beta$ by $\vec{\mathbf{r}} \cdot \vec{\rho}_1$, we calculate r_1 to be

$$r_1 = \sqrt{r^2 + \rho_1^2 - 2\,\vec{\rho}_1 \cdot \vec{\mathbf{r}}} \tag{9.30}$$

and with $\vec{\mathbf{r}} = r\vec{\mathbf{u}}_r$ to

$$r_1 = r\sqrt{1 + \frac{\rho_1^2}{r^2} - \frac{2\,\vec{\rho}_1 \cdot \vec{\mathbf{u}}_r}{r}} \; . \tag{9.31}$$

The root expression is developed into a Taylor series:

$$r_1 = r\left\{1 + \frac{1}{2}\left(\frac{\rho_1^2}{r^2} - \frac{2\,\vec{\rho}_1 \cdot \vec{\mathbf{u}}_r}{r}\right) - \frac{1}{8}\left(\frac{\rho_1^2}{r^2} - \frac{2\,\vec{\rho}_1 \cdot \vec{\mathbf{u}}_r}{r}\right)^2 + \ldots\right\} . \tag{9.32}$$

In the far field, $\rho_1/r \ll 1$ applies. Because r_1 occurs in the phase term of the electric field, all links with ρ_1^2/r^2 are nevertheless initially taken into account to calculate r_1:

$$r_1 = r - \vec{\rho}_1 \cdot \vec{\mathbf{u}}_r + \frac{1}{2}\frac{1}{r}\left(\rho_1^2 - \left(\vec{\rho}_1 \cdot \vec{\mathbf{u}}_r\right)^2\right) . \tag{9.33}$$

Apart from the phase quantities $\varphi_i = k_0\, r_i$ of the exponential functions in Eq. (9.28), the coarser approximation can be introduced for the quantities r_i and $\vec{\mathbf{u}}_{r_i}$:

$$r_i \approx r \qquad \text{and} \qquad \vec{\mathbf{u}}_{r_i} \approx \vec{\mathbf{u}}_r \; .$$

The phase error can be estimated using the third element, which is proportional $(\rho_1^2 - (\vec{\rho}_1 \cdot \vec{\mathbf{u}}_r)^2)$. The phase error caused by the third element is maximum when $\vec{\rho}_1 \cdot \vec{\mathbf{u}}_r = 0$. If you use this relationship and set a more or less arbitrary error limit of $\pi/8$ for the phase error $\Delta\varphi$, then you obtain the third element of the phase $\varphi_1 = k_0 r_1$ as a measure of the phase error:

$$\begin{aligned} \Delta\varphi &\approx k_0 \frac{1}{2}\frac{1}{r}\rho_1^2 < \frac{\pi}{8}, \\ \Delta\varphi &\approx \frac{\pi\,\rho_1^2}{\lambda_0\, r} < \frac{\pi}{8}. \end{aligned} \tag{9.34}$$

Let $D = |\vec{\rho}_1 - \vec{\rho}_2|$ denote the distance between the two source points and assume that $\vec{\rho}_1 = -\vec{\rho}_2$ and thus $D = 2\,\rho_1$, then the inequality (9.34) is satisfied if

$$r > \frac{2\,D^2}{\lambda_0} \qquad \textbf{(Far field conditions)} \tag{9.35}$$

are chosen.

The spatial region defined by the inequality (9.35) is called the far-field zone or `Fraunhofer zone` and the field in this region is called far field. If the first three elements in the series are used to approximate the phase, then one speaks of the `Fresnel zone`. Taking into account the approximations for the far field zone discussed above, the far field of the two elementary emitters is obtained if the order in the vector product is reversed:

$$\vec{\mathbf{E}}(\vec{\mathbf{r}}) = \frac{e^{-jk_0 r}}{r}\,\vec{\mathbf{E}}_0(\vec{\mathbf{u}}_r) \qquad \text{with}$$

$$\vec{\mathbf{E}}_0(\vec{\mathbf{u}}_r) = \frac{j\,Z_0}{2\lambda_0}\,\vec{\mathbf{u}}_r \times \left\{ \vec{\mathbf{u}}_r \times \left[I_1\,\vec{\ell}_1\, e^{(jk_0\,\vec{\mathbf{u}}_r \cdot \vec{\rho}_1)} + I_2\,\vec{\ell}_2\, e^{(jk_0\,\vec{\mathbf{u}}_r \cdot \vec{\rho}_2)} \right] \right\}. \tag{9.36}$$

This procedure, shown using the example of two electric elementary emitters, can be transferred to the calculation of the field of any spatially distributed surface current density. The current element $I\,\vec{\ell}$ is replaced by the surface current element $\vec{\mathbf{J}}\,ds$ with the surface current density $\vec{\mathbf{J}}$ (Fig. 9.21).

$$\vec{\mathbf{J}}\,ds = \vec{\mathbf{J}}\,d\ell_1\,d\ell_2 \tag{9.37}$$

Similar considerations apply to the magnetic surface current density $\vec{\mathbf{M}}$ as were made for the electric surface current density $\vec{\mathbf{J}}$.

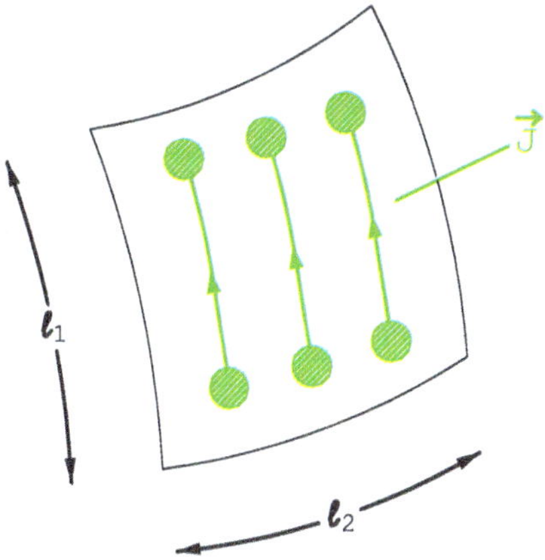

Fig. 9.21 Representation of the surface current element

To calculate the field of an extended current distribution (Fig. 9.21), the sum is replaced by an integral over all sources. This gives the far field:

$$\vec{\mathbf{E}}(\vec{\mathbf{r}}) = \frac{e^{-jk_0 r}}{r} \vec{\mathbf{E}}_0(\vec{\mathbf{u}}_r) \tag{9.38}$$

with

$$\vec{\mathbf{E}}_0(\vec{\mathbf{u}}_r) = \frac{j}{2\lambda_0} \iint_S \left\{ \vec{\mathbf{u}}_r \times \left[Z_0 \left(\vec{\mathbf{u}}_r \times \vec{\mathbf{J}}(\vec{\rho}) \right) + \vec{\mathbf{M}}(\vec{\rho}) \right] e^{(jk_0 \vec{\mathbf{u}}_r \cdot \vec{\rho})} \right\} ds \tag{9.39}$$

resp.

$$\begin{aligned} \vec{\mathbf{E}}_0(\vec{\mathbf{u}}_r) = \frac{-j}{2\lambda_0} \iint_S \Big\{ \Big[Z_0 \Big(\vec{\mathbf{J}}(\vec{\rho}) - \left(\vec{\mathbf{J}}(\vec{\rho}) \cdot \vec{\mathbf{u}}_r \right) \cdot \vec{\mathbf{u}}_r \Big) \\ + \vec{\mathbf{M}}(\vec{\rho}) \times \vec{\mathbf{u}}_r \Big] e^{(jk_0 \vec{\mathbf{u}}_r \cdot \vec{\rho})} \Big\} ds \end{aligned} \tag{9.40}$$

and

$$\vec{\mathbf{H}}(\vec{\mathbf{r}}) = \frac{1}{Z_0} \left(\vec{\mathbf{u}}_r \times \vec{\mathbf{E}}(\vec{\mathbf{r}}) \right) . \tag{9.41}$$

Fig. 9.22 serves to explain the transformation from Eq. (9.39) to Eq. (9.40).

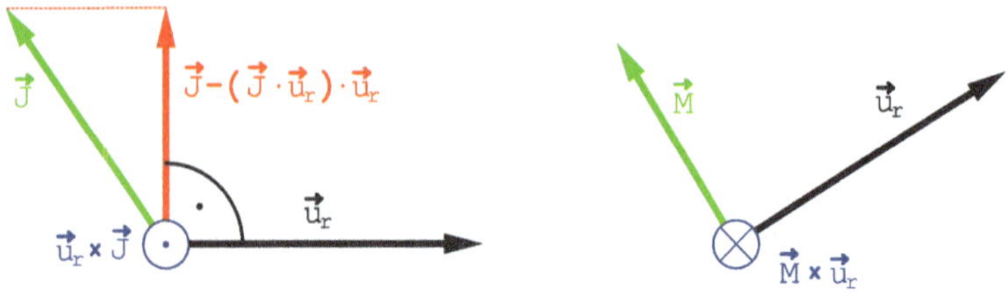

Fig. 9.22 To explain a vector transformation

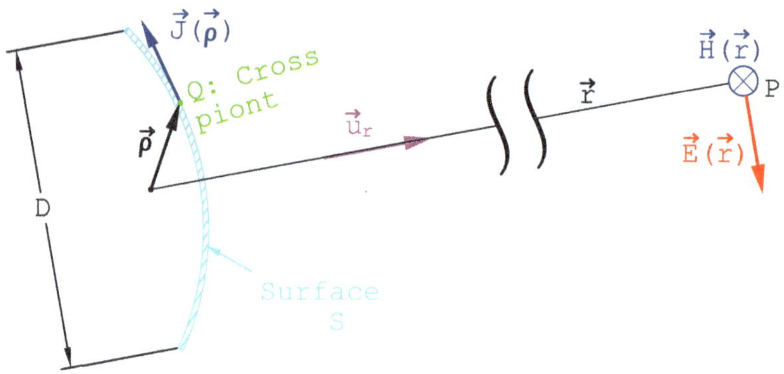

Fig. 9.23 To explain the integral over the surface current density

The cross product $\vec{\mathbf{u}}_r \times \vec{\mathbf{J}}$ is perpendicular to the two vectors $\vec{\mathbf{u}}_r$ and $\vec{\mathbf{J}}$ and the double cross product $\vec{\mathbf{u}}_r \times \vec{\mathbf{u}}_r \times \vec{\mathbf{J}}$ now perpendicular to $\vec{\mathbf{u}}_r$ and $\vec{\mathbf{u}}_r \times \vec{\mathbf{J}}$[4]. The direction of the vectors are assigned in the sense of a right-hand screw.

The application of Eqs. (9.38) to (9.41) will be explained using the Fig. 9.23 for the case that only one surface current density $\vec{\mathbf{J}}(\vec{\rho})$ is present.

The integration must be carried out taking into account the drawn vectors over the area S drawn in the section.

The dimension D of the far field condition (Eq. (9.35)) now has the meaning of the largest linear dimension of the current distribution.

Eqs. (9.38) to (9.41) represent the result of this section. The properties of the far field of an arbitrary current distribution resulting from these equations can be summarized as follows:

1. The spatial dependence of the electric field strength $\vec{\mathbf{E}}$ is given by the product of two functions. The first factor, namely $\frac{1}{r}\, e^{-jk_0 r}$ is independent of the specific shape of the current distribution and describes the amplitude decay and the phase of the field as a function of the distance r. The second factor $\vec{\mathbf{E}}_0(\vec{\mathbf{u}}_r) = \vec{\mathbf{E}}_0(\vartheta, \varphi)$ is only dependent on the direction $\vec{\mathbf{u}}_r = \vec{\mathbf{u}}_r(\vartheta, \varphi)$, i.e. independent of r. Here,

[4] The magnitude of the cross products is equal to the area of a parallelogram.

ϑ and φ are the two coordinate angles in spherical coordinates. This is called $\vec{\mathbf{E}}_0(\vec{\mathbf{u}}_r) = \vec{\mathbf{E}}_0(\vartheta, \varphi)$ the **absolute** directional characteristic of the current distribution.

2. Since according to Eq. (9.40) the vectors $(\vec{\mathbf{J}} - (\vec{\mathbf{J}} \cdot \vec{\mathbf{u}}_r) \cdot \vec{\mathbf{u}}_r)$ and $\vec{\mathbf{M}} \times \vec{\mathbf{u}}_r$ are perpendicular to $\vec{\mathbf{u}}_r$ in the integrand, is also $\vec{\mathbf{E}}(\vec{\mathbf{r}})$ perpendicular to $\vec{\mathbf{u}}_r$, i.e. tangential to the spherical surface around the zero point.
3. According to Eq. (9.41), $\vec{\mathbf{H}}$ is perpendicular to $\vec{\mathbf{E}}$ and $\vec{\mathbf{u}}_r$. The following applies to the amounts:

$$\left|\vec{\mathbf{H}}\right| = \frac{\left|\vec{\mathbf{E}}\right|}{Z_0} . \tag{9.42}$$

The electric and magnetic fields are in phase.

9.5 The Equivalence Principle (Huygen's Principle)

The design of a radar system requires knowledge of the radiation properties such as directional characteristics, gain, half-width, sidelobe attenuation of the antennas in question on the one hand and knowledge of the backscattering properties such as scattering matrix, radar cross-section of possible radar targets on the other.

The determination of these quantities generally represents a complicated electromagnetic boundary value problem, the strict solution of which generally requires a very high computational effort. As a rule, approximation methods are used whose results are practically sufficient for the purpose at hand. In addition, these approximation methods have the advantage that they provide a good insight into the principle mode of action of antennas.

From the previous section, the relationship between the distribution of the electric and magnetic currents (sources) and the associated electromagnetic field is known under the assumption of free space (i.e. homogeneous space with $\varepsilon_r = \mu_r = 1$ for all points). These results cannot be applied directly to the problems at hand (antenna, radar target). This is because the condition of homogeneous space is violated here, as the antennas and radar targets are made of material with $\varepsilon_r \neq 1$ (mostly metal). In addition, the actual sources of electromagnetic fields are located in the tubes and semiconductor components used. However, the equivalence principle (also known as Huygen's principle) can be used to describe the problem at hand using sources in free space.

The statements of the equivalence principle required here can be summarized as follows (Fig. 9.24):

i) Volume regions V_i, $(i = 1, 2, \ldots)$ of arbitrary geometric shape are introduced, which include those structures that disturb the homogeneity of the space. In the images 9.25 and 9.25 a radar system including antenna is included in V_1 and the radar target in V_2.

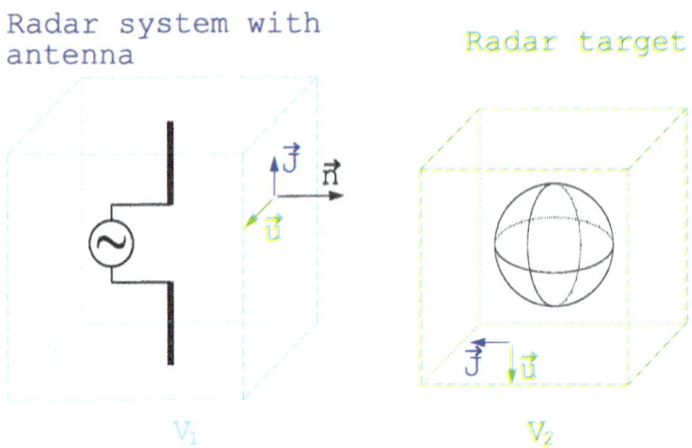

Fig. 9.24 The radar system with dipole antenna and the radar target are included in the volume V_1 and V_2

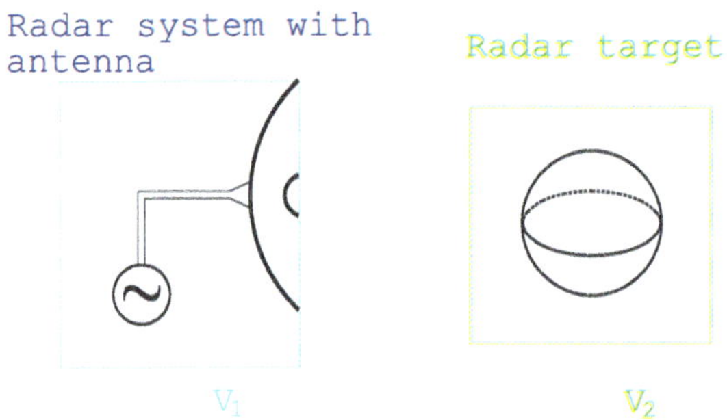

Fig. 9.25 The radar system with parabolic antenna and the radar target are integrated into the volume V_1 and V_2

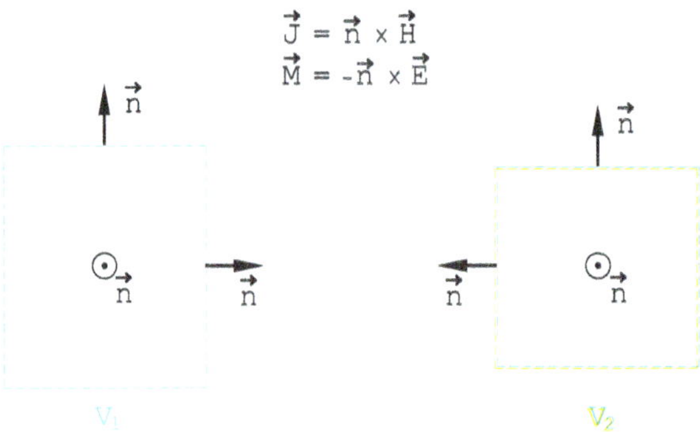

Fig. 9.26 Area normal vector for the radar system and the radar target replaced

ii) In all points on the surface of these volume areas "equivalent current densities" $\vec{\mathbf{J}}$ and $\vec{\mathbf{M}}$, which are defined from the field strengths prevailing at field strengths prevailing at these points $\vec{\mathbf{E}}$ and $\vec{\mathbf{H}}$ via

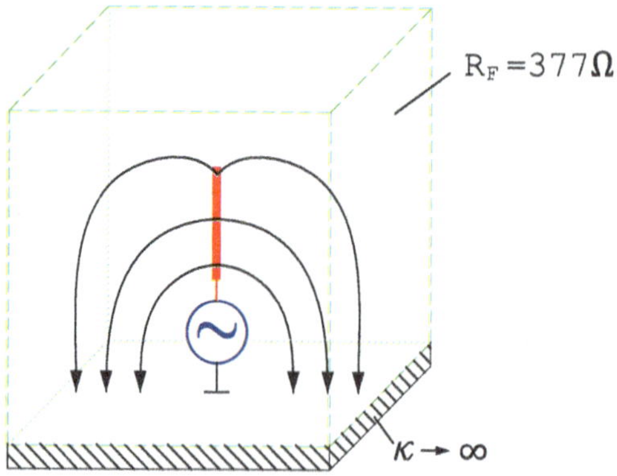

Fig. 9.27 Representation of the electric field of a monopole in the volume

$$\vec{\mathbf{J}} = \vec{\mathbf{n}} \times \vec{\mathbf{H}} \tag{9.43}$$

and

$$\vec{\mathbf{M}} = -\vec{\mathbf{n}} \times \vec{\mathbf{E}} \tag{9.44}$$

yield. Where $\vec{\mathbf{n}}$ is the outward-pointing surface normal vector, Fig. 9.26.

iii) If one imagines the equivalent current densities defined under **ii)** to be located in free space (Fig 9.26), then they generate an electromagnetic field outside V_1 and V_2 which corresponds to that of the original arrangement. Within these volume ranges, the field generated by the equivalent currents becomes zero. Thus configuration (Fig. 9.25) can be replaced by an arrangement consisting of electrical and magnetic sources in free space (Fig. 9.26) and is equivalent with regard to the field outside V_1 and V_2 is equivalent. This only introduces a new description method. Since the equivalent currents according to Eq. (9.43) and (9.44) depend on the tangential part of the electric and magnetic field in the surface, but this field is unknown, the boundary value problem remains unsolved. However, as will be explained below, there are useful approximations for the tangential components of the fields in the aperture of surface radiators and on the surface of metallic scatterers (radar targets).

In practical applications, a distinction is made between planar antennas (antennas with one aperture) and the rest of the antennas. The procedure described here must also be used in the numerical field simulation!

If you do not have planar antennas, you must include the signal source in the calculation. In field simulation, a gap source is often used here. The selected volume can already end in the near range, as shown in Fig. 9.27.

With flat antennas, the volume already ends in the surface. If a waveguide is used as the input, an inner source is not even necessary, see Fig. 9.28.

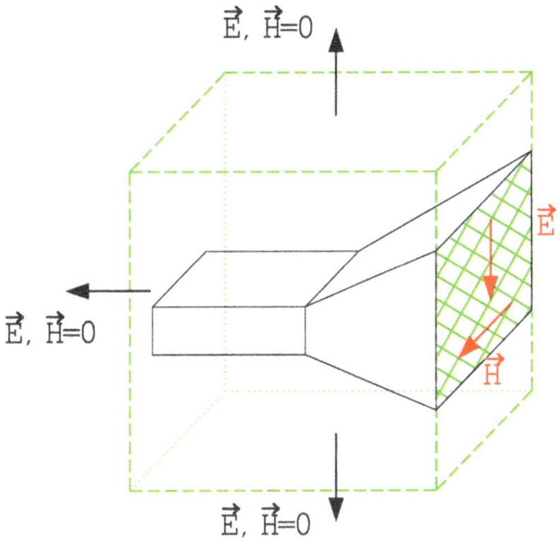

Fig. 9.28 Application of Huygen's principle for the horn antenna

9.6 Relationship Between Aperture Assignment and Directional Characteristic for Area Antennas

Radar devices in ship or flight surveillance require antennas that have a sharp directional characteristic in the horizontal plane (azimuth) and a relative low bundling in the vertical plane (elevation).

The linear antennas presented so far are only used in radar technology as elements of so-called array antennas, which are discussed in detail in the next chapter. In addition to the electronically steerable array antennas, mechanically steerable flat panel antennas. Surface-mounted antennas are divided into two classes, transmitter and reflector antennas. With waveguide radiator, very good broadband matching to the characteristic impedance of the free space is achieved with a funnel-shaped transition. The electromagnetic field in the aperture area (aperture area) corresponds to the field of the excited wave type: H_{10} in the rectangular or H_{11} in the circular waveguide. Because with rectangular horns the directional can be calculated very precisely, they are also used as calibration antennas with a known directivity factor.

Some examples of horn or funnel antennas can be seen in Fig. 9.29 to.

Horn antennas are generally used as exciters in mirror antennas.

Simple mirror antennas are dimensioned using the methods of geometrical optics: The path of all beams is from the wave center in the focal point F via the reflection on the parabolically shaped mirror surface to the aperture and therefore the phase assignment is constant in the aperture. However, it must be clearly pointed out that the horn antenna does not have a constant phase and amplitude assignment over the elevation and azimuth angles. Particularly with precise range measurements in radar applications, the phase changes have a noticeable effect as measurement errors in the range.

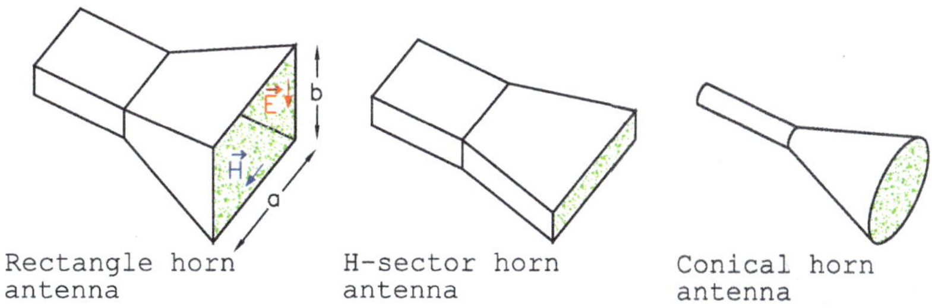

Fig. 9.29 Sketch of various waveguide antennas

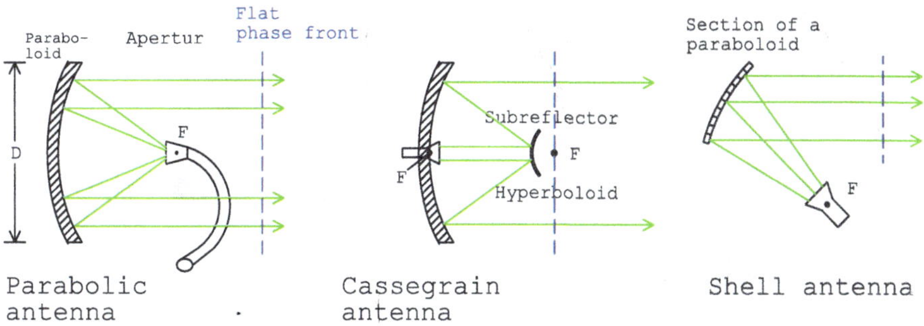

Fig. 9.30 Examples for panel radiators

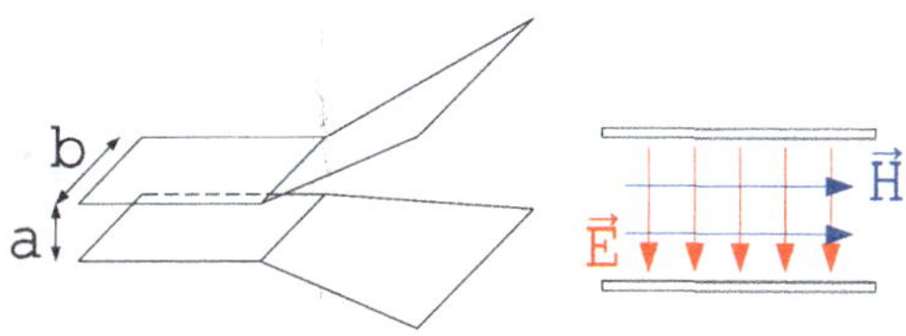

Fig. 9.31 Example: TEM horn antenna

Fig. 9.30 illustrates three frequently used mirror antennas. Fig. 9.31 shows a TEM horn antenna based on the expansion of a TEM band line. Both types of panel antennas have in common that

a) whose geometric dimensions are a multiple of the wavelength λ_0.
b) whose main beam direction is perpendicular to the main extension of the antenna.
c) a flat surface (aperture) of limited dimensions located in the immediate vicinity of the antenna, through which by far the largest part of the radiation passes.

According to Section 2.3, to calculate the radiation field, the antenna can be replaced by an equivalent current distribution in free space (Eq. (9.43)). For an exact calculation, one would have to the equivalent current distribution on a closed surface surrounding the entire antenna. If we think of this surface in such a way that the aperture introduced above is part of this surface, then due to the definition of the aperture (= plane through which by far the largest part of the radiation passes),

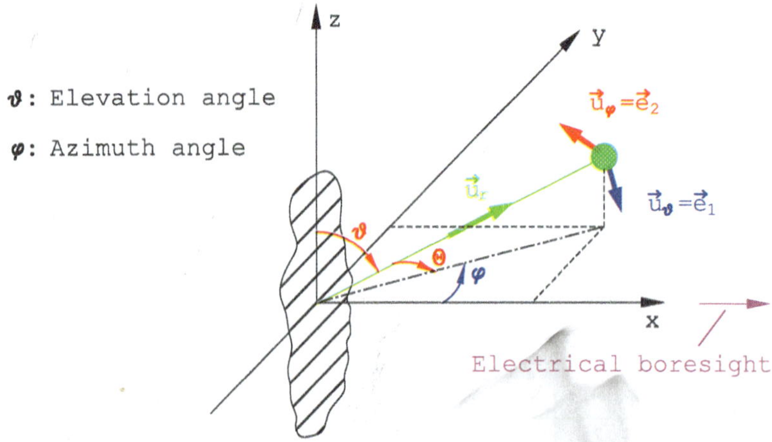

Fig. 9.32 Aperture in the plane $x = 0$ and corresponding coordinate system

the contribution of the equivalent currents on the remaining part of the surface is negligible, especially with respect to the radiation field on the front of the antenna.

So instead of integrating over the closed surface, one only integrates approximately over the flat aperture surface. In the remaining area of the closed surface, the surface currents are assumed to be zero. In principle, the aperture can be chosen more or less arbitrarily for a given planar antenna. In general, however, it is chosen so that a simple approximation can be given with sufficient accuracy for the course of the equivalent current densities (proportional to the transverse components of $\vec{\mathbf{E}}$ and $\vec{\mathbf{H}}$) in the aperture. In the case of a parabolic mirror symmetrical to the axis with a circular edge, it coincides with the mirror aperture, in the case of a horn radiator with the horn aperture. Fig. 9.32 shows an aperture placed in the plane $x = 0$.

Eq. 9.44 gives an equivalent magnetic surface current density for the aperture of

$$\vec{\mathbf{M}}(y,z) = -\vec{\mathbf{n}} \times \vec{\mathbf{E}}(0,y,z) = -\vec{\mathbf{u}}_x \times \vec{\mathbf{E}}_t(y,z)\ . \tag{9.45}$$

Here it $\vec{\mathbf{E}}_t(y,z) = \vec{\mathbf{E}} - E_x \cdot \vec{\mathbf{u}}_x$ the portion of the electric field oriented parallel to the aperture plane (transverse portion).

If we assume in a further approximation that the relationship between the transverse component of the electric and magnetic field in the aperture is approximately the same as for TEM waves, namely

$$\vec{\mathbf{H}}_t(y,z) \approx \frac{1}{Z_0}\vec{\mathbf{u}}_x \times \vec{\mathbf{E}}_t(y,z)\ . \tag{9.46}$$

then the electric surface current density can also be expressed by $\vec{\mathbf{E}}_t(y,z)$:

$$\begin{aligned}\vec{\mathbf{J}}(y,z) &= \vec{\mathbf{n}} \times \vec{\mathbf{H}}_t(0,y,z) \;=\; \vec{\mathbf{u}}_x \times \vec{\mathbf{H}}_t(y,z)\,,\\ &\approx \vec{\mathbf{u}}_x \times \left(\frac{1}{Z_0}\,\vec{\mathbf{u}}_x \times \vec{\mathbf{E}}_t(y,z)\right)\,,\\ &= -\frac{1}{Z_0}\,\vec{\mathbf{E}}_t(y,z)\,.\end{aligned} \tag{9.47}$$

After inserting this special expression for $\vec{\mathbf{M}}$ and $\vec{\mathbf{J}}$ in Eq. (9.39)

$$\vec{\mathbf{E}}_0(\vec{\mathbf{u}}_r) = -\frac{j}{2\,\lambda_0}\int\!\!\int_S \vec{\mathbf{u}}_r \times \left[\left(\vec{\mathbf{u}}_x + \vec{\mathbf{u}}_r\right) \times \vec{\mathbf{E}}_t(y,z)\right] e^{\,jk_0\vec{\mathbf{u}}_r\vec{\rho}}\;dy\,dz\,, \tag{9.48}$$

with $\qquad \vec{\rho} = \left(y \cdot \vec{\mathbf{u}}_y + z \cdot \vec{\mathbf{u}}_z\right)\,.$

The integral is to be extended over the plane aperture surface S. The integral expression can be rewritten using the coordinates introduced in Fig. 9.32. Note that the angle θ is oriented to the main beam direction and does not correspond to the angle ϑ according to Fig. 9.13, which is calculated from the z-axis. In detail:

a) the transverse field $\vec{\mathbf{E}}_t(y,z)$ in the aperture is broken down into its components with respect to the z- and y-axis.

$$\vec{\mathbf{E}}_t(y,z) \;=\; E_y(y,z)\,\vec{\mathbf{u}}_y + E_z(y,z)\,\vec{\mathbf{u}}_z \tag{9.49}$$

b) the direction from the coordinate zero point to the point at which the far field is to be determined (receptor or emission point) is described by the angles θ and φ:

$$\vec{\mathbf{u}}_r \;=\; \sin\theta \cdot \vec{\mathbf{u}}_z + \cos\theta \cdot \left(\cos\varphi \cdot \vec{\mathbf{u}}_x + \sin\varphi \cdot \vec{\mathbf{u}}_y\right)\,. \tag{9.50}$$

c) the vector $\vec{\mathbf{E}}_0(\theta,\varphi)$ is divided into components with respect to the unit vectors $\vec{\mathbf{e}}_1$ and $\vec{\mathbf{e}}_2$, which are perpendicular to each other and tangential to the surface of the sphere:

$$\vec{\mathbf{E}}_0(\theta,\varphi) \;=\; E_{01} \cdot \vec{\mathbf{e}}_1 \;+\; E_{02} \cdot \vec{\mathbf{e}}_2\,, \tag{9.51}$$

with $\quad \vec{\mathbf{e}}_1 = -\sin\varphi \cdot \vec{\mathbf{u}}_x + \cos\varphi \cdot \vec{\mathbf{u}}_y \quad$ and $\quad \vec{\mathbf{e}}_2 = \vec{\mathbf{e}}_1 \times \vec{\mathbf{u}}_r\,.$

If Eq. (9.48) is calculated using Eqs. (9.49) and (9.50) in the components $\vec{\mathbf{u}}_x$, $\vec{\mathbf{u}}_y$, $\vec{\mathbf{u}}_z$, $\vec{\mathbf{E}}_y$, $\vec{\mathbf{E}}_z$, θ, φ and then express $\vec{\mathbf{E}}_0$ from Eq. (9.48) into two components in the direction of $\vec{\mathbf{e}}_1$ and $\vec{\mathbf{e}}_2$ by forming the scalar product:

$$\vec{\mathbf{E}}_0(\theta,\varphi) = \frac{j}{2\lambda_0} \Bigg\{ \left[\vec{\mathbf{e}}_1 \cdot (\cos\theta + \cos\varphi) + \vec{\mathbf{e}}_2 \cdot \sin\varphi \cdot \sin\theta\right] \cdot \iint E_y(y,z)\, e^{jk_0\,(z\sin\theta + y\cos\theta\sin\varphi)}\, dy\,dz + \left[\vec{\mathbf{e}}_1 \cdot \sin\varphi \cdot \sin\theta - \vec{\mathbf{e}}_2 \cdot (\cos\theta + \cos\varphi)\right] \cdot \iint E_z(y,z)\, e^{jk_0\,(z\sin\theta + y\cos\theta\sin\varphi)}\, dy\,dz \Bigg\}. \tag{9.52}$$

For emission points in the horizontal plane $\theta = 0$ (plane $z = 0$), Eq. (9.52) simplifies to

$$\vec{\mathbf{E}}_0(0,\varphi) = \frac{j(1+\cos\varphi)}{2\lambda_0} \Bigg\{ \vec{\mathbf{e}}_1 \cdot \iint E_y(y,z)\, e^{jk_0 y\sin\varphi}\, dy\,dz - \vec{\mathbf{e}}_2 \cdot \iint E_z(y,z)\, e^{jk_0 y\sin\varphi}\, dy\,dz \Bigg\}, \tag{9.53}$$

and for emission points in the vertical plane $\varphi = 0$ (plane $y = 0$) to

$$\vec{\mathbf{E}}_0(\theta,0) = \frac{j(1+\cos\theta)}{2\lambda_0} \Bigg\{ \vec{\mathbf{e}}_1 \cdot \iint E_y(y,z)\, e^{jk_0 z\sin\theta}\, dy\,dz - \vec{\mathbf{e}}_2 \cdot \iint E_z(y,z)\, e^{jk_0 z\sin\theta}\, dy\,dz \Bigg\}. \tag{9.54}$$

The directivity of the individual components of $\vec{\mathbf{E}}_0(\theta,\varphi)$ is calculated according to Eqs. (9.53) and (9.54) in the main planes $\theta = 0$ and $\varphi = 0$ is determined by the product of the respective integral expressions with the function $(1+\cos\varphi)$ or $(1+\cos\theta)$. For the angular range in the vicinity of the main beam direction $\theta = \varphi = 0$, which is of interest for strongly converging antennas, $1+\cos\varphi \approx 2$ and $1+\cos\theta \approx 2$ can be set so that the directional dependence is completely given by the integral expressions. As the various integral expressions are obtained by swapping components and coordinates, one of the integral expressions will be used here as an example, for example

$$W(\sin\varphi) = \frac{1}{\lambda_0} \iint E_y(y,z)\, e^{jk_0 y\sin\varphi}\, dy\,dz\,. \tag{9.55}$$

Since the e- function in the integrand does not depend on z, $E_y(y,z)$ (aperture assignment with respect to the y- component) can first be integrated over z, so that a function $\tilde{w}(y)$ dependent only on y remains. In this way you get

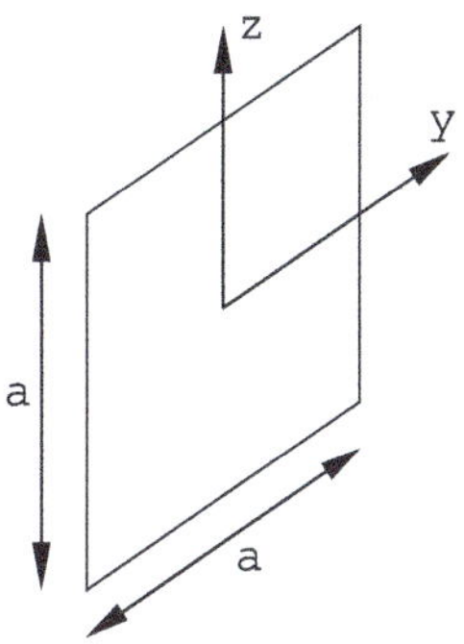

Fig. 9.33 Geometric shape for constant distribution

$$W(\sin\varphi) = \frac{1}{2\pi} \int_{-\infty}^{+\infty} w(k_0 y)\, e^{j k_0 y \sin\varphi}\, d(k_0 y) \tag{9.56}$$

with the occupancy function (here with regard to the y- component and the horizontal plane)

$$\begin{aligned} \tilde{w}(y) = w(k_0 y) &= \int_{z_{min}(y)}^{z_{max}(y)} E_y(y,z)\, dz \quad \text{für } y_{min} < y < y_{max};\ y \in S_a \\ &= 0 \qquad \text{otherwise}\,. \end{aligned} \tag{9.57}$$

This means that the distribution function depends both on the geometric shape and on the spatial dependence of the components of $\vec{\mathbf{E}}$ (here $\vec{\mathbf{E}}_y$) within the aperture. Eq. (9.56) shows that the distribution function $w(k_0 y)$ and the function $W(\sin\varphi)$, which is decisive for the directional characteristic, are related to each other via a Fourier transformation (Fig. 9.33).

$$\begin{aligned} W(\sin\varphi) &= \mathscr{F}^{-1}\{w(k_0 y)\} \\ \text{and} \quad w(k_0 y) &= \mathscr{F}\{W(\sin\varphi)\} \end{aligned} \tag{9.58}$$

$$\text{resp.} \quad W(\sin\varphi) \quad \bullet\!\!-\!\!\circ \quad w(k_0 y) \tag{9.59}$$

The Figs. 9.34 and 9.35 as well as Eq. (9.60) are intended to illustrate the analogy to signal theory, whereby the directional characteristic corresponds to the time course $f(t)$ of a signal and the distribution function corresponds to the spectrum of this signal $F(\omega)$.

$$f(t) = \frac{1}{2\pi} \int_{-\infty}^{+\infty} F(\omega)\, e^{j\omega t}\, d(\omega) \qquad \text{resp.} \qquad f(t) \quad \bullet\!\!-\!\!\circ \quad F(\omega) \tag{9.60}$$

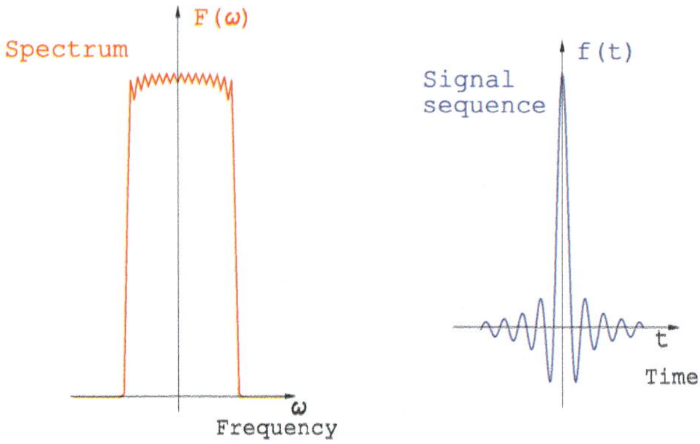

Fig. 9.34 The analogy between signal and antenna theory I

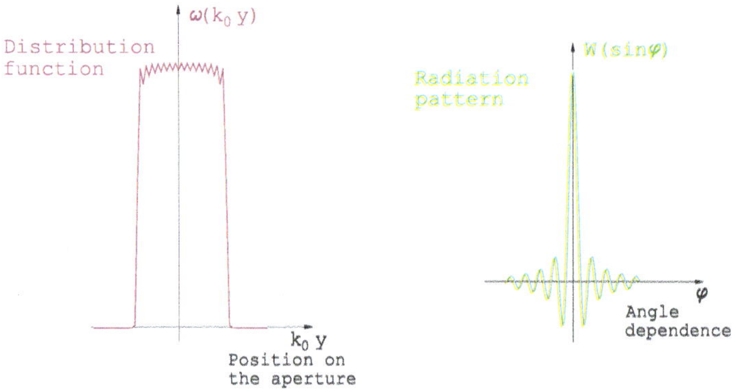

Fig. 9.35 The analogy between signal and antenna theory II

For the most important special case of a constant distribution function of width a, i. h.

$$w(k_0 y) = \begin{cases} 1 : & \text{for } |y| \leq a/2 \\ 0 : & \text{otherwise} \end{cases} \tag{9.61}$$

the directional characteristic $W(\sin\varphi)$ is given by an si function (also known as a sinc function),

$$\begin{aligned} W(\sin\varphi) &= \frac{a}{\lambda_0} \cdot \frac{\sin(\sin\varphi\, k_0\, a/2)}{\sin\varphi\, k_0\, a/2} \\ &= \frac{a}{\lambda_0} \cdot \mathrm{si}(\sin\varphi\, k_0\, a/2) \\ &\simeq \frac{a}{\lambda_0} \cdot \mathrm{si}(\varphi\, k_0\, a/2) \qquad \text{for} \quad \varphi \ll \frac{\pi}{2} \end{aligned} \tag{9.62}$$

the magnitude of which is shown in Fig. 9.36.

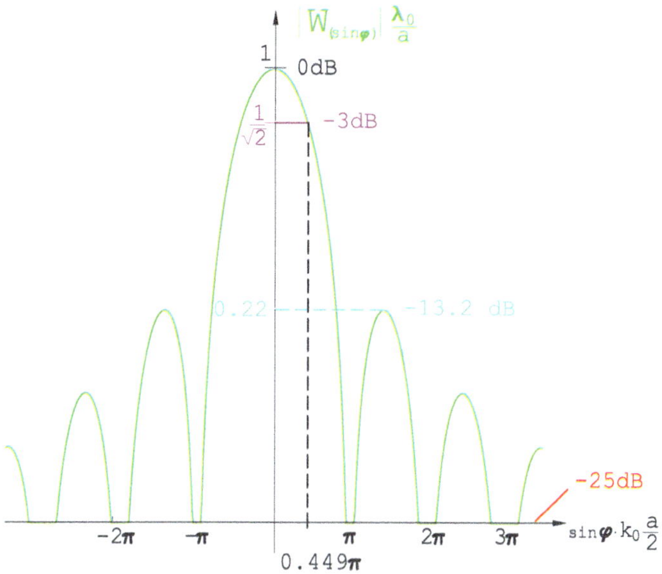

Fig. 9.36 Diagram of the directional characteristic for a rectangular distribution function

It can be seen that the 3 dB limits of the main lobe are at $\sin\varphi = \pm 0.44\,\lambda_0/a$. For approximately $\lambda_0/a < 1/3$ a good approximation for the 3 dB bandwidth (half-width band) $\Delta\varphi$ can thus be written:

$$\begin{aligned} \Delta\varphi &= 2\cdot 0.443\cdot\frac{\lambda_0}{a} = 0.886\cdot\frac{\lambda_0}{a} \\ \text{or}\qquad \Delta\varphi/^\circ &\approx 51\cdot\frac{\lambda_0}{a}\,. \end{aligned} \tag{9.63}$$

The larger the ratio of the aperture width a to the wavelength λ_0, the narrower the main lobe. An example: $a = 50\,\lambda_0$, therefore $\Delta\varphi = 1^\circ$ (Fig. 9.37, 9.38, 9.39, 9.40, and 9.41).

It can also be seen that, in addition to the main lobe, further maxima occur in the so-called side lobes. In the case of the rectangular distribution function considered here, the maximum of the first side lobe is $20\,\log(0.22) = -13.2\,\text{dB}$ below the maximum value at $\varphi = 0$. This side lobe attenuation of 13.2 dB is too low for most applications. It is known from the theory of the Fourier transform that the side lobe maxima can be reduced can be reduced if a function that decreases steadily towards the edge is used instead of the rectangular function ("taper"). The effect of a taper is to be demonstrated using various occupancy functions with the normalized length of $a/2 = 1/k_0$ will be demonstrated:

$$w(k_0 y) = \begin{cases} \cos^n\left(\frac{\pi}{2}k_0 y\right) & : \quad \text{for } |y| \le 1/k_0 \\ 0 & : \quad \text{otherwise} \end{cases} \tag{9.64}$$

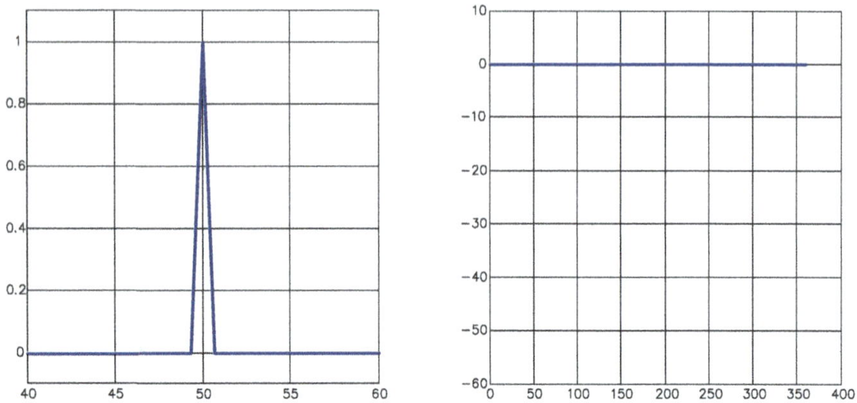

Fig. 9.37 Distribution function and directional characteristic of a dirac-shaped distribution function

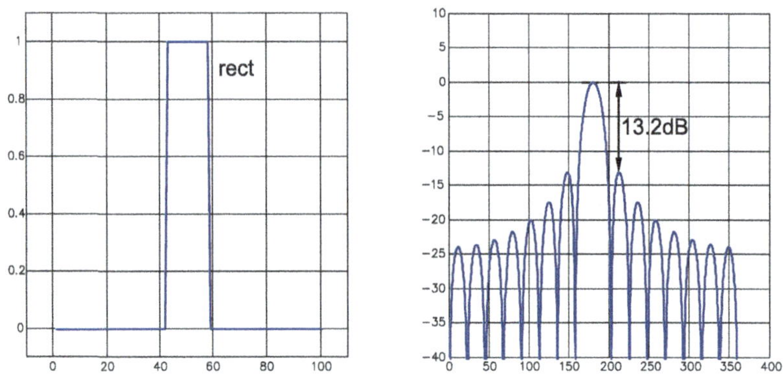

Fig. 9.38 Distribution function and directional characteristic of a rectangular shaped distribution function

with $n = 0, 1, 2$.

The following applies to the resulting directional characteristic:

$$W(\sin\varphi) = \frac{1}{2\pi} \int_{-1}^{1} \cos^n\left(\frac{\pi}{2} k_0 y\right) e^{j k_0 y \sin\varphi}\, d(k_0 y) \,. \tag{9.65}$$

For $n = 0$ the rectangular function already considered is obtained. A comparison of the Fourier transforms of the cos-distribution function ($n = 1$) and the $\cos^2$-distribution function ($n = 2$) with the rectangular function with regard to the height of the first secondary maximum and the 3 dB bandwidth is shown in the following table (Tab. 9.1 and Fig. 9.42):

Both the table and the directional characteristics in Fig. 9.43 can be determined directly from the results for the three cases.

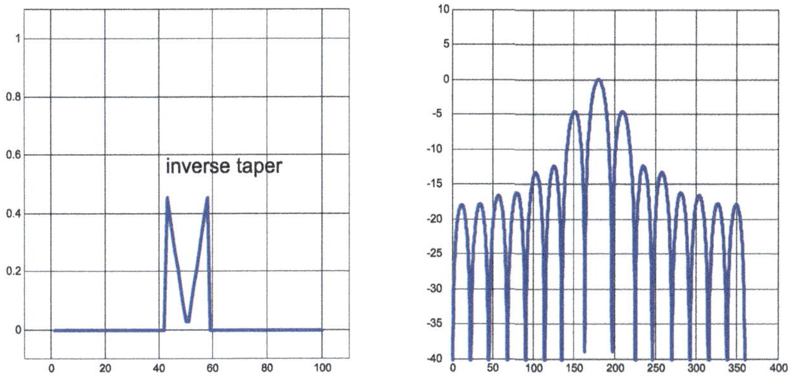

Fig. 9.39 Distribution function and directional characteristic of a inverse tapered distribution function

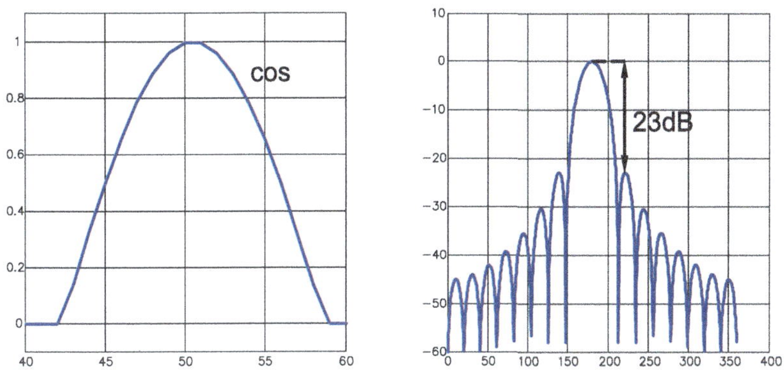

Fig. 9.40 Distribution function and directional characteristic of a cosinus-shaped distribution function

Tab. 9.1 Side lobe attenuation and 3 dB bandwidth of an area antenna in dependence on different distribution functions

Distribution function	*Side lobe attenuation*	*3 dB bandwidth/°*
Rectangle $w(k_0y) = 1$ für $\lvert y\rvert \le 1/k_0$	13.2 dB	$50\ \lambda_0$
Cosine $w(k_0y) = \cos\left(\frac{\pi}{2}k_0y\right)$ for $\lvert y\rvert \le 1/k_0$	23 dB	$69\ \lambda_0/a$
Cosine–square $w(k_0y) = \cos^2\left(\frac{\pi}{2}k_0y\right)$ for $\lvert y\rvert \le 1/k_0$	32 dB	$83\ \lambda_0/a$

$$n = 0: \qquad W(\sin\varphi) = \frac{1}{\pi}\,\frac{\sin(\sin\varphi)}{\sin\varphi} \tag{9.66}$$

$$n = 1: \qquad W(\sin\varphi) = \frac{2}{\pi^2}\,\frac{\cos(\sin\varphi)}{1-\left(\frac{2\sin\varphi}{\pi}\right)^2} \tag{9.67}$$

$$n = 2: \qquad W(\sin\varphi) = \frac{\pi}{2\pi^2 - 2\sin^2\varphi}\,\frac{\sin(\sin\varphi)}{\sin\varphi} \tag{9.68}$$

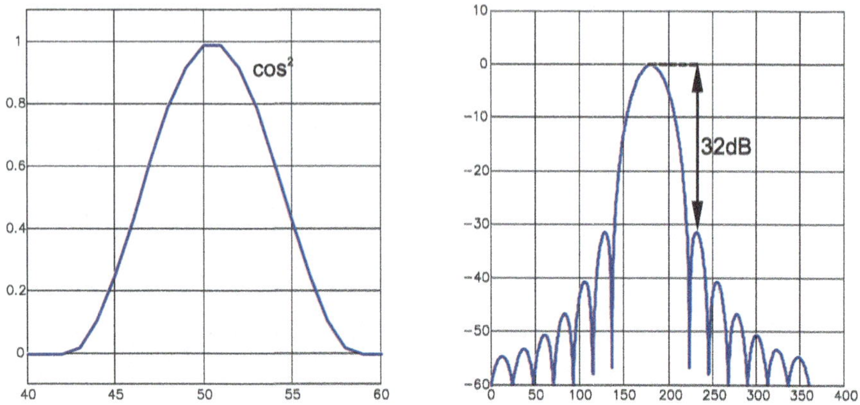

Fig. 9.41 Distribution function and directional characteristic of a cosinus quadrat distribution function

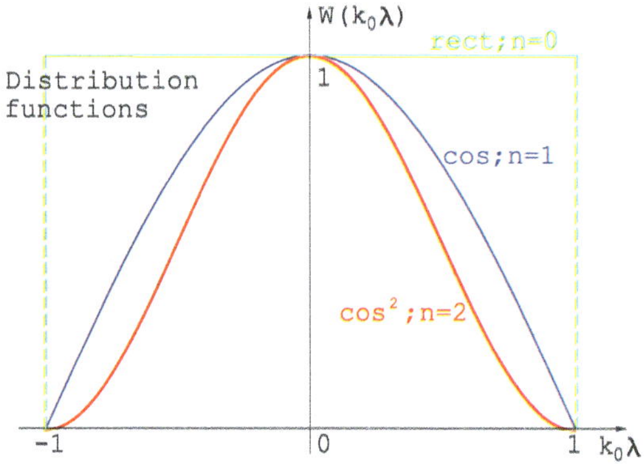

Fig. 9.42 Comparison of different distribution functions I

The possibility of a significant increase in side lobe damping by tapering the distribution function can be seen. However, an increase in the half-width must be accepted. For a given antenna width, the requirements for high sidelobe attenuation and low half-width are contradictory. A compromise must therefore be reached between these two requirements.

The above considerations were based on the ideal case that the phase of the field within the aperture is constant. In practice, however, small phase deviations are unavoidable. These phase errors mean that an increase in the sidelobe attenuation beyond the value of around 35 to 40 dB can only be achieved at great expense. The contributions of the electric and magnetic surface current density to the field in the main beam direction ($\theta = 0,\ \varphi = 0$) are equal and add up if the approximation applies that the ratio of electric and magnetic field in the aperture is equal to the field characteristic impedance Z_0 (Eq. (9.46)). This fact can be seen from Eq. (9.48). On

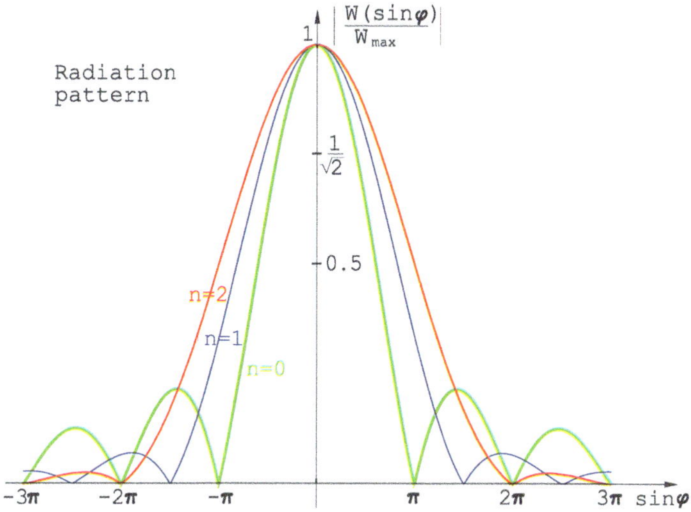

Fig. 9.43 Comparison of different distribution functions II

the other hand, under the same conditions, the field in the reverse direction, i.e. opposite to the main beam direction, is zero because the contributions from the electric and magnetic surface current density then just cancel each other out. This result can be seen, for example, in the Eq. (9.40).

The fact that the distribution function and the directional characteristic are linked via a Fourier transformation allows the application of some theorems known from signal theory. For example, the following applies:

a) *Similarity theorem:*

$$F(\alpha\,\omega) \quad \circ\!\!-\!\!\bullet \quad \frac{1}{|\alpha|}\, f\Big(\frac{t}{\alpha}\Big)$$

$$w(\alpha\, k_0 y) \quad \circ\!\!-\!\!\bullet \quad \frac{1}{|\alpha|}\, W\Big(\frac{\sin\varphi}{\alpha}\Big) \tag{9.69}$$

A "stretching" ($\alpha < 1$) or "compression" ($\alpha > 1$) of the distribution function leads to a "compression" ($\alpha < 1$) or "stretching" ($\alpha > 1$) of the directional characteristic as a function of $\sin\varphi$.

b) *Displacement theorem:*

$$F(\omega)\, e^{-j\omega t_0} \quad \circ\!\!-\!\!\bullet \quad f(t-t_0)$$

$$w(k_0 y)\, e^{-j k_0 y \sin\varphi_0} \quad \circ\!\!-\!\!\bullet \quad W(\sin\varphi - \sin\varphi_0) \tag{9.70}$$

A linearly increasing phase of the occupancy function leads to a beampivot (Fig. 9.44 and Fig. 9.45).

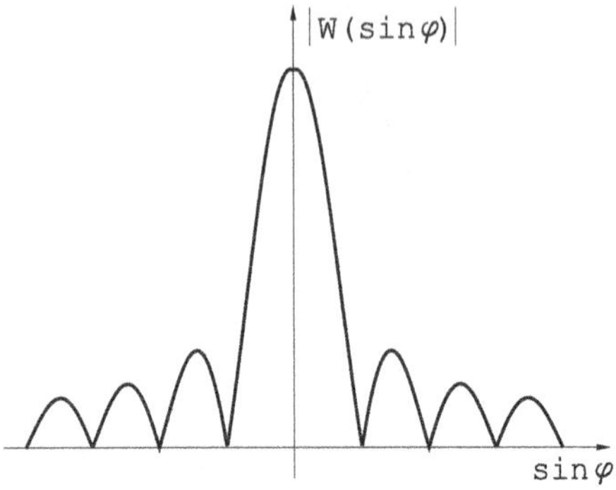

Fig. 9.44 Beam deflection by linearly increasing phase of the distribution function I

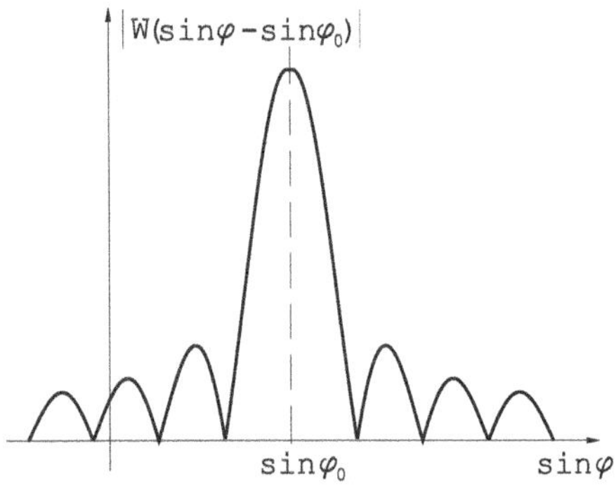

Fig. 9.45 Beam deflection by linearly increasing phase of the distribution function II

As we will see, this is the basic idea for the realization of antennas with electronically pivoting directional characteristics.

c) *Displacement theorem, applied in the opposite direction:*

$$F(\omega-\omega_0) \quad \circ\!\!-\!\!\bullet \quad f(t)\, e^{j\omega_0 t}$$

$$w(k_0(y-y_0)) \quad \circ\!\!-\!\!\bullet \quad W(\sin\varphi)\, e^{j k_0 y_0 \sin\varphi} \tag{9.71}$$

A spatial shift of the distribution function changes the phase of the directional characteristic.

9.7 Characteristics of an Antenna

The properties of an antenna when transmitting and receiving electromagnetic waves are characterized by parameters. The reciprocity theorem can be used to de-

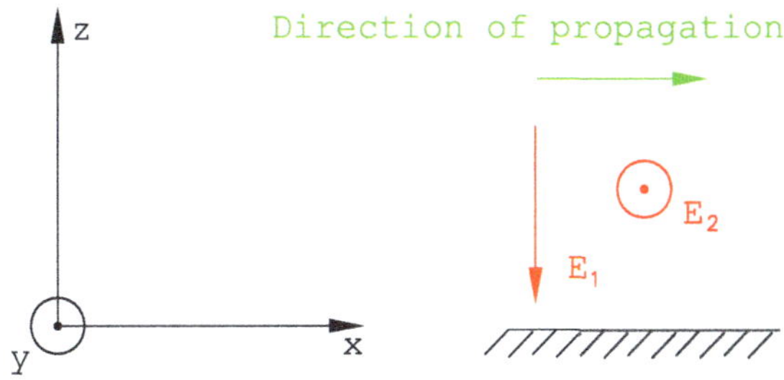

Fig. 9.46 Representation of the two linear polarizations

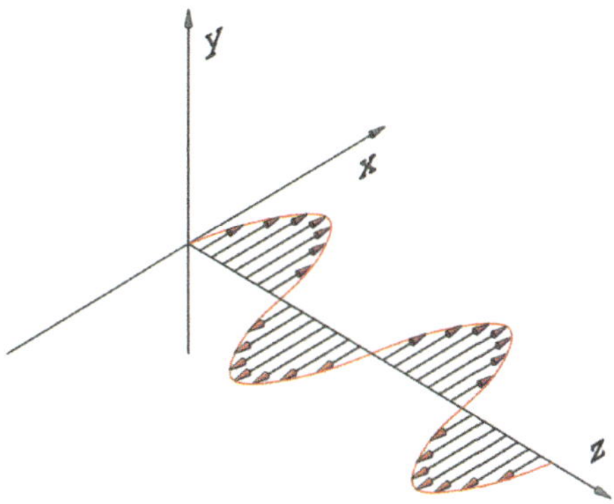

Fig. 9.47 Electrical field image in the space of the horizontally polarized wave

rive a relationship between the characteristics in the case of reception and transmission.

9.7.1 Polarization of the Free-Space Wave

Linear Polarization

To understand the fundamentals of free-space transmission polarization is an important point. The polarization is named after the orientation of the electric field.

The most common cases are

vertical polarisation: $E_1 \cdot \vec{u}_z$ and the
horizontal polarisation: $E_2 \cdot \vec{u}_y$,

which are also referred to as linear polarization (of the field or the antenna) (Fig. 9.46).

Fig. 9.47 shows a "snapshot" of the horizontally polarized E-field.

In addition to linear polarization with two modes (horizontal and vertical), there is also circular polarization with two possible modes (left and right circular).

Circular Polarization

The generation of a circularly polarized wave using a GHz signal, which is split in phase and amplitude via a signal splitter and fed to 50 % via a 90° phase shifter offset to two dipole antennas (turnstile antenna), is shown in Fig. 9.48.

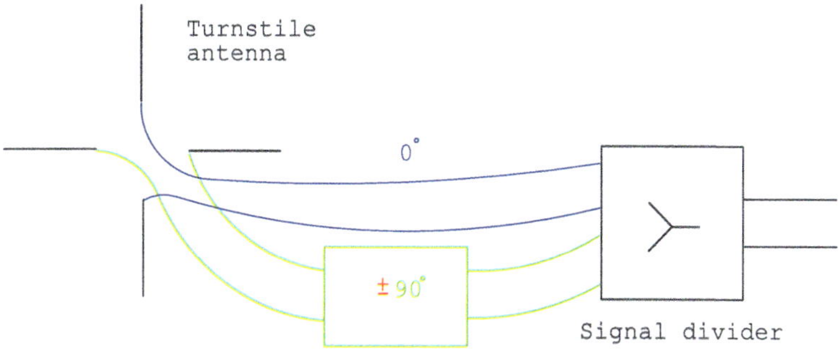

Fig. 9.48 Generation of circular polarization using two dipole antennas

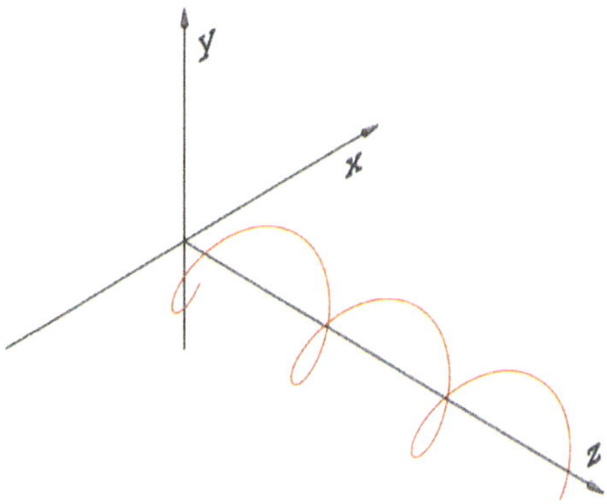

Fig. 9.49 Representation of the left-hand circular wave

Tab. 9.2 Representation of the left- and right-circular wave with regard to the energy distribution on the horizontally and vertically polarized mode

$\varphi = 90$	Vertical	Horizontal	$\varphi = -90$	Vertical	Horizontal
$\omega t = 0$	0	1	$\omega t = 0$	0	−1
$\omega t = 45$	1/2	1/2	$\omega t = 45$	−1/2	−1/2
$\omega t = 90$	1	0	$\omega t = 90$	−1	0
$\omega t = 135$	1/2	−1/2	$\omega t = 135$	−1/2	1/2
$\omega t = 180$	0	−1	$\omega t = 180$	0	1
$\omega t = 225$	−1/2	−1/2	$\omega t = 225$	1/2	1/2
$\omega t = 270$	−1	0	$\omega t = 270$	1	0
$\omega t = 315$	−1/2	1/2	$\omega t = 315$	1/2	−1/2

The Fig. 9.49 illustrates a left-circular (= left-hand circular) wave propagating in space.

The table shows the formation of the circular wave based on the two linearly polarized waves (Tab. 9.2).

9.7.2 Characteristics of an Antenna for the Transmission Case

As explained in the previous sections, the following applies to any antenna in the far field

$$\vec{\mathbf{E}}(r) = \vec{\mathbf{E}}_0(\theta,\varphi)\ \frac{e^{-jk_0 r}}{r}\ . \tag{9.72}$$

Here, $\vec{\mathbf{E}}_0(\theta,\varphi)$ depends on the amplitude and phase of the time-harmonic wave a fed into the antenna input. It is therefore useful to define the `absolute` directional characteristic $\vec{\mathbf{C}}(\theta,\varphi)$ of the antenna so that it is a pure antenna characteristic. This can, for example, be in the form

$$\vec{\mathbf{E}}_0(\theta,\varphi) = \sqrt{(Z_0/(2\pi))}\cdot\vec{\mathbf{C}}(\theta,\varphi)\cdot a\,, \tag{9.73}$$

$$\text{with}\quad \vec{\mathbf{C}}(\theta,\varphi) = \left(C_1(\theta,\varphi)\vec{\mathbf{e}}_1 + C_2(\theta,\varphi)\vec{\mathbf{e}}_2\right) \tag{9.74}$$

happen. Here, $C_1(\theta,\varphi)$ and $C_2(\theta,\varphi)$ are the directional characteristics for polarization 1 or 2 in the direction $\vec{\mathbf{e}}_1$ or $\vec{\mathbf{e}}_2$. The unit vectors $\vec{\mathbf{e}}_1$ and $\vec{\mathbf{e}}_2$ are perpendicular to each other and are tangential to the spherical surface in the center of which the antenna is located. By normalizing the absolute directional characteristic to the generally largest value in the main beam direction, the `relative` directional characteristic is obtained.

$$\frac{C(\theta,\varphi)}{C_{max}}:\quad \text{relative directional characteristic}$$

As a rule, the following applies $C_{max} = C(\theta = 0, \varphi = 0)$.

A directional diagram (or pattern) is the graphical representation of a section through the relative directional characteristic. Because of

$$\vec{\mathbf{H}}(\vec{\mathbf{r}}) = \frac{1}{Z_0}\cdot\left(\vec{\mathbf{u}}_r\times\vec{\mathbf{E}}\right) = \frac{1}{Z_0}\cdot\left[\vec{\mathbf{u}}_r\times\vec{\mathbf{E}}_0(\theta,\varphi)\right]\cdot\frac{e^{-jk_0 r}}{r} \tag{9.75}$$

applies to the radiation density S in the far field (with $[S] = W/m^2$)

$$S(\theta,\varphi) = \frac{1}{2}\ \mathrm{Re}\left\{\left(\vec{\mathbf{E}}\times\vec{\mathbf{H}}^*\right)\cdot\vec{\mathbf{u}}_r\right\} = \frac{1}{2}\cdot\frac{1}{Z_0}\cdot\left|\vec{\mathbf{E}}\right|^2, \tag{9.76}$$

$$S(\theta,\varphi) = \frac{|a|^2}{4\pi r^2}\cdot\left\{|C_1(\theta,\varphi)|^2 + |C_2(\theta,\varphi)|^2\right\}.$$

Fig. 9.50 shows the power distribution in a transmitting antenna. In the same way, however, there is also a phase dependency and thus runtime differences over the various angles of non-isotrope antennas. Of the absorbed active power $P = |a|^2$ a (usually small) proportion P_v is converted into heat as power loss within the antenna. It was assumed that the antenna is matched. The radiated power

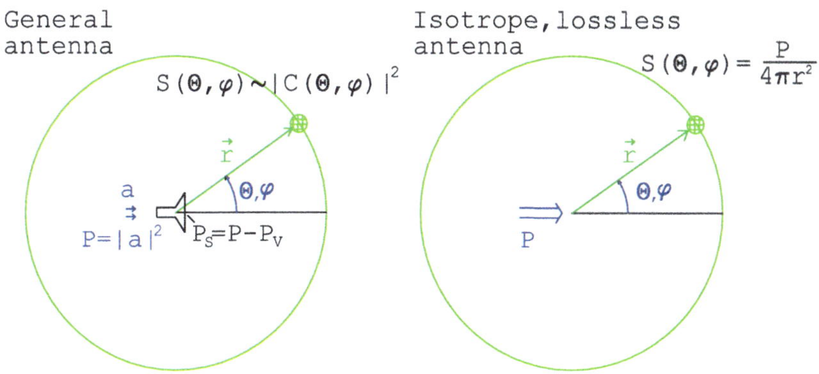

Fig. 9.50 Radiation density of a transmitting antenna

$$P_s = P - P_v = \eta_a P \qquad (\eta_a : \text{antenna efficiency}) \tag{9.77}$$

is radiated into free space and the following applies

$$P_s = \iint_A S(\theta, \varphi)\, dA \,. \tag{9.78}$$

In order to optimize the antenna efficiency, the metallic and/or dielectric losses can be reduced.

Reduction of the metallic losses:

- low surface roughness,
- "a lot" of metal and/or
- high electrical conductivity.

Reduction of dielectric losses:

- $DK = 1$ (air) or dielectrics,
- lower $\tan\delta \to 0$ and/or
- better circuit board material, instead of FR4 e.g. Teflon or Al_2O_3 .

To describe the directivity of an antenna, its radiation density can be related to the average radiation density that would occur if the entire absorbed power were to be radiated uniformly into the entire room, as is the case with a hypothetical isotropic radiator. The following then applies:

$$\tilde{G}(\theta, \psi) = \frac{S(\theta, \varphi)}{P/(4\pi r^2)} = |C_1(\theta, \varphi)|^2 + |C_2(\theta, \varphi)|^2 \,. \tag{9.79}$$

One calls $\tilde{G}(\theta, \varphi)$ the gain function of an antenna. The value of the gain function in the main beam direction is referred to as the antenna gain G.

$$G = \tilde{G}(\theta = 0, \varphi = 0) \tag{9.80}$$

The gain of a Hertzian dipole results in

$$G_{Hertz} = 1.5 \mathrel{\hat{=}} 1.76\,\text{dB}$$

and the gain of a $\lambda/2$ – dipol is

$$G_{\lambda/2} = 1.64 \mathrel{\hat{=}} 2.15\,\text{dB} .$$

For a lossless antenna with homogeneous field coverage (i.e. the surface current within the aperture is constant), the following applies

$$G_{hom.Apert.} = 4\pi \, \frac{A}{\lambda_0^2} . \tag{9.81}$$

Example Astra satellite antenna:
Given: $\varnothing = 30\,\text{cm}$, $f = 10\,\text{GHz} \rightarrow \lambda_0 = 3\,\text{cm}$, $A = \Pi r^2 = \Pi \cdot 15^2\,\text{cm}^2$

$$G_{hom.Apert.} = \frac{4\Pi^2 \cdot 15^2\,\text{cm}^2}{3^2\,\text{cm}^2} \approx 1000 \tag{9.82}$$

$$G^{dB}_{hom.Apert.} = 30\,\text{dB} \tag{9.83}$$

Example Effelsberg (Fig. 9.51), in operation since 1972:
Given: $\varnothing = 100\,\text{m}$, $f = 100\,\text{GHz} \rightarrow \lambda_0 = 0.003\,\text{m}$, $A = \Pi r^2 = \Pi \cdot 50^2\,\text{m}^2$

$$G_{hom.Apert.} = \frac{4\Pi^2 \cdot 50^2\,\text{m}^2}{0.003^2\,\text{m}^2} \approx 11 \cdot 10^9 \tag{9.84}$$

$$G^{dB}_{hom.Apert.} = 100.4\,\text{dB} \tag{9.85}$$

Where A is the geometric aperture area. The `half-width` $\Delta\theta$ or $\Delta\varphi$ is the angular range within which the radiance drops to half of the maximum value. According to Eq. (9.63), the following applies to the 3 dB-bandwidth of the homogeneously illuminated rectangular aperture

$$\Delta\varphi = 0.88 \, \frac{\lambda_0}{L} , \tag{9.86}$$

where L is the width measured parallel to the plane under consideration.

With non-homogeneous, tapered illumination, the half-width increases to about twice the value. This gives an estimate valid for any plane antenna:

$$\Delta\varphi/^\circ \approx (50\ldots100) \, \frac{\lambda_0}{L} , \tag{9.87}$$

where the smallest value applies to the case of homogeneous illumination.

For a rough estimate of the correlation between the half-power widths with the gain G for antennas with so-called "pencil lobes" can be arrived at by assuming

Fig. 9.51 Radioteleskope Effelsberg in Germany, fully rotatable and with a diameter of 100 m, an aperture area of 785 m^2 and for a frequency range from 30 MHz to 90 GHz

that the total energy in a range defined by the half-widths $\Delta\theta$ and $\Delta\varphi$ given by the rectangular beam is uniformly distributed:

$$G \approx \frac{4\pi r^2}{r^2\,\Delta\theta\,\Delta\varphi} = \frac{4\pi}{\Delta\theta\,\Delta\varphi}\,, \tag{9.88}$$

or as a numerical value equation with the half-value widths (3 dB-widths) in degrees

$$G \approx 41.000 \cdot \frac{1}{\Delta\theta^\circ\,\Delta\varphi^\circ}\,. \tag{9.89}$$

9.7.3 Characteristics of an Antenna for the Receiving Case

To describe the receiving behavior of an antenna, a plane homogeneous wave `incident` on the antenna in the direction $-\vec{\mathbf{u}}_r$ (angle θ, φ) is assumed.

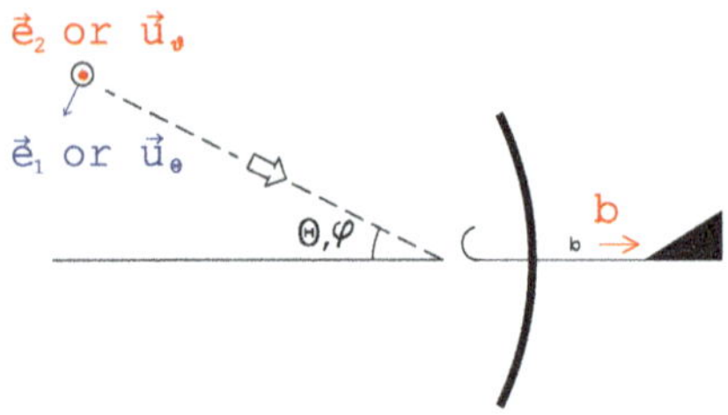

Fig. 9.52 Explanation of a receiving antenna

$$\vec{\mathbf{E}}(\vec{\mathbf{r}}) = (E_1 \cdot \vec{\mathbf{e}}_1 + E_2 \cdot \vec{\mathbf{e}}_2)\, e^{-jk_0\,(-\vec{\mathbf{u}}_r \cdot \vec{\mathbf{r}})} \tag{9.90}$$

Insert the radiation density S into the Eq. (9.90):

$$S = \frac{1}{2Z_0}\,|\vec{\mathbf{E}}|^2\,, \tag{9.91}$$

this equation can also be written in the following form:

$$\vec{\mathbf{E}}(\vec{\mathbf{r}}) = \sqrt{2\,Z_0\,S}\,\left(k \cdot \vec{\mathbf{e}}_1 + \sqrt{(1-k^2)}\,e^{j\psi} \cdot \vec{\mathbf{e}}_2\right)\,e^{jk_0\,\underbrace{\vec{\mathbf{u}}_r \cdot \vec{\mathbf{r}}}_{=r}}\,. \tag{9.92}$$

Here, k and ψ are real numbers that describe the polarization of the wave. In general, $k \geq 0$ and $-\pi \leq \psi \leq \pi$ and for special polarizations:

$k = 1$: $\vec{\mathbf{E}}$ is a linearly polarized wave in the direction $\vec{\mathbf{e}}_1$,
$k = 0$; $\psi = 0$: $\vec{\mathbf{E}}$ is a linearly polarized wave in the direction $\vec{\mathbf{e}}_2$,
$k = \frac{1}{\sqrt{2}}$; $\psi = \pm\frac{\pi}{2}$: $\vec{\mathbf{E}}$ is a circularly polarized wave (right or left).

An appropriately polarized incident wave causes the active power $P = |b|^2$ to be emitted at a matched load (Fig. 9.52) of an antenna.

Since b is a linear function of the field components E_1 and E_2 an expression of the form

$$b = \frac{1}{\sqrt{2\,Z_0}}\,\left(D_1(\theta,\varphi) \cdot E_1 \,+\, D_2(\theta,\varphi) \cdot E_2\right)\,. \tag{9.93}$$

Here, $D_1(\theta,\varphi)$ and $D_2(\theta,\varphi)$ are to be designated as the absolute **reception directional characteristics** for polarization 1 and 2 respectively.

The relative receive directional characteristics are obtained by normalizing the magnitudes of D_1 and D_2 so that the maximum value is just one.

If the field components of the incident plane homogeneous wave are represented by the radiation density S as in Eq. (9.91) and inserted into Eq. (9.93), then we obtain

$$b = \sqrt{S}\left(k \cdot D_1(\theta,\varphi) + \sqrt{(1-k^2)}\, e^{j\psi} \cdot D_2(\theta,\varphi)\right) . \tag{9.94}$$

This means that the active power delivered to the load of the receiver $P = |b|^2$:

$$P = bb^* = S \cdot \left\{k^2|D_1|^2 + (1-k^2)|D_2|^2 + 2k\sqrt{(1-k^2)}\ \mathrm{Re}\left\{D_1^*\, D_2\, e^{j\psi}\right\}\right\}. \tag{9.95}$$

The active power P is therefore dependent on the power given by

a) θ and φ determined by the direction of radio receiving,
b) k and ψ defined polarization.

The expression for P in Eq. (9.95) assumes its extreme values as a function of ψ when $D_1^* \cdot D_2 \cdot e^{j\psi}$ becomes real. The expression for P assumes its maximum as a function of ψ when $D_1^* \cdot D_2 \cdot e^{j\psi}$ becomes positive real. This is the case for the angle ψ_{opt}

$$\psi = \psi_{opt} = \arg D_1 - \arg D_2 . \tag{9.96}$$

Then

$$P = S \cdot \left(k \cdot |D_1| + \sqrt{(1-k^2)} \cdot |D_2|\right)^2 . \tag{9.97}$$

This power is used for a maximum of

$$k = k_{opt} = \frac{|D_1|}{\sqrt{(|D_1|^2 + |D_2|^2)}} , \tag{9.98}$$

with the value $P = P_{max}$

$$P = P_{max} = S \cdot \left(|D_1|^2 + |D_2|^2\right) . \tag{9.99}$$

On the other hand

$$\psi = \psi_{block} = \psi_{opt} + \pi \qquad \text{and} \tag{9.100}$$

$$k = k_{block} = \frac{|D_2|}{\sqrt{(|D_1|^2 + |D_2|^2)}} \tag{9.101}$$

the power delivered, i.e. $P = 0$.

For each receiving direction, there is therefore an optimal polarization determined by Eqs. (9.96) and (9.98), which leads to a maximum power output, and a blocking polarization given by Eqs. (9.100) and (9.101), at which no power is output.

The power P_{max} absorbed for a certain receiving direction with optimum polarization according to Eq. (9.99) corresponds to the power that the undisturbed plane homogeneous wave of radiation density S passes through a surface of size

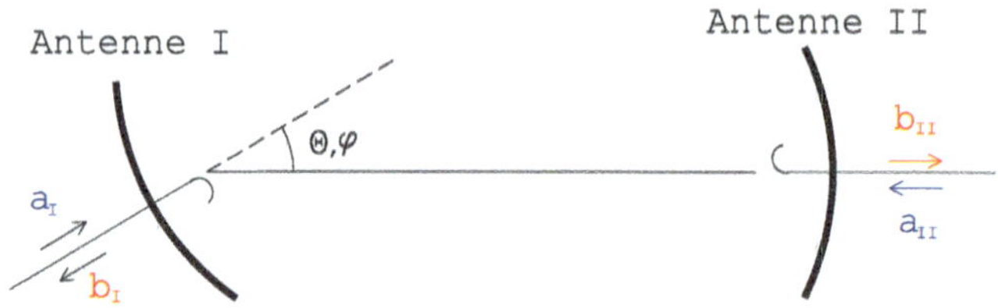

Fig. 9.53 To explain the arrangement of the transmitting and receiving antenna

$$\tilde{A}_w(\theta,\varphi) = \frac{P_{max}}{S} = |D_1(\theta,\varphi)|^2 + |D_2(\theta,\varphi)|^2 \tag{9.102}$$

and which can be picked up by the receiving antenna. One refers to $\tilde{A}_w(\theta,\varphi)$ as **active area function**. The maximum value of $\tilde{A}_w(\theta,\varphi)$, which lies here in the main beam direction at $\theta = \varphi = 0$, is referred to as `antenna effective area` A_w.

$$A_w = \tilde{A}_w(\theta = 0, \varphi = 0) \tag{9.103}$$

9.7.4 The Reciprocity Theorem

The **reciprocity theorem** provides an relationship between the directional characteristics $C_1(\theta,\varphi)$ and $C_2(\theta,\varphi)$ for the transmit case and $D_1(\theta,\varphi)$ and $D_2(\theta,\varphi)$ for the receive case. For this purpose, an antenna arrangement is considered in which the antenna I to be examined can be swiveled (Fig. 9.53).

Let the antenna II be aligned with its main beam direction along the axis towards the antenna I. In the event that the antenna I transmits and the antenna II receives, the following applies according to Eqs. (9.73) and (9.93)

$$\frac{b_{II}}{a_I} = S_{21}(\theta,\varphi) = \frac{e^{-jk_0 r}}{2\sqrt{\pi}\,r}\left\{C_1^I(\theta,\varphi)\cdot D_1^{II}(0,0) + C_2^I(\theta,\varphi)\cdot D_2^{II}(0,0)\right\}. \tag{9.104}$$

For the reverse case, i.e. antenna II as transmitting antenna and antenna I as receiving antenna with the same spatial arrangement, the following applies:

$$\frac{b_I}{a_{II}} = S_{12}(\theta,\varphi) = \frac{e^{-jk_0 r}}{2\sqrt{\pi}\,r}\left\{C_1^{II}(0,0)\cdot D_1^I(\theta,\varphi) + C_2^{II}(0,0)\cdot D_2^I(\theta,\varphi)\right\}. \tag{9.105}$$

If the antenna path is constructed from isotropic materials and does not contain any active components, the reciprocity theorem is valid and applies:

$$S_{21} = S_{12} \tag{9.106}$$

and therefore also

$$C_1^I(\theta,\varphi)\cdot D_1^{II}(0,0)+C_2^I(\theta,\varphi)\cdot D_2^{II}(0,0)=\\ C_1^{II}(0,0)\cdot D_1^I(\theta,\varphi)+C_2^{II}(0,0)\cdot D_2^I(\theta,\varphi)\,. \tag{9.107}$$

If we assume that the antenna II only transmits and receives polarization 1 in the main beam direction, i.e.

$$C_2^{II}(0,0)=D_2^{II}(0,0)=0$$

applies, then Eq. (9.107) reduces to:

$$\frac{D_1^I(\theta,\varphi)}{C_1^I(\theta,\varphi)}=\frac{D_1^{II}(0,0)}{C_1^{II}(0,0)}=\text{const.}=\varphi_0\,. \tag{9.108}$$

This gives the following important result:

The ratio of the directional characteristics for the receiving case $D_{1,2}(\theta,\varphi)$ to the directional characteristics for the transmitting case $C_{1,2}(\theta,\varphi)$ is independent of the direction and also independent of the antenna type. So if you know the ratio for the specific antenna type and for a specific direction, you know it for all antenna types and directions.

The following applies to the Hertzian dipole:

$$\varphi_0=\frac{\lambda_0}{2\sqrt{\pi}}\,. \tag{9.109}$$

So in general

$$D_{1,2}(\theta,\varphi)=\frac{\lambda_0}{2\sqrt{\pi}}\,C_{1,2}(\theta,\varphi)\,. \tag{9.110}$$

From Eq. (9.110), the relationship between the effective antenna area A_w and the antenna gain via Eq. (9.79) and (9.102) is also given by the relationship

$$A_w=\frac{\lambda_0^2}{4\,\pi}\,G\,. \tag{9.111}$$

A comparison of Eq. (9.111) with Eq. (9.81) allows the conclusion that the geometric area A and the effective area A_w are equal for the area antenna with homogeneous aperture assignment. An optimally aligned receiving antenna absorbs a power that results from the product of the radiation density S and the antenna effective area A_w. The antenna effective area can be calculated from the gain of the antenna using Eq. (9.111).

The proportionality between the transmit and receive characteristics means that the sidelobe attenuation of the transmit and receive antennas is multiplied. With the same transmit and receive antenna and rectangular occupancy function, the effective sidelobe attenuation is therefore $2\cdot 13.2\,\text{dB}=26.4\,\text{dB}$.

Chapter 10
Phase-Controlled Antennas, Patch Antennas, Measurement Technology

Electronically controlled or phased array antennas are increasingly being used in radar technology. With these, the antenna pattern is not swiveled mechanically, but by means of a phase control of a group of individual radio emitters. The advantage of such antennas is that the antenna beam can be swiveled much faster compared to a mechanically swiveling antenna. This chapter also introduces the frequently used patch antennas. It also explains how a chaotic design can reduce sidelobes immensely and briefly describes the basic antenna measurement technology.

10.1 Basic Procedure for Phased Array Antennas

In radar technology, electronically steered or phased array antennas are increasingly being used. are being used. With these antennas, the antenna pattern is not steered mechanically, but by phase control of a group of individual radio emitters. Such antennas have the advantage that the antenna beam can be swiveled much faster compared to a mechanically swiveling antenna. Because no heavy mass has to be accelerated, the beam can be deflected in any sequence. This makes it possible, for example, to scan a room and tracking of several objects almost simultaneously.

A group of individual radiators, e.g. electric dipoles, horn radiators or so-called slot radiators, which are approximately dual to electric dipoles, are individually equipped with electronically switchable phase shifters. This allows the phase position of each individual radiator to be adjusted. In general, however, not every arbitrary phase value can be set; for reasons of cost, the phase shifter only has a finite number of discrete stages. For example, a 3-bit phase shifter is a phase shifter that has a 45°, 90° and 180° phase switch in series and therefore allows the values $0°, 45°, 90°, 135°, 180°, 225°, 270°, 315°$ to be set in $2^3 = 8$ steps. The values 360° and 0° are identical.

The phase shifters are generally controlled digitally with the support of computers. The 3-bit phase shifter therefore has a discretization error of $\pm 22.5°$.

We will discuss the effects of such a discretization error later. Ferrite phase shifters and phase shifters controlled by semiconductor components such as PIN diodes or FET transistors are mainly used. However, the design of switchable phase shifters will not be discussed in detail here. The principle of an electronically slewable antenna is explained using a linear radiator arrangement, which should have N elements (Fig. 10.1).

H. Heuermann, *Microwave Technology*,
https://doi.org/10.1007/978-3-658-45686-3_10

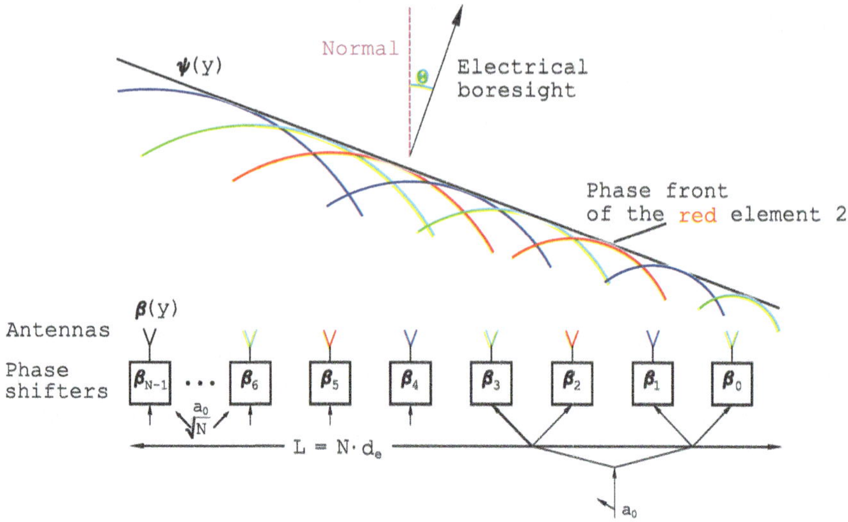

Fig. 10.1 Linear group of radiator elements

As shown in Fig. 10.1, in the following $\beta(y)$ describes the phase in the radiator plane or radiator line and $\Psi(y)$ the phase in a plane or line perpendicular to the considered propagation direction with the direction θ. The differential phase values $\Delta\beta$ and $\Delta\Psi$ denote the corresponding phase changes from radiating element to radiating element.

It should be assumed that the wave $a_0/\sqrt{N}$ is incident on the matched radiating elements of the group (transmission case). In addition, no coupling should occur between the individual radiators and all radiators should have the same directional characteristic $C_e(\theta)$. The wave a_0 is to be introduced into the feed network, which is assumed to be lossless.

10.2 The Directional Characteristic of a Phased Array Antenna

The individual radiators should have the same polarization 1. According to Eq. (9.73), the following applies to the single radiator for the electric field in the far field:

$$\vec{\mathbf{E}}(r,\theta) = \sqrt{\frac{Z_0}{2\pi}}\ \frac{1}{r}\ \frac{1}{\sqrt{N}}\ C_e(\theta)\ e^{-jk_0 r}\ \vec{\mathbf{e}}_1\ a_0\,. \tag{10.1}$$

While we assume that the contribution to the far field by the factor $1/r$ is the same for all radiating elements, as with the flat antenna, we must take into account that the phase factor $e^{-jk_0 r}$ is not the same for the different radiators when summing up the field contributions of the individual radiators.

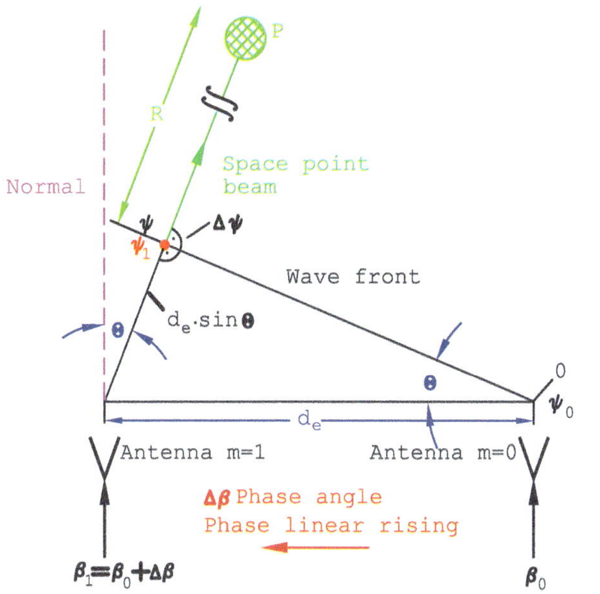

Fig. 10.2 Phase differences in the direction of radiation

We want to assume that there are no discretization errors in the phase shifters and that the phase $\beta(m)$, $m = 0, 1, 2, \ldots, N$, of the phase shifters increases linearly by $\Delta\beta$ from phase shifter to phase shifter.

The phase change to a neighboring radiator $\Delta\Psi$ transverse to the considered propagation direction θ is taken from Fig. 10.2.

The phase change is

$$\Delta\Psi = \Psi_1 - \Psi_0 \,,$$

$$\Delta\Psi = \Delta\beta - k_0 \; d_e \; \sin\theta \,,$$

$$\Delta\Psi = -k_0 \; d_e \; (\sin\theta - \sin\theta_0) \,, \tag{10.2}$$

$$\text{with} \qquad \Delta\beta = \beta_1 - \beta_0 = k_0 \; d_e \; \sin\theta_0 \,. \tag{10.3}$$

We will see that θ_0 denotes the main beam direction. Starting at zero, the phase contribution to a receptor point in the far field increases by $\Delta\Psi$ from radiator to radiator to $N\Delta\Psi$. For the total field $\vec{\mathbf{E}}_i(R,\theta)$, which is set at the space point P, the contributions of all N individual radiators assumed to be equal here must be summed up with a field contribution according to Eq. (10.1).

$$\vec{\mathbf{E}}(R,\theta) = \sqrt{\frac{Z_0}{2\pi}} \; \frac{e^{-jk_0R}}{R} \; C_e(\theta) \left(\frac{1}{\sqrt{N}} \sum_{m=0}^{N-1} e^{\,jm\Delta\Psi} \right) a_0 \, \vec{\mathbf{e}}_1 \tag{10.4}$$

$(m = \text{Running index})$

The factor $1/\sqrt{N}$ in Eq. (10.4) takes into account that $|a_0|^2$ should be the total power fed into all individual radiators. The sum in Eq. (10.4) should be designated as the directivity of the antenna array $C_g(\theta)$ with

$$C_g(\theta) = \frac{1}{\sqrt{N}} \sum_{m=0}^{N-1} e^{jm\Delta\Psi} . \tag{10.5}$$

The total characteristic C_{res} is the product of the individual radiator characteristic C_e and the group characteristic C_g.

The sum in Eq. (10.5) can be understood as a geometric series and calculated explicitly. This gives the following result:

$$C_g(\theta) = \frac{1}{\sqrt{N}} \frac{1 - e^{(jN\Delta\Psi)}}{1 - e^{(j\Delta\Psi)}} ,$$

$$C_g(\theta) = \frac{e^{\left(j(N-1)\frac{\Delta\Psi}{2}\right)}}{\sqrt{N}} \frac{e^{\left(-jN\frac{\Delta\Psi}{2}\right)} - e^{\left(jN\frac{\Delta\Psi}{2}\right)}}{e^{\left(-j\frac{\Delta\Psi}{2}\right)} - e^{\left(j\frac{\Delta\Psi}{2}\right)}} ,$$

$$C_g(\theta) = e^{\left(j(N-1)\frac{\Delta\Psi}{2}\right)} \frac{1}{\sqrt{N}} \frac{\sin\left(N\frac{\Delta\Psi}{2}\right)}{\sin\left(\frac{\Delta\Psi}{2}\right)} ,$$

$$C_g(\theta) = e^{\left(j(N-1)\frac{\Delta\Psi}{2}\right)} \frac{1}{\sqrt{N}} \frac{\sin\left(\frac{N\pi d_e}{\lambda_0}(\sin\theta - \sin\theta_0)\right)}{\sin\left(\frac{\pi d_e}{\lambda_0}(\sin\theta - \sin\theta_0)\right)} . \tag{10.6}$$

Finally, for the gain function $\tilde{G}_g(\theta)$ of the group, the insignificant phase pre-factor in Eq. (10.6) is omitted and the result is obtained:

$$\tilde{G}_g(\theta) = \left| C_g(\theta) \right|^2 = \frac{\sin^2\left(\frac{N\pi d_e}{\lambda_0}(\sin\theta - \sin\theta_0)\right)}{N \sin^2\left(\frac{\pi d_e}{\lambda_0}(\sin\theta - \sin\theta_0)\right)} . \tag{10.7}$$

For the favorable choice $d_e = \lambda_0/2$ and a sufficiently large number of elements N, this diagram is similar to that of a homogeneous occupied surface antenna (Fig. 10.1). The main beam direction is at $\sin\theta = \sin\theta_0$ or $\theta = \theta_0$ as already described by Eq. (9.70). The first side lobe is about 13.2 dB below the main lobe at large N.

As long as the element spacing $\lambda_0/2$ or smaller, the side lobes remain small compared to the main lobe. If however, the element spacing is greater than $\lambda_0/2$, additional lobes can appear in the radiation pattern with a gain comparable gain to the main lobe. Such secondary lobes (or "grating lobes") occur when the numerator and denominator in Eq. (10.7) simultaneously become zero but the quotient in the limit remains finite. This is the case if the following applies:

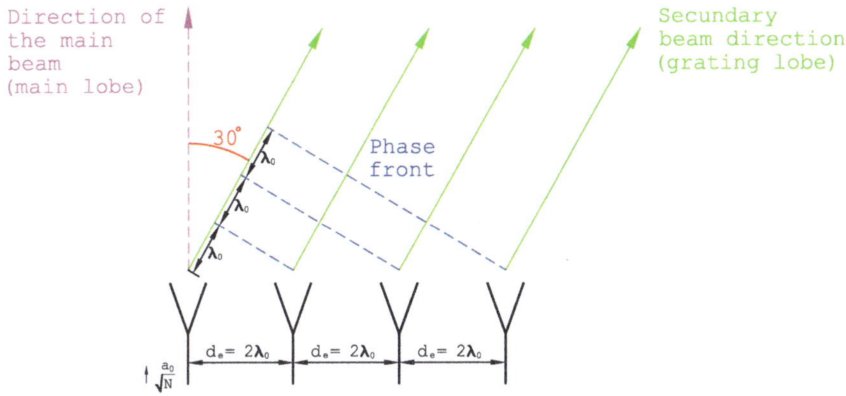

Fig. 10.3 Graphic explanation for a secondary lobe for $\theta = 30°$

$$\frac{\pi d_e}{\lambda_0}\,(\sin\theta - \sin\theta_0) = \pm\mu\,\pi \qquad (\mu \text{ whole number}),$$

$$\text{or} \qquad \left|\,\sin\theta - \sin\theta_0\,\right| = \mu\,\frac{\lambda_0}{d_e}, \qquad m = 1,2,3,\ldots\,. \tag{10.8}$$

For example, if $\theta_0 = 0$ and the element spacing $d_e = 2\lambda_0$, then unwanted secondary lobes occur at θ_s with

$$\theta_s = \pm 30°;\ \pm 90°\,.$$

The side lobes at $\pm 90°$ are generally not critical because the gain of the individual radiators is usually low at this angle.

Fig. 10.3 explains graphically that at $d_e = 2\,\lambda_0$ and an angle of 30° there is again a main beam direction, i.e. a common phase front of all individual lobes is formed.

As can be seen from Fig. 10.3, for $\theta_s = 30°$ the condition is fulfilled that the path differences to the individual radiators are λ_0, i.e. the contributions of the individual radiators are summed up. Algebraically, the condition for such constructive interference is when Ψ_0 is an arbitrary initial phase:

$$\Psi_0 + \Delta\beta - k_0\,d_e\,\sin\theta = \Psi_0 \ \pm\ \mu\,2\,\pi \qquad \text{with} \qquad \mu = 0,1,2,\ldots,$$

$$\text{or} \qquad \sin\theta = \frac{\Delta\beta}{k_0\,d_e} \pm \frac{\mu\,2\,\pi}{k_0\,d_e}\,. \tag{10.9}$$

Because of Eq. (10.3), however, this relationship is identical to Eq. (10.8). For $\mu = 1,2,3,\ldots$, unwanted secondary lobes can form if Eq. (10.8) can be fulfilled by a real $\theta = \theta_s$. This is most likely possible with $\mu = 1$.

To reliably avoid secondary lobes, the distance d_e between the individual radiators must only slightly exceed $\lambda_0/2$. Secondary lobes can generally be permitted at $\theta_s = \pm 90°$ because the individual radiators radiate little in this direction due to their

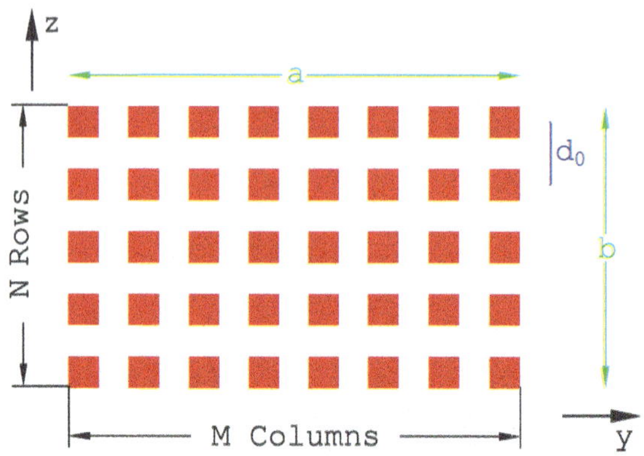

Fig. 10.4 Grid arrangement of the elementary radiators with M columns and N rows

inherent characteristics. If the maximum deflection angle is also restricted, e.g. to $\pm\theta_m$, then $\theta_s = \pm 90°$ follows from Eq. (10.8) for the distance d_e to be maintained:

$$\frac{\lambda_0}{d_e} \geq 1 + \sin\theta_m \,. \tag{10.10}$$

For $\theta_m = \pm 60°$ follows from this $d_e \leq 0.536\,\lambda_0$. For $\theta_m \to 0$ the permissible distance between the radiating elements increases to λ_0.

If we consider a rectangular planar group arrangement of individual radiators, namely M columns and N rows (Fig. 10.4), then the gain function of the group in the horizontal and vertical planes is obtained:

$$\tilde{G}_g(\theta,0) = M \left. \frac{\sin^2\left(\frac{N\pi d_e}{\lambda_0}(\sin\theta - \sin\theta_0)\right)}{N\sin^2\left(\frac{\pi d_e}{\lambda_0}(\sin\theta - \sin\theta_0)\right)} \right|_{\varphi_0 = 0} , \tag{10.11}$$

$$\tilde{G}_g(0,\varphi) = N \left. \frac{\sin^2\left(\frac{M\pi d_e}{\lambda_0}(\sin\varphi - \sin\varphi_0)\right)}{M\sin^2\left(\frac{\pi d_e}{\lambda_0}(\sin\varphi - \sin\varphi_0)\right)} \right|_{\theta_0 = 0} . \tag{10.12}$$

The phase assignment $\beta(y)$ must be selected so that a plane of constant phase is perpendicular to the main beam direction. If the main beam direction does not coincide with either the vertical or horizontal plane, the plane of constant phase must also be skewed accordingly. This makes the phase assignment β a function of y and z, i. h. $\beta = \beta(y,z)$.

10.3 Characteristics of a Phased Array Antenna

From Eqs. (10.11) and (10.12), the gain G_g of the antenna array first in the main beam direction with $\theta = \theta_0$ and $\varphi = \varphi_0$ results in

$$G_g = \tilde{G}(\theta = \theta_0, \varphi = \varphi_0) = N\,M\,. \tag{10.13}$$

For a distance of the individual radiators of $d_e = \lambda_0/2$ and the side lengths of the group a and b (Fig. 10.4) and the area $A = a\,b$ the Eqs. (10.11) and (10.12) can also be written:

$$G_g = N\,M = \frac{N\,\frac{\lambda_0}{2}\,M\,\frac{\lambda_0}{2}}{\frac{\lambda_0^2}{4}} = \frac{4\,A}{\lambda_0^2}\,. \tag{10.14}$$

If we compare this expression with the gain of an area antenna with homogeneous occupancy, Eq. (9.81), and further assume that the individual radiators have at least the gain of 1.5 of a Hertzian dipole, then we find that the gain G of the array antenna is approximately a factor of $2\pi/3$ less than that of an area radiator of the same area and homogeneous occupancy. We can obtain an estimate of the beam half-width by the following consideration. The gain functions of the array $\tilde{G}_g$ according to Eq. (10.11) and (10.12) have the term $\mathrm{si}^2 u$ except for a constant factor, whereby the abbreviation

$$u = \frac{N\,k_0\,d_e}{2}\;(\sin\theta - \sin\theta_0)\,,$$

respectively

$$u = \frac{M\;k_0\,d_e}{2}\;(\sin\varphi - \sin\varphi_0)\,, \tag{10.15}$$

was introduced. For this purpose, $\sin u \approx u$ was set in the denominator of Eqs. (10.11) and (10.12) because of $N, M \gg 1$.

The gain is reduced by 3 dB when u is increased from zero to the value

$$u = \pm\,0.443\;\pi \tag{10.16}$$

has increased (see also Eq. (9.63)).

For further consideration we want to use the following identity:

$$\sin\theta - \sin\theta_0 = \sin(\theta - \theta_0)\;\cos\theta_0 - \overbrace{\Big[\overbrace{1\;-\;\cos(\theta - \theta_0)}^{\approx 0}\Big]\;\sin\theta_0}^{\approx 0}\,. \tag{10.17}$$

The second summand on the right-hand side of Eq. (10.17) can be neglected if θ_0 is not too large and θ does not deviate too much from θ_0. Then the following applies approximately:

$$\sin\theta - \sin\theta_0 = \sin(\theta - \theta_0)\ \cos\theta_0\ . \tag{10.18}$$

This gives the half-width of the group profit:

$$\Delta\theta_b = \frac{2\cdot 0.443\ \lambda_0}{N\ d_e\ \cos\theta_0} = 0.886\ \frac{\lambda_0}{N\ d_e\ \cos\theta_0}\ ,$$
$$\Delta\varphi_b = 0.886\ \frac{\lambda_0}{M\ d_e\ \cos\varphi_0}\ . \tag{10.19}$$

The corresponding equations in degrees are as follows:

$$\Delta\theta_b = \frac{51^\circ\ \lambda_0}{N\ d_e\ \cos\theta_0}\ ,$$
$$\Delta\varphi_b = \frac{51^\circ\ \lambda_0}{M\ d_e\ \cos\varphi_0}\ . \tag{10.20}$$

The factor $\cos\theta_0$ in the denominator of Eq. (10.20) means that approximately only the projection of the aperture area in the main beam direction is decisive for the half-width.

It should be mentioned that, as with the flat radiator, better suppression of the side lobes can be achieved if the amplitude of the distribution function is tapered. This then also leads to an increase in the half-width.

10.4 Discretization Error with a Phased Array Antenna

The phase of the individual radiating elements can generally only assume discrete values. For a n- bit phase shifter, the step size is $\delta\Psi$:

$$\delta\Psi = \frac{2\ \pi}{2^n}\ . \tag{10.21}$$

This results in a maximum quantization error β_m of

$$\beta_m = \frac{\delta\Psi}{2} = \frac{\pi}{2^n}\ . \tag{10.22}$$

Fig. 10.5 shows the ideal phase progression $\Gamma(y)$ and the actual phase progression $\beta(y)$, which has a staircase effect due to the quantization. However, only individual discrete values $\beta(y_m)$ are actually assumed at the points y_m, $m = 0,1,2,\ldots,N$, which are shown as points in Fig. 10.5. However, we want to carry out the following considerations, which are only estimates, with $\beta(y)$. Fig. 10.5 also shows the progression of the error phase

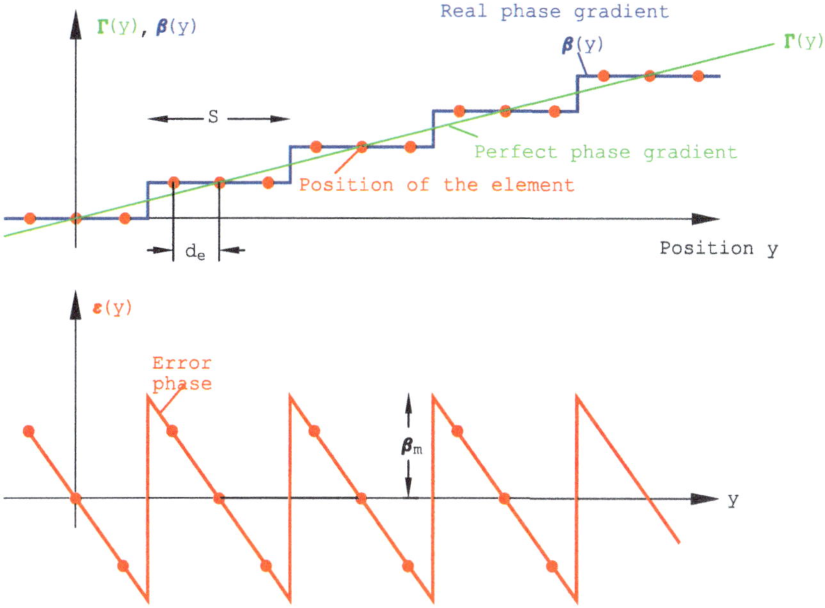

Fig. 10.5 Progression of the ideal phase $\Gamma(y)$, the actual phase $\beta(y)$ and the error phase $\varepsilon(y) = \Gamma(y) - \beta(y)$ as a function of the location y with discrete phase adjustment

$$\varepsilon(y) = \beta(y) - \Gamma(y) \tag{10.23}$$

as a function of the location y. In this representation, the error phase $\varepsilon(y)$ is a triangular function with the vertex value β_m. The ideal phase progression $\Gamma(y)$ is represented by the expression

$$\Gamma(y) = k_0 \ \sin\theta_0 \ y \tag{10.24}$$

is described. This means that for the period S of the error phase $\varepsilon(y)$ from

$$2\,\beta_m = k_0 \ \sin\theta_0 \ S$$

respectively

$$S = \frac{2\,\beta_m}{k_0 \ \sin\theta_0} \,. \tag{10.25}$$

Due to the triangular shape of the error phase $\varepsilon(y)$, it has a rectangular probability density $p(\varepsilon)$, as shown in Fig. 10.6.

For the directional characteristic of the group C_g in the main beam direction, i. h. $\theta = \theta_0$, the following applies according to Eq. (10.5) with $\Delta\Psi = 0$

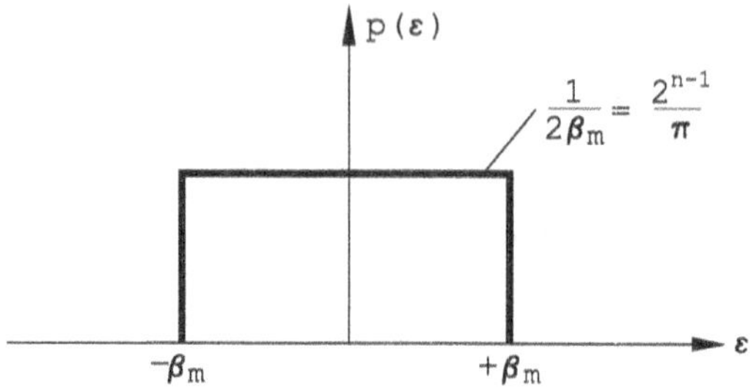

Fig. 10.6 Probability density of the error phase

Tab. 10.1 Gain loss due to discretization error

n	2	3	4
G_e/G in dB	1.0	0.23	0.06

$$C_g(\theta_0) = \frac{1}{\sqrt{N}} \sum_{m=0}^{N-1} e^{\,jm\Delta\Psi} = \frac{1}{\sqrt{N}}\, N\,. \tag{10.26}$$

The expected value of the directional characteristic of the group in the main beam direction C_e with an assumed rectangular probability density as in Fig. 10.6:

$$C_e(\theta_0) = \mathrm{E}\Big(C_g(\theta_0)\Big) = \frac{1}{\sqrt{N}}\, \mathrm{E}\left(\sum_{m=0}^{N-1} e^{\,j\varepsilon_m}\right) = \frac{1}{\sqrt{N}} \sum_{m=0}^{N-1} \mathrm{E}\left(e^{\,j\varepsilon}\right),$$

$$C_e(\theta_0) = \frac{1}{\sqrt{N}}\, N\, \mathrm{E}\left(e^{\,j\varepsilon}\right) = \frac{1}{\sqrt{N}}\, N \int_{-\infty}^{+\infty} p(\varepsilon)\, e^{\,j\varepsilon}\, d\varepsilon\,,$$

$$C_e(\theta_0) = \frac{N}{\sqrt{N}}\, \frac{1}{2\,\beta_m} \int_{-\beta_m}^{+\beta_m} e^{\,j\varepsilon}\, d\varepsilon = \frac{N}{\sqrt{N}}\, \frac{\sin\beta_m}{\beta_m}\,,$$

$$C_e(\theta_0) \approx \sqrt{N}\left(1 - \frac{1}{6}\,\beta_m^2\right) \quad \text{(the first two links in the series for si}(\beta_m))\,. \tag{10.27}$$

This gives the ratio of the profit with the error phase $G_e(\theta_0)$ to the profit without error phase $G(\theta_0)$ for $\beta_m \ll 1$:

$$\frac{G_e(\theta_0)}{G(\theta_0)} = \left|\frac{C_e(\theta_0)}{C_g(\theta_0)}\right|^2 = 1 - \frac{1}{3}\, \frac{\pi^2}{2^{2n}}\,. \tag{10.28}$$

Tab. 10.1 shows the gain reduction in dB as a function of the number of bits n of the phase shifters.

Furthermore, the phase quantization causes additional side lobes, so-called quantization lobes. For a simplified mathematical treatment, we want to assume that the

Tab. 10.2 Suppression of the quantization lobes

n	2	3	4	5
$\beta_m^2/4$ in dB	−8	−14	−20	−26

error phase has a sinusoidal shape over the location y with the period S as shown in Fig. 10.6. The sinusoidal curve can be imagined as the first element of a Fourier series development of the sawtooth curve of Fig. 10.5. We set the amplitude of the sinusoidal phase perturbation equal to β_m (more precisely: with an ideal sawtooth curve equal to $\tilde{\beta}_m = 2\beta_m/\pi$). The distribution function then has the following curve:

$$w(y) = e^{-j\beta(y)} = e^{\left(-j\left[k_0\,\sin(\theta_0)\,y\,+\,\tilde{\beta}_m\,\sin\left(\frac{2\pi y}{S}\right)\right]\right)}\,. \tag{10.29}$$

Assuming that $\tilde{\beta}_m \ll is1$, the following approximations apply:

$$w(y) \approx e^{\left(-jk_0\sin(\theta_0)y\right)}\left[1 - j\,\tilde{\beta}_m\,\sin\left(\frac{2\pi y}{S}\right)\right],$$

$$w(y) \approx e^{\left(-jk_0\sin(\theta_0)y\right)}\left[1 - \frac{\tilde{\beta}_m}{2}\,e^{\left(j\frac{2\pi y}{S}\right)} + \frac{\tilde{\beta}_m}{2}\,e^{\left(-j\frac{2\pi y}{S}\right)}\right],$$

$$w(y) \approx e^{\left(-jk_0\sin(\theta_0)y\right)} - \frac{\tilde{\beta}_m}{2}\,e^{\left(-jk_0\left[\sin(\theta_0) - \frac{\lambda_0}{S}\right]y\right)} + \frac{\tilde{\beta}_m}{2}\,e^{\left(-jk_0\left[\sin(\theta_0) + \frac{\lambda_0}{S}\right]y\right)}\,. \tag{10.30}$$

The spatial modulation of the phase assignment function thus leads to two side lobes at the angles $\theta_{1,2}$ with

$$\sin\left(\theta_{1,2}\right) = \sin\left(\theta_0\right) \pm \frac{\lambda_0}{S}\,. \tag{10.31}$$

The relative amplitude of the side lobes is $\tilde{\beta}_m/2 = \pi/2^{n+1}$. Table 10.2 shows the suppression of these quantization lobes as a function of the number of bits.

It can be seen from Tab. 10.2 that sufficient suppression of the quantization lobes will probably determine the selection of the required number of bits more than the loss of gain.

10.5 Patch Antennas

Patch antennas manufactured on microstrip line substrates (also known as stripline antennas) are used especially in the high frequency range from 100 MHz to 100 GHz. They offer the possibility of a standardized design of the microwave circuit, the feed network and the antenna on a common substrate.

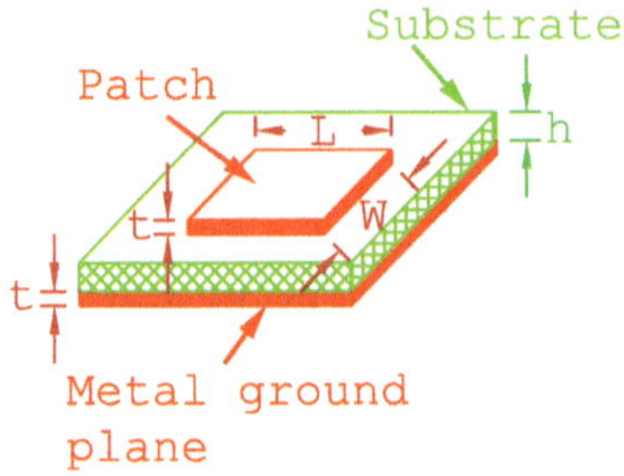

Fig. 10.7 Basic design of a rectangular patch antenna

Advantages

- High degree of miniaturization,
- reproducibility and automated mass production (low costs),
- mechanical resilience due to vibration and shock and high reliability.

Disadvantages

- Low efficiency due to losses in the substrate,
- which limits radiant power and gain and
- small relative bandwidth (a few percent).

A closed-form solution for wave propagation cannot be given. Therefore, empirical approximation formulas are used, which are derived from the static fields and generalized for higher frequencies.

10.5.1 Design of the Patch Antenna

As can be seen in Fig. 10.7, a patch antenna consists of individual basic elements. A dielectric substrate between a metallic structure, the patch element and a metallic base plate of thickness $t \to 0$.

The patch element can be animated in various ways:

- a) with a coaxial cable from below through the base plate,
- b) direct feed with a strip line,
- c) electrodynamic coupling to reduce parasitic radiation or
- d) aperture coupling through slots in an intermediate metallization.

A good match to the feed line can often be achieved by selecting a suitable feed point (x_s, y_s), which means that an additional matching network is often no longer required (Fig. 10.8).

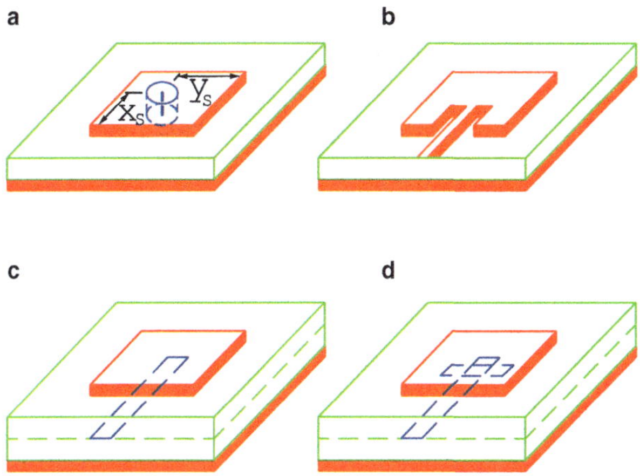

Fig. 10.8 Different ways to excite patch antennas

A rectangular patch element can be regarded as a stripline of length L and width W that is open on all four sides. The stray fields arising at $y = 0, W$ are approximately taken into account with $u = W/h$ by a relative permittivity, which is smaller ε_r, since the field lines run both in the substrate and in the outer space:

$$\varepsilon_{r,eff}^{(0)} \cong \frac{\varepsilon_r + 1}{2} + \frac{\varepsilon_r - 1}{2}\left(1 + \frac{10}{u}\right)^{1/2} . \tag{10.32}$$

The approximation (10.32) assumes a quasi-TEM wave on the stripline. The line impedance of this wave is :

$$Z_L^{(0)} \cong \frac{60\,\Omega}{\sqrt{\varepsilon_{r,eff}^{(0)}}} ln\left[\frac{F(u)}{u} + \sqrt{1 + \frac{4}{u^2}}\right] \quad \text{with}$$
$$F(u) = 6 + (2\pi - 6)\, exp\left[-\left(\frac{30.666}{u}\right)^{0.7528}\right] . \tag{10.33}$$

When the frequency is increased, the fields become more concentrated in the substrate, which leads to an increase in the effective relative permittivity:

$$\varepsilon_{r,eff} \cong \varepsilon_r - \frac{\varepsilon_r - \varepsilon_{r,eff}^{(0)}}{1 + G(f/f_p)^2} \Rightarrow \lambda_{eff} = \frac{\lambda_0}{\sqrt{\varepsilon_{eff}}} \tag{10.34}$$

with the auxiliary variables

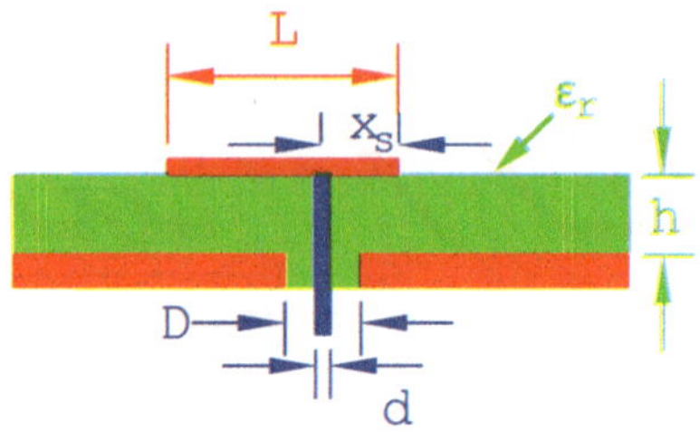

Fig. 10.9 Coaxial feed through the base plate with feed point at x_s

$$G = 0.6 + 0.009 \frac{Z_L^{(0)}}{\Omega} \; sowie \; f_p = \frac{Z_L^{(0)}}{2\,\mu_0 h} \,. \tag{10.35}$$

The stray fields at $x = 0, L$ make the line appear electrically longer by ΔL:

$$\Delta L \cong 0.412h \frac{\varepsilon_{r,eff} + 0.300}{\varepsilon_{r,eff} - 0.258} \frac{u + 0.262}{u + 0.813} \,. \tag{10.36}$$

With (10.34), (10.36) and $L_{eff} = L + 2\Delta L = \lambda_{eff}/2$ you will find the geometric patch length L:

$$L = \frac{\lambda_{eff}}{2} - 2\Delta L \,. \tag{10.37}$$

The aim is therefore to operate in half-wave resonance, in which a shortening due to capacitive end loading is effective. The width W of the patch antenna results from:

$$W = \sqrt{\frac{h\lambda_0}{\sqrt{\varepsilon_r}}\left[ln\left(\frac{\lambda_0}{h\sqrt{\varepsilon_r}}\right) - 1\right]} \,. \tag{10.38}$$

The feed-in point (x_s, y_s) is calculated from (Fig. 10.9):

$$x_s \cong \frac{\lambda_{eff}}{2\pi} \arccos\sqrt{\frac{R_E}{R_S}} \quad \text{and} \qquad y_s = \frac{W}{2} \tag{10.39}$$

with the radiation resistance R_S according to Eq. (10.59).

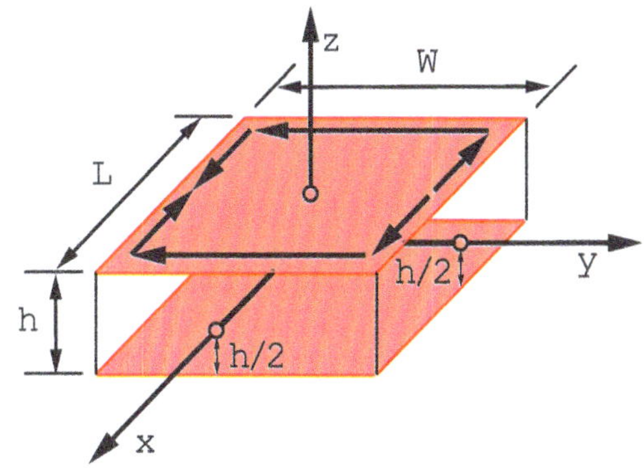

Fig. 10.10 Air-filled cavity resonator with two electrical and four magnetic walls

10.5.2 Radiation Fields of the Patch Antenna

To calculate the radiation of a rectangular patch, we use a Cartesian coordinate system whose origin lies in the center under the rectangular volume of the patch $V = WL_{eff}h$ (Fig. 10.10).

The cuboid cavity under the patch radiates laterally from the 4 flat slits of height h, which are located at $x = \pm L_{eff}/2$ and $y = \pm W/2$. Due to the low substrate height $h << \lambda_0$ we can assume that the fields of the cavity are not dependent on z. Furthermore, we first assume that $\varepsilon_r = 1$ and $W = L_{eff}$.

The vertical electric field strength is obtained by solving the Helmholtz equation, assuming a magnetic wall as boundary condition in the four slit planes:

$$\vec{E}_z = \vec{E}_0 \cos\frac{m\pi(x + L_{eff}/2)}{L_{eff}} \cos\frac{n\pi(y + W/2)}{W} \quad \text{with} \tag{10.40}$$

$$\frac{\delta \vec{E}_z}{\delta x} = 0 \text{ für } x = \pm L_{eff}/2 \text{ and } \frac{\delta \vec{E}_z}{\delta y} = 0 \text{ für } y = \pm W/2\,. \tag{10.41}$$

Es gilt: $L_{eff} = \lambda_{eff}/2$.

At half-wave resonance, the simplest waveform, the E10 fundamental wave, occurs if $L_{eff} \geq W$ applies. With $m - 1$ and $n = 0$ we obtain from (10.41):

$$\vec{E}_z = -\vec{E}_0 \sin\frac{\pi x}{L_{eff}}\,. \tag{10.42}$$

We replace the electric aperture fields according to Huygens' principle by equivalent magnetic surface current densities $\vec{M}_F = \vec{E} \times n$ and obtain in all four slits, where n is to be taken as the outer surface normal in each case:

$$
\begin{aligned}
\vec{M}_F(x=-L_{eff}/2) &= \vec{E}_0 u_z \times (-u_x) = -\vec{E}_0 u_y \\
\vec{M}_F(x=L_{eff}/2) &= -\vec{E}_0 u_z \times u_x = -\vec{E}_0 u_y \\
\vec{M}_F(y=-W/2) &= -\vec{E}_0 u_z \times (-u_y)\sin\frac{\pi x}{L_{eff}} = -\vec{E}_0 u_x \sin\frac{\pi x}{L_{eff}} \\
\vec{M}_F(y=W/2) &= -\vec{E}_0 u_z \times u_y \sin\frac{\pi x}{L_{eff}} = \vec{E}_0 u_x \sin\frac{\pi x}{L_{eff}} \,.
\end{aligned}
\tag{10.43}
$$

The slits at $x = \pm L_{eff}/2$ radiate in phase, while the slits at $y = \pm W/2$ radiate in antiphase (10.10). We can therefore group the four slots into two groups of two elements each and take their combined effect into account using two group factors. The electric vector potential for the single slot at $x = L_{eff}/2$ follows from

$$
\vec{F}_y = -\vec{E}_0 \frac{e^{-jk_0 r}}{4\pi r} \int_{z'=-W/2}^{h/2} e^{jk_0 y' \sin\vartheta \sin\varphi} dy' \int_{z'=-h/2}^{h/2} e^{jk_0 z' \cos\vartheta} dz', \tag{10.44}
$$

while for the other slot at $y = -W/2$:

$$
\vec{F}_x = \vec{E}_0 \frac{e^{-jk_0 r}}{4\pi r} \int_{z'=-L_{eff}/2}^{L_{eff}/2} \sin\frac{\pi x'}{L_{eff}} e^{jk_0 y' \sin\vartheta \cos\varphi} dx' \int_{z'=-h/2}^{h/2} e^{jk_0 z' \cos\vartheta} dz'. \tag{10.45}
$$

The evaluation of the integrals (10.44) and (10.45) leads to

$$
\begin{aligned}
\vec{F}_y &= -\vec{E}_0 \frac{e^{-jk_0 r}}{4\pi r} W h \frac{\sin Y}{Y} \frac{\sin Z}{Z} \\
\vec{F}_x &= \vec{E}_0 \frac{e^{-jk_0 r}}{4\pi r} 4jLh \frac{X\cos X}{\pi^2 - (2X)^2} \frac{\sin Z}{Z}
\end{aligned}
\tag{10.46}
$$

with the abbreviations

$$
\begin{aligned}
X = 0.5 \cdot k_0 L_{eff} \sin\vartheta \cos\varphi,\ Y &= 0.5 \cdot k_0 W \sin\vartheta \sin\varphi \\
\text{and\_}Z &= 0.5 \cdot k_0 h \cos\vartheta \,.
\end{aligned}
\tag{10.47}
$$

Two opposing slits are combined to form a group of two. For the in-phase group along the x-axis at a distance L_{eff} and the anti-phase group along the y-axis at a distance W, we obtain the group factors

$$
\begin{aligned}
F_1 &= 2\cos[0.5 \cdot k_0 L_{eff} \sin\vartheta \cos\varphi] = 2\cos X \\
F_2 &= 2\cos\frac{-\pi + k_0 W \sin\vartheta \sin\varphi}{2} = 2\sin Y \,.
\end{aligned}
\tag{10.48}
$$

Now we multiply (10.46) by the respective group factor (10.48)

$$
\begin{aligned}
\vec{F}_y F_1 &= -\vec{E}_0 \frac{e^{-jk_0 r}}{2\pi r} W h \cos X \frac{\sin Y}{Y} \frac{\sin Z}{Z}\,, \\
\vec{F}_x F_2 &= j\vec{E}_0 \frac{e^{-jk_0 r}}{2\pi r} L_{eff}/h \frac{4X\cos X}{\pi^2 - (2X)^2} \sin Y \frac{\sin Z}{Z} \,.
\end{aligned}
\tag{10.49}
$$

and obtain the far fields of a rectangular stripline antenna (for $\varepsilon_r = 1$):

$$\begin{aligned}\vec{E}_\vartheta &= Z_0\vec{H}_\varphi = -jk_0(-\vec{F}_xF_2\sin\varphi + \vec{F}_yF_1\cos\varphi)\,,\\ \vec{E}_\varphi &= -Z_0\vec{H}_\vartheta = -jk_0\cos\vartheta\,(\vec{F}_xF_2\cos\varphi + \vec{F}_yF_1\sin\varphi)\,.\end{aligned} \tag{10.50}$$

After inserting (10.49) and (10.50), the following finally follows

$$\begin{aligned}\vec{E}_\vartheta &= jk_0h\vec{E}_0\frac{e^{-jk_0r}}{2\pi r}\left(\frac{4jL_{eff}X}{\pi^2-(2X)^2}\sin\varphi + \frac{W}{Y}\cos\varphi\right)\cos X\sin Y\frac{\sin Z}{Z}\,,\\ \vec{E}_\varphi &= jk_0h\vec{E}_0\frac{e^{-jk_0r}}{2\pi r}\cos\vartheta\left(\frac{-4jL_{eff}X}{\pi^2-(2X)^2}\cos\varphi + \frac{W}{Y}\sin\varphi\right)\cos X\sin Y\frac{\sin Z}{Z}\,.\end{aligned} \tag{10.51}$$

In the vertical section in the E-plane at $\varphi = 0$ with $Y = 0$ and $X = 0.5\cdot k_0L_{eff}\sin\vartheta$ we get:

$$\begin{aligned}\vec{E}_\vartheta &= jWh\vec{E}_0\frac{e^{-jk_0r}}{\lambda_0 r}\cos[0.5\cdot k_0L_{eff}\sin\vartheta]\frac{\sin[0.5\cdot k_0h\cos\vartheta]}{0.5\cdot k_0h\cos\vartheta}\,,\\ \vec{E}_\varphi &= 0\,,\end{aligned} \tag{10.52}$$

while in the H-plane at $\varphi = \pi/2$ mit $X = 0$ and $Y = 0.5\cdot k_0W\sin\vartheta$ applies:

$$\begin{aligned}\vec{E}_\vartheta &= 0\,,\\ \vec{E}_\varphi &= jWh\vec{E}_0\frac{e^{-jk_0r}}{\lambda_0 r}\cos\vartheta\frac{\sin[0.5\cdot k_0W\sin\vartheta]}{0.5\cdot k_0W\sin\vartheta}\frac{\sin[0.5\cdot k_0h\cos\vartheta]}{0.5\cdot k_0h\cos\vartheta}\,,\end{aligned} \tag{10.53}$$

The main directions of radiation are perpendicular to the surface of the patch element at $\vartheta = 0$ and $\vartheta = \pi$. It is noticeable that the antiphase slits at $y = \pm W/2$ in the main sections do not contribute any radiation. This contribution also remains relatively small in other φ planes in the area around the main lobes, as $|X| << \pi L/\lambda_0 \cong 1$ applies there. This is why it is usually not dealt with in the literature. As the radiation field is proportional to h, it can be increased by increasing the substrate thickness. However, the secondary condition

$$h \leq \frac{0.3\lambda_0}{2\pi\sqrt{\varepsilon_r}} \tag{10.54}$$

must be taken into account. With greater thicknesses, the energy loss due to surface waves along the dielectric increases noticeably. At the cut-off frequency

$$f_c = \frac{nc_0}{4h\sqrt{\varepsilon_r-1}} \text{ with } n = 0, 1, 2, \ldots \tag{10.55}$$

surface waves become capable of propagation. (10.54) ensures that the lowest surface wave with $n = 0$ only has a low energy and no higher waves can propagate. At 10 GHz and $\varepsilon_r = 2.2$, the substrate height h should therefore be at most 0.97 mm.

So far, we have neglected the influence of the extended ground plane and the dielectric substrate. These effects can be taken into account by using an in-phase mirror image. Better accuracy can be achieved by multiplying the following correction factors by $\vec{E}_\vartheta$ or $\vec{E}_\varphi$ in (10.51):

$$\vec{F}_3 = \frac{2\cos\vartheta\sqrt{\varepsilon_r - \sin^2\vartheta}}{\sqrt{\varepsilon_r - \sin^2\vartheta} - j\varepsilon_r\cos\vartheta\cot\left(k_0 h\sqrt{\varepsilon_r - \sin^2\vartheta}\right)} \quad (\text{for } \vec{E}_\vartheta) \tag{10.56}$$

$$\vec{F}_4 = \frac{2\cos\vartheta}{\cos\vartheta - j\sqrt{\varepsilon_r - \sin^2\vartheta}\cot\left(k_0 h\sqrt{\varepsilon_r - \sin^2\vartheta}\right)} \quad (\text{for } \vec{E}_\varphi)\,. \tag{10.57}$$

We obtain the radiated power of the entire stripline antenna from $\vec{U}_0 = h\vec{E}_0$:

$$P_S = \frac{4}{2Z_0}\int_{\varphi=0}^{\pi/2}\int_{\vartheta=0}^{\pi/2}\left(\left|\vec{F}_3\vec{E}_\vartheta\right|^2 + \left|\vec{F}_4\vec{E}_\varphi\right|^2\right) r^2 \sin\vartheta\, d\vartheta\, d\varphi = \frac{1}{2}\left|\vec{U}_0\right|^2 G_S\,. \tag{10.58}$$

The following approximation applies to the radiation conductance G_S in A/V for $h/\lambda_0 \leq 0.02$:

$$G_S = \frac{1}{R_S} \cong \begin{cases} W^2/(45\lambda_0^2) & \text{for } W < 0.35\lambda_0 \\ W/(60\lambda_0) - 1/(30\pi^2) & \text{for } 0.35\lambda_0 < W < 2\lambda_0 \\ W/(60\lambda_0 & \text{for } 2\lambda_0 < W\,. \end{cases} \tag{10.59}$$

10.6 Array of Patch Antennas

An antenna array offers the possibility of reducing the aperture angle and increasing the antenna gain. Here, several possible array antennas were simulated on the basis of a single radiator (Fig. 10.11 and Fig. 10.12).

It can be seen that the antenna gain of 8.6 dBi in the main main beam direction is very good for a single element. However the aperture angle of 61° is quite large.

The number of elements in the array was set at 16 using the rule of thumb of 3 dB gain per doubling of the elements. This theoretically results in an increase in antenna gain of 12 dB.

In the following, a linear patch array with different distances between the elements was simulated.

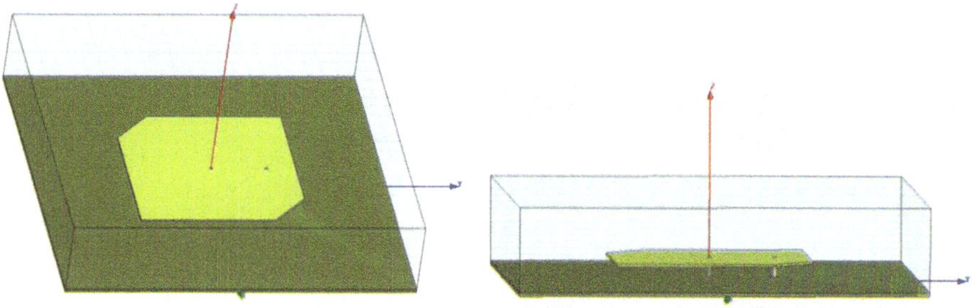

Fig. 10.11 3D model of the patch antenna: top view on the left, side view on the right with an edge length of approx. lambda-effective 2

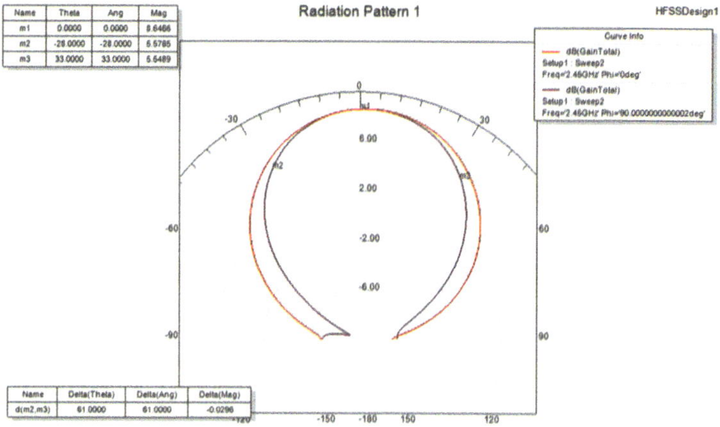

Fig. 10.12 Directional diagram of the patch antenna

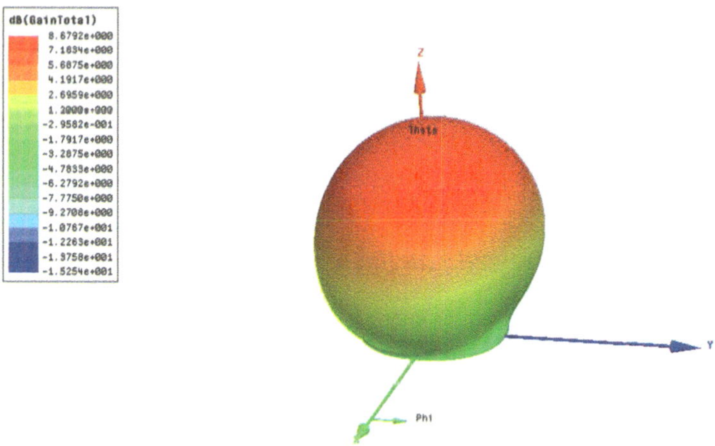

Fig. 10.13 3D plot of the directional diagram of the patch antenna

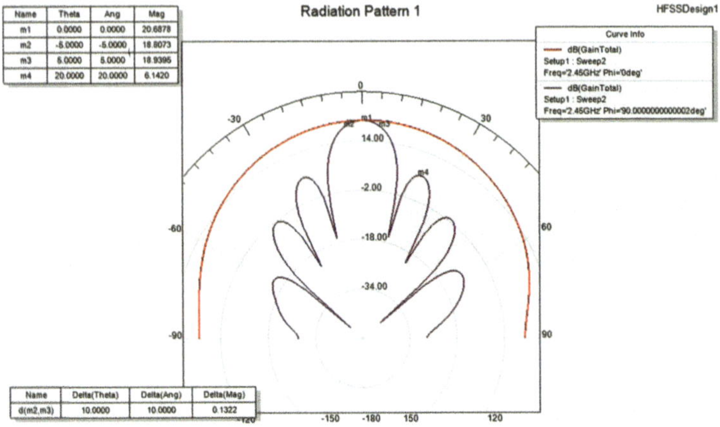

Fig. 10.14 Directional diagram of the linear patch array with lambda-quarter spacing and 16 elements

10.6.1 Linear Patch Array with Lambda-Quarter Spacing

HFSS can easily calculate a linear array from a simulated model of an antenna. The spacing of the radiating elements in this simulation is

$$\lambda/4 = \frac{3\cdot 10^8\,\text{m/s}}{4\cdot 2.45\,\text{GHz}} = 30.6\,\text{mm}\,. \tag{10.60}$$

The matching remains unchanged in the simulation, which is why only the plots for the directional diagram are given.

As can be clearly seen, an antenna gain of 20.6 dBi is obtained as expected. The aperture angle has also significantly narrower, it is now 10°. The emergence of the side lobes can be seen very clearly. The first sidelobe appears at 20° next to the main lobe and the sidelobe attenuation is approx. 14.5 dB. This means that the sidelobes can be classified as non-interfering (Fig. 10.13 and Fig. 10.14).

10.6.2 Linearly Arranged Patch Array with Lambda-Half Spacing

The spacing of the radiating elements in this simulation is

$$\lambda/2 = \frac{3\cdot 10^8\,\text{m/s}}{2\cdot 2.45\,\text{GHz}} = 61.2\,\text{mm}\,. \tag{10.61}$$

It is easy to see that the number of sidelobes has increased significantly compared to the linear patch array with $\lambda/4$ spacing, and the sidelobe suppression has also decreased, now amounting to approx. 13.5 dB. Furthermore, the first sidelobe is

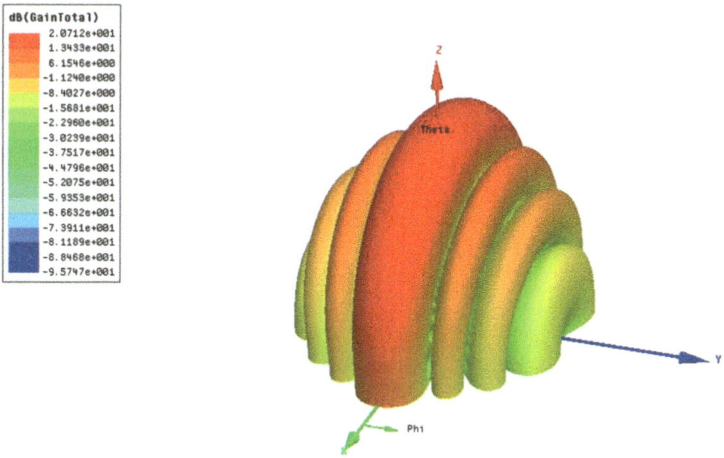

Fig. 10.15 3D plot of the directional diagram of the linear patch array with $\lambda/4$ spacing and 16 elements

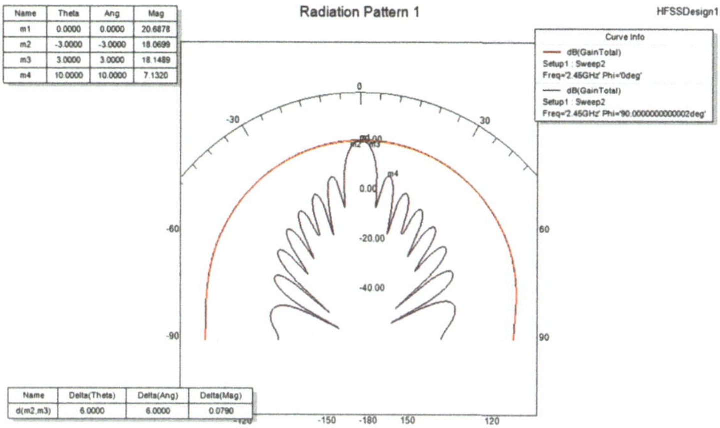

Fig. 10.16 Directional diagram of the linear patch array with $\lambda/2$ distances and 16 elements

now only 10° away from the main lobe. As expected, the opening angle has become narrower and is now 6°. The antenna gain is unchanged, as the number of elements has remained the same (Fig. 10.15, Fig. 10.16 and Fig. 10.17).

10.6.3 Linearly Arranged Patch Array with Three-Lambda Quarter Spacing

The spacing of the radiating elements in this simulation is

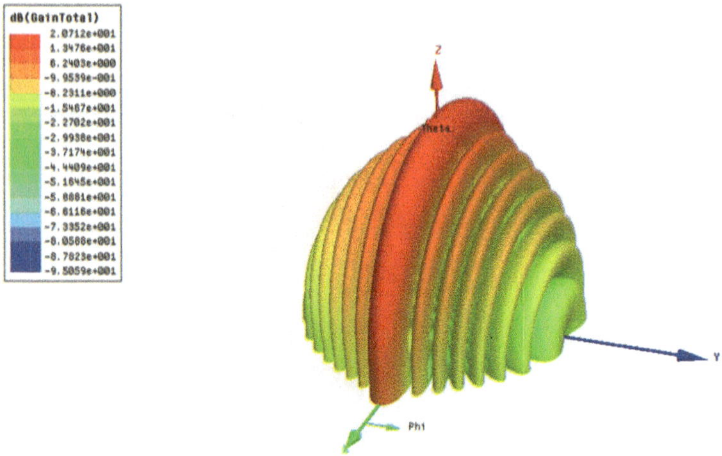

Fig. 10.17 3D plot of the directional diagram of the linear patch array with $\lambda/2$ spacing and 16 elements

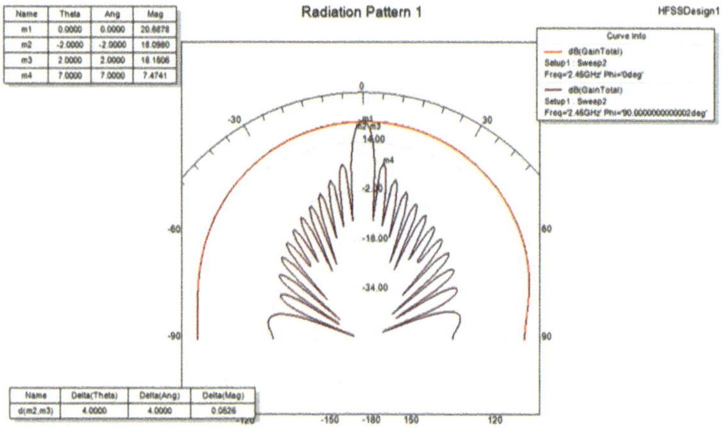

Fig. 10.18 Directional diagram of the linear patch array with $3\lambda/4$ spacing and 16 elements

$$3\lambda/4 = \frac{3 \cdot 3 \cdot 10^8\,\mathrm{m/s}}{4 \cdot 2.45\,\mathrm{GHz}} = 91.8\,\mathrm{mm}\,. \tag{10.62}$$

The sidelobe attenuation has become lower again and is now approx. 13.2 dB. As before, the first side lobe is now closer to the main lobe, it is only 7° away, which can lead to problems in distinguishing between main lobe and side lobe. The opening angle is now only 4° (Fig. 10.18 and Fig. 10.19).

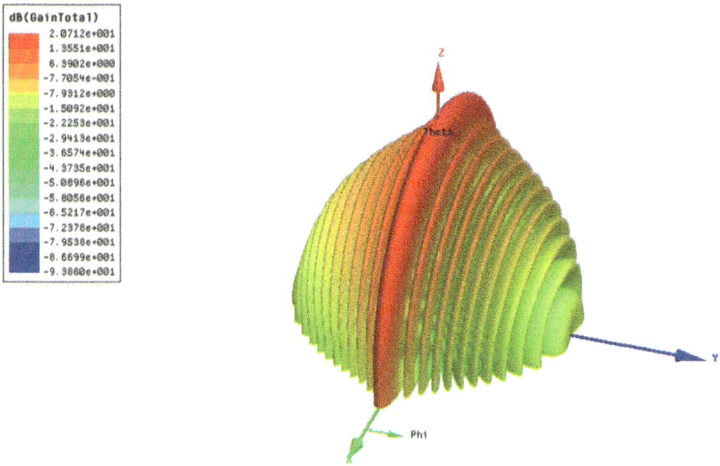

Fig. 10.19 3D plot of the directional diagram of the linear patch array with $3\lambda/4$ distances and 16 elements

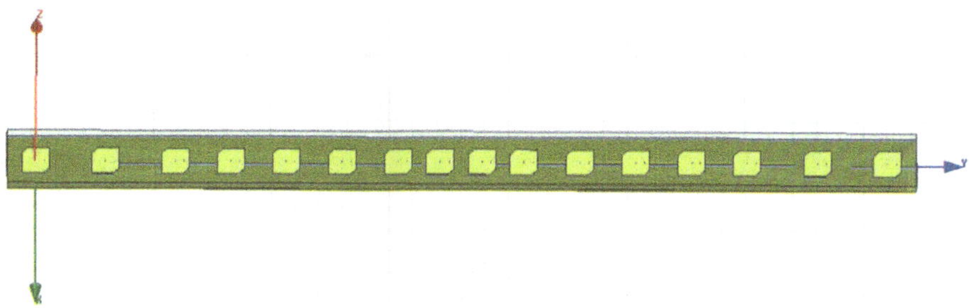

Fig. 10.20 Tapped patch array with different distances between the 16 elements

10.6.4 Tapped Arranged Patch Array

The greater the distance between the elements, the narrower the opening angle. However, this results in significantly more side lobes, which are also closer to the lie closer to the main lobe.

The aim is now to achieve a very narrow main lobe despite good sidelobe suppression. To achieve this, the elements are no longer distributed linearly, but are tapered. Namely in the range from $3\lambda/4$ to $\lambda/4$ (Fig. 10.20).

As can be seen, the spacing of the elements increases from the center of the antenna to the edge. The distances are $\lambda/4$, $\lambda/2$ and $3\lambda/4$.

The narrow main lobe and the well-suppressed secondary lobes quickly catch the eye. The aperture angle is still only 4°, the sidelobe attenuation reaches a good 17.5 dB and the antenna gain is 20.2 dBi (Figs. 10.21 and 10.22).

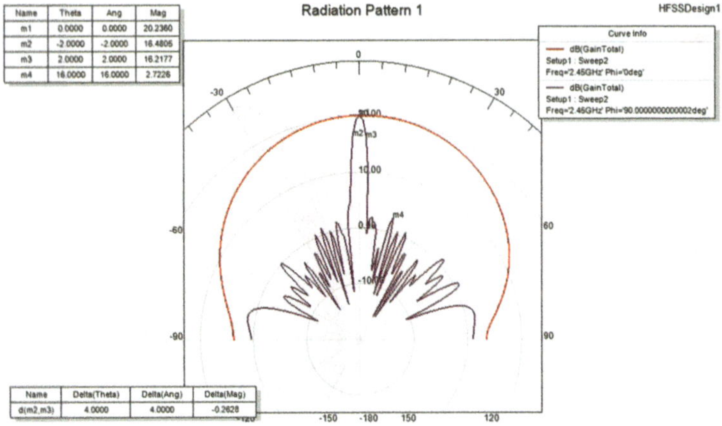

Fig. 10.21 Directional diagram of the tapered patch array

Fig. 10.22 3D plot of the directional diagram of the tapered patch array

10.7 Phase-Controlled Receiving Antennas

In principle, a phased array antenna works in the same way when receiving as when transmitting, provided that the phase shifters used are reciprocal. If amplifiers are used, they must of course have the correct gain sense. PIN diode phase shifters are practically always reciprocal, but ferrite phase shifters are often not.

If, in the case of reception, an arrangement is placed at the output of each individual radiating element that allows the received signal to be measured according to magnitude and phase (i.e. in principle network analyzers) and digitized, then the phase shift and subsequent summation can also be carried out purely mathematically. According to the state of the art, such a procedure is not yet fast enough for pulse radars and is therefore more suitable for frequency-modulated continuous wave radars.

10.8 Antenna Measurement Technology

The measurements of antennas help to make relative statements about the antenna parameters with regard to the opening angle, the gain gain, the forward and backward ratio and the directivity factor. Furthermore the measurements to compare and verify the antenna directivity values from the simulation. check.

10.8.1 The Measuring Device

The measurement to determine the aperture angle and the gain of the antenna is carried out using two antennas carried out. A first antenna is placed on the transmitting side and a second antenna is placed on the receiving side. The distance between the transmitting antenna and the receiving antenna can vary between 3 m and 10 m. As the antennas are reciprocal, the transmitting or receiving antenna can be freely selected. A third antenna is used as a reference antenna in order to obtain more accurate statement about the gain of the measured antenna

In order to minimize the reflection of the waves by the building walls and the ceiling, all measurements are carried out outdoors, e.g. on the roof of the university of applied sciences building see Fig. 10.23, or in absorption rooms.

There are two ways to measure the antenna parameters.

- **1. Transmission measurement:** The transmission parameter S_{21} is measured in dB using a vector network analyzer (e. g. Rohde and Schwarz, 300 kHz to 8.0 GHz, ZVM), the level difference between the two gates is read off the display.
- **2nd level measurement:** A signal is sent on one side using a signal generator (e.g. Rohde and Schwarz, 5 kHz to 3.0 GHz, SME 03). On the other side, the received signal is read with the aid of a spectrum analyzer (e. g. Rohde and Schwarz, 9 kHz to 3.0 GHz, FSP). The scalar transmission parameter S_{21} is determined here, with the level in dBm.

The radio field attenuation is measured. Radio field attenuation is the attenuation between the transmitting and receiving antenna (Fig. 10.24).

When transmitting data, the radiated power is expressed as effective radiated power. The effective radiated power (ERP) required when transmitting data is the transmission power emitted by a transmitting antenna in a specific direction compared to a half-wave dipole antenna. The gain of a half-wave dipole antenna refers to the gain of an isotropic antenna. The effective isotropic radiated power (EIRP) is determined from the isotropic antenna. The EIRP indicates the transmission power

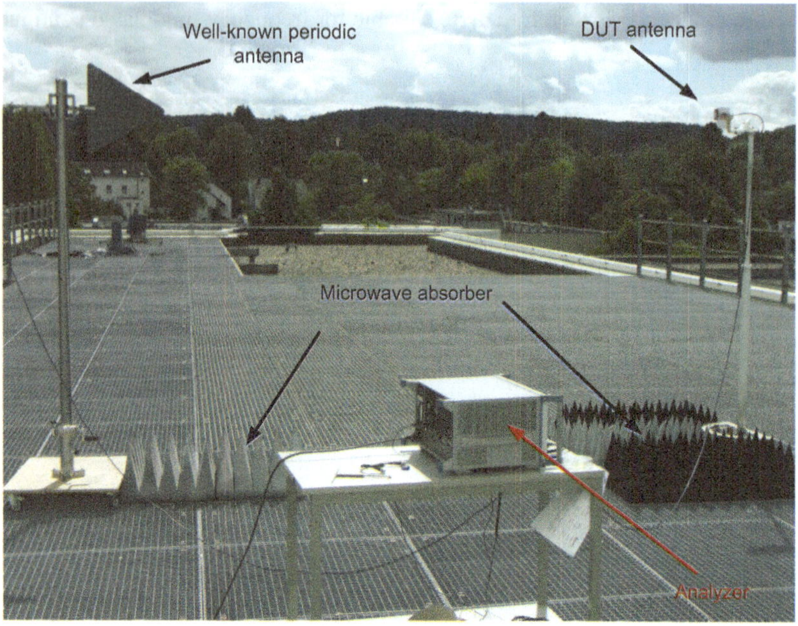

Fig. 10.23 Measuring device for the antennas on the roof of the FH Aachen building

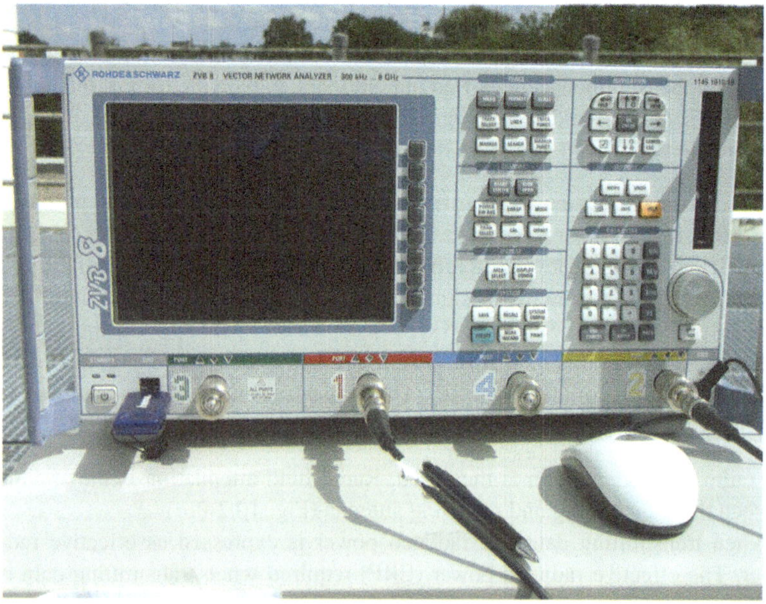

Fig. 10.24 Network analyzer (Rohde and Schwarz, 300 kHz to 8.0 GHz, ZVB 8) with four ports

that would have to be supplied to an isotropic antenna in order to generate the same field strength in the far field as a directional antenna in its main beam direction.

$$ERP = \frac{P_S \cdot 10^{\frac{ga}{10dBi}}}{L} = EIRP \cdot 0.610 \tag{10.63}$$

Calculation in dBm:

$$ERP = 10 \cdot log \frac{P_S}{1\,mW}[dBm] + Ga[dB] - 10 \cdot logL[dB] \tag{10.64}$$

with P_S as transmit power in watts, G_a as antenna gain of the measuring antenna in dBi and L as system losses (cable attenuation, connectors, passages, etc.).

Since ERP refers to a half-wave dipole, EIRP is greater than ERP by a factor of 1.64 (2.15 dB):

$$EIRP = \frac{P_t \cdot G_t}{L} = P_t \cdot G_{max} = ERP \cdot 1.64 \tag{10.65}$$

with P_t as transmit power in watts, G_t as antenna gain related to an isotropic antenna and G_{max} as maximum possible gain of the receiving antenna related to an isotropic antenna.

The regulations according to [ETSI] for antenna measurement can be summarized as follows

- The distance between the transmitting and receiving antennas should be 3 m or 10 m. The transmitter should be set to RF.
- The distance between the measuring device and the antenna should be at least 3 m.
- The measuring devices must be set to the measured frequency, the cable attenuation should be measured and taken into account before the measurement.
- The measurements should be carried out in a measuring chamber or in an anechoic plane.
- For the first measurement, the measured antenna is aligned horizontally and rotated by $\pm 180°$ with any step according to the azimuth angle φ in relation to the main direction of transmission. For the second measurement, the measured antenna is aligned vertically and the recording of the values is repeated as for the first measurement.
- During the measurement, the height must be varied so that the measurement at the receiver has the strongest signal.
- The transmitting antenna should be aligned vertically. To determine the determine the aperture angle, the measuring antenna is rotated by the azimuth angle (horizontal) until the half-value width (3 dB drop compared to the maximum level) is reached.

For all measurements carried out, the values are given with a measurement inaccuracy of typically $\pm 2dB$.

Chapter 11
Scattering of Electromagnetic Waves at Radar Targets

In general, the exact analytical description of the backscattering behavior of radar targets, such as ships and airplanes, is an intractable problem. However, many radar targets, such as raindrops or the level in a tank, can be analytically captured quite well by idealized targets, such as spheres or infinitely extended metallic planes. The aim of this chapter is to give a description of idealized radar targets.

For a description of the electromagnetic backscattering at radar targets, it is advisable to it is advisable to make a number of case distinctions. The first case (case I in Fig. 11.1), which will be considered here, concerns radar targets whose linear dimensions L are small compared to the width of the "radar beam", i.e.

$$L \ll \Delta\phi \, R \,. \tag{11.1}$$

Here $\Delta\phi$ is the half-width of the antenna lobe and R is the distance to the radar target (Fig. 11.1).

In the other limiting case, which will be considered later (case II), the radar target is large compared to the width of the antenna lobe.

11.1 General Consideration of the Scattering on a Single Radar Target

In the absence of the scattering body (radar target), the far field of an antenna in the vicinity of a point O_R is a plane homogeneous wave (Fig. 11.2).

The following should apply to the general consideration of the radar scattering body:

$$\lambda_0 \ll L \ll \Delta\phi \, R \,. \tag{11.2}$$

In the absence of the scattering body, this incident plane wave has the form (the index i stands for „incident“):

$$\vec{\mathbf{E}}_i = \left(E_{i1}\,\vec{\mathbf{e}}_1 + E_{i2}\,\vec{\mathbf{e}}_2\right)\, e^{-jk_0 x_i} \qquad \text{with}$$

$$E_{i1,2} = \frac{1}{R} e^{-jk_0 R} \sqrt{\left(\frac{Z_0}{2\pi}\right)}\, C_{1,2}(\theta,\varphi)\, a_0 \,. \tag{11.3}$$

H. Heuermann, *Microwave Technology*,
https://doi.org/10.1007/978-3-658-45686-3_11

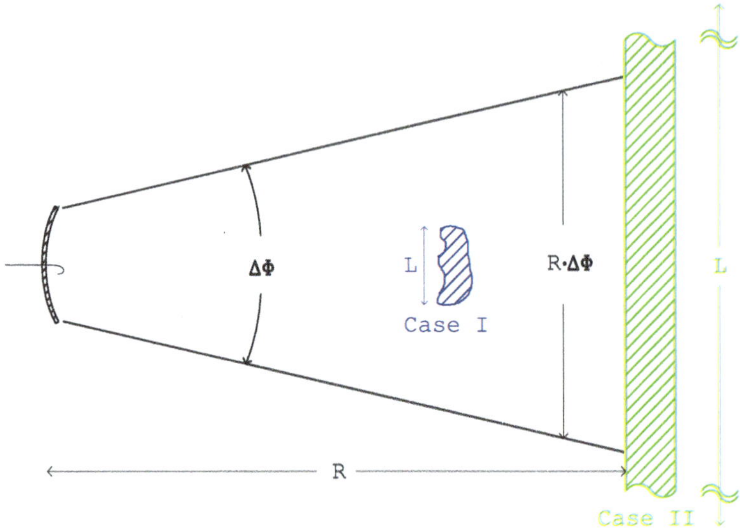

Fig. 11.1 Linear dimension of the radar target small (case I) or large (case II) compared to the width of the antenna lobe

Here $C_{1,2}(\theta, \varphi)$ is the transmission characteristic for polarization 1 or 2 and a_0 is the wave fed into the antenna.

If the scattering body is present, its backscattering behavior can be replaced by the induced equivalent electric and magnetic surface currents $\vec{\mathbf{J}}_s$ and $\vec{\mathbf{M}}_s$ in free space. The field $\vec{\mathbf{E}}_s$ (stray field) generated by these currents is superimposed on the radiation field of the antenna $\vec{\mathbf{E}}_i$ (incident field).

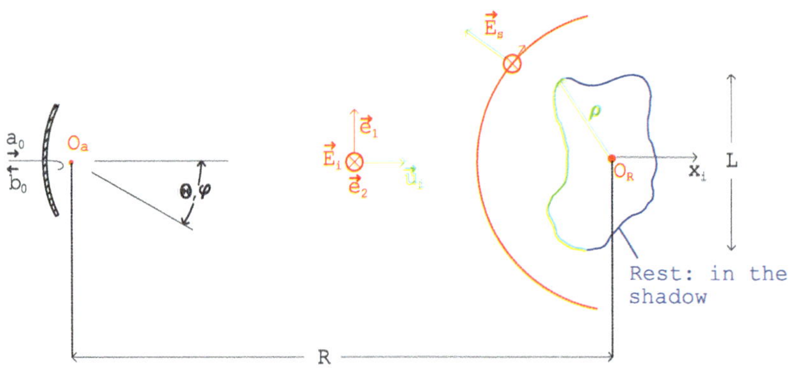

Fig. 11.2 Detailed representation of a radar target that is small compared to the incident plane homogeneous wavefront

The course of the induced currents as a function of the location on the surface of the scattering body is just such that the total field

$$\vec{\mathbf{E}}_{ges} = \vec{\mathbf{E}}_i + \vec{\mathbf{E}}_s \tag{11.4}$$

fulfills all boundary and continuity conditions. For example, for an ideally conducting metallic scattering body, the tangential component of $\vec{\mathbf{E}}_{ges}$ must disappear on the surface of the scattering body.

If the antenna is also located in the scatterer field of the radar target, i.e.

$$R > \frac{2\,L^2}{\lambda_0} \tag{11.5}$$

(no identical far-field conditions for the transmitting and scattering case), where L is the largest linear dimension of the scattering body, then the scattered field in the vicinity of the point O_a (see Fig. 11.2) can also be interpreted as a plane homogeneous wave in the absence of the antenna.

$$\vec{\mathbf{E}}_s = \left(E_{s1}\,\vec{\mathbf{e}}_1 + E_{s2}\,\vec{\mathbf{e}}_2\right) e^{-jk_0x_i} \tag{11.6}$$

Here, E_{s1} and E_{s2} result from the equivalent surface currents $\vec{\mathbf{J}}(\rho')$ and $\vec{\mathbf{M}}(\rho')$ according to Eqs. (9.38) and (9.39). Since the components E_{s1} and E_{s2} depend linearly on the fields E_{i1} and E_{i2}, it must be possible to formulate the following four-pole equation in general:

$$\begin{pmatrix} E_{s1} \\ E_{s2} \end{pmatrix} = \frac{e^{-jk_0R}}{2\,\sqrt{\pi}\,R} \begin{bmatrix} \gamma_{11} & \gamma_{12} \\ \gamma_{21} & \gamma_{22} \end{bmatrix} \begin{pmatrix} E_{i1} \\ E_{i2} \end{pmatrix} . \tag{11.7}$$

The elements γ_{11} to γ_{22} are referred to as scattering amplitudes and the corresponding matrix (marked with []) as radar scattering matrix. The choice of the prefactor in Eq. (11.7) will become plausible later.

It follows from the reciprocity theorem that $\gamma_{12} = \gamma_{21}$ must apply. The three complex scattering amplitudes γ_{11}, γ_{12} and γ_{22} contain the complete information about the backscattering at a radar target for a given frequency and a fixed aspect angle. They are independent of the distance R and the directivity $C_{1,2}$ of the antenna.

The first index of the scattering amplitudes characterizes the polarization of the incident wave and the second index that of the scattered wave.

Once this scattering matrix has been determined for two orthogonal polarizations (e.g. horizontal and vertical linear), the radar scattering matrix for any two other orthogonal polarizations (e.g. left-handed and right-handed circular) can be determined from the measurement result.

For the wave b_0 received by the antenna and incident on a matched load, Eqs. (9.93) and (9.110):

$$b_0 = \frac{\lambda_0}{2\,\sqrt{\pi}\,\sqrt{2Z_0}} \left(C_1(\theta,\varphi)\,E_{s1} + C_2(\theta,\varphi)\,E_{s2}\right) . \tag{11.8}$$

The directional characteristics for the transmitting case were also used for the receiving case, which are meant in the following when only the directional characteristic is mentioned. If E_{s1} and E_{s2} from Eq. (11.7) are inserted into this equation and then the expression from Eq. (11.3) is used for E_{i1} and E_{i2}, the result is (transmit and receive antenna should be identical):

$$b_0 = \frac{\lambda_0\, e^{(-j2k_0R)}}{8\,\pi\, R^2\,\sqrt{\pi}} \left(C_1^2\,\gamma_{11} + C_2^2\,\gamma_{22} + 2\,C_1\,C_2\,\gamma_{12}\right) a_0\,. \tag{11.9}$$

For the ratio of the received active power $|b_0|^2$ to the fed-in power $|a_0|^2$, i. h. for the insertion loss, the expression results:

$$\frac{|b_0|^2}{|a_0|^2} = \frac{\lambda_0^2}{(4\pi)^3\,R^4} \left|C_1^2\,\gamma_{11} + C_2^2\,\gamma_{22} + 2\,C_1\,C_2\,\gamma_{12}\right|^2. \tag{11.10}$$

After introducing the gain function $\tilde{G}(\theta,\varphi)$ from Eq. (9.79) and the radar cross-section σ one obtains the expression:

$$\frac{|b_0|^2}{|a_0|^2} = \frac{\lambda_0^2}{(4\pi)^3\,R^4}\,\tilde{G}^2(\theta,\varphi)\;\sigma \tag{11.11}$$

with the backscatter cross-section:

$$\sigma = \frac{\left|C_1^2\,\gamma_{11} + C_2^2\,\gamma_{22} + 2\,C_1\,C_2\,\gamma_{12}\right|^2}{\left(|C_1|^2 + |C_2|^2\right)^2}\,,$$

$$\text{resp.}\qquad \sigma = \frac{\left|\gamma_{11} + \left(\frac{C_2}{C_1}\right)^2\,\gamma_{22} + 2\,\frac{C_2}{C_1}\,\gamma_{12}\right|^2}{\left(1 + \left|\frac{C_2}{C_1}\right|^2\right)^2}\,. \tag{11.12}$$

The complex quantity C_2/C_1 characterizes the polarization of the radiated electromagnetic field. Thus, the quantity σ, which is referred to as radar cross section, backscattering area or backscattering cross section, depends both on the scattering amplitudes γ_{11} to γ_{22} and on the polarization. The expression in Eq. (11.12) is simplified for the important special cases of linear and circular polarization. For a linear polarization in the direction of the unit vector $\vec{\mathbf{e}}_1$, i. h. $C_2/C_1 = 0$, the radar cross-section σ is obtained:

$$\sigma = \left|\gamma_{11}\right|^2 \qquad \text{(linear polarization in the direction of } \vec{\mathbf{e}}_1\text{)}\,. \tag{11.13}$$

The same applies to linear polarization in the direction of $\vec{\mathbf{e}}_2$, and with this $C_2/C_1 \rightarrow \infty$,

$$\sigma = \left|\gamma_{22}\right|^2 \qquad \text{(linear polarization in the direction of } \vec{\mathbf{e}}_2) \,. \tag{11.14}$$

For a circular polarization, $C_2/C_1 = \pm j$ applies according to the clockwise or counterclockwise polarization, and the following is obtained for the backscatter cross-section

$$\sigma = \frac{1}{4}\left|\gamma_{11} - \gamma_{22} \pm j\,2\,\gamma_{12}\right|^2 \qquad \text{(circular polarization)}. \tag{11.15}$$

In order to arrive at a clear interpretation of the radar cross-section, the Eq. (11.11) can be rewritten with the help of the Equ. (9.111) and the radar equation is obtained:

$$\left|\,b_0\,\right|^2 = \frac{|a_0|^2}{4\pi R^2}\;\tilde{G}(\theta,\varphi)\;\;\sigma\;\;\frac{1}{4\,\pi\,R^2}\;\;\tilde{A}_w \,. \tag{11.16}$$

This equation can be interpreted as follows: The transmitted power $|a_0|^2$, divided over the spherical surface $4\pi R^2$ and increased by the gain $\tilde{G}$, results in a radiation density S_R at the location of the scattering body. According to the radar cross-section or the backscattering area σ, a power $S_R\sigma$ is scattered back isotropically in all spatial directions. This results in a radiation density $S_e = (S_R\sigma)/(4\pi R^2)$ at the location of the receiving antenna, which according to the effective area $\tilde{A}_w$ leads to an input power $|b_0|^2 = S_e\tilde{A}_w$ in the receiving antenna. It should be emphasized once again that a scattering body does not scatter back isotropically. This is taken into account in Eq. (11.16) by the fact that the scattering cross-section σ is a function of the aspect angle and the polarization. In addition, σ is generally frequency-dependent.

11.2 Calculation of the Radar Cross-Section of Metallic Objects

In this section, the previous generally treated `backscatter cross-section` σ for special metallic objects with a homogeneous incident wave will be calculated.

In order to make statements about the homogeneous incident wave

$$\vec{\mathbf{E}}_i = \left(E_{i1}\,\vec{\mathbf{e}}_1 + E_{i2}\,\vec{\mathbf{e}}_2\right)\,e^{-jk_0x_i} \tag{11.17}$$

$$\text{with} \qquad x_i = \vec{\mathbf{u}}_i \cdot \vec{\rho}$$

to find the radar backscattering caused by the radar, Maxwell's equations would have to be solved taking into account the boundary conditions of an ideally conducting metallic scattering body, i.e. under the boundary condition that the tangential component of the electric field must disappear. This means

$$\vec{\mathbf{E}} \times \vec{\mathbf{n}} = \vec{\mathbf{0}} \qquad \text{(perfect conductor)} \,. \tag{11.18}$$

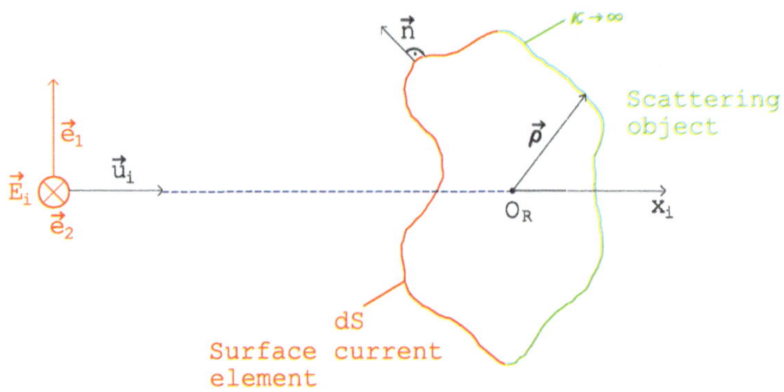

Fig. 11.3 Coordinates for the incident wave and the scattering body

If the solution of the electromagnetic field including backscattering is known, the stray body can be replaced by equivalent surface currents in free space (see section 2.3).

The following applies to the equivalent magnetic surface currents according to Eq. (9.44) $\vec{\mathbf{M}}_s = -\vec{\mathbf{n}} \times \vec{\mathbf{E}}$ and therefore due to Eq. (11.18) $\vec{\mathbf{M}}_s = \vec{\mathbf{0}}$. This therefore leaves the electric surface current according to Eq. (9.43) with

$$\vec{\mathbf{J}} = \vec{\mathbf{n}} \times \vec{\mathbf{H}}\,. \tag{11.19}$$

Fig. 11.3 explains the choice of the two polarization directions $\vec{\mathbf{e}}_1$ and $\vec{\mathbf{e}}_2$ and the coordinates.

For the electric stray field components E_{s1} and E_{s2} from Eq. (11.6), one can use Eq. (9.40) taking into account that no components occur in the direction of propagation $(\vec{\mathbf{J}}(\vec{\rho}) \cdot \vec{\mathbf{u}}_r)\, \vec{\mathbf{u}}_r = \vec{\mathbf{0}}$ and that the boundary condition of an ideal metallic surface must be fulfilled ($\vec{\mathbf{M}}(\vec{\rho}) = \vec{\mathbf{0}}$), also write:

$$E_{s1,2} = \frac{e^{-jk_0R}}{2\sqrt{\pi}\,R} \left\{ \frac{-j\sqrt{\pi}\,Z_0}{\lambda_0} \iint_S \vec{\mathbf{J}}(\vec{\rho}) \cdot \vec{\mathbf{e}}_{1,2}\; e^{-jk_0x_i}\, dS \right\} \tag{11.20}$$

$$\text{with} \qquad x_i = -\vec{\mathbf{u}}_r \cdot \vec{\rho}\,.$$

A comparison with the definition equation, Eq. (11.7), provides for the scattering amplitudes γ_{11} to γ_{22}:

$$\gamma_{11} = -j\,\frac{\sqrt{\pi}}{\lambda_0}\, Z_0 \iint_S \left(\vec{\mathbf{J}} \cdot \vec{\mathbf{e}}_1\right) e^{-jk_0x_i}\, dS\,,$$

$$\gamma_{21} = -j\,\frac{\sqrt{\pi}}{\lambda_0}\, Z_0 \iint_S \left(\vec{\mathbf{J}} \cdot \vec{\mathbf{e}}_2\right) e^{-jk_0x_i}\, dS\,,$$

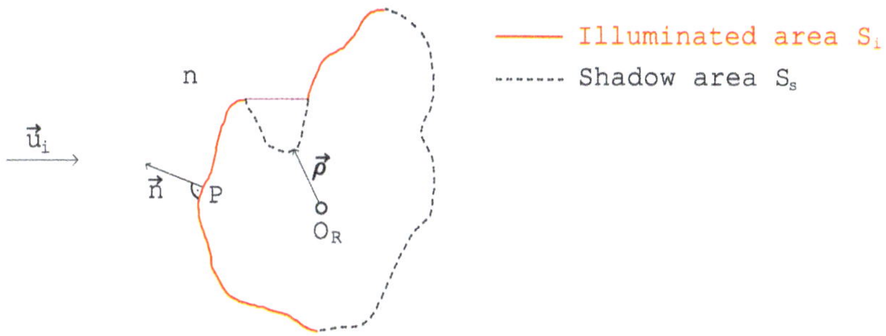

Fig. 11.4 Explanation of the distribution of space flows

with $\vec{\mathbf{J}}$ for $E_{i1} = 1$ and $E_{i2} = 0$,

$$\gamma_{12} = -j\frac{\sqrt{\pi}}{\lambda_0} Z_0 \iint_S \left(\vec{\mathbf{J}} \cdot \vec{\mathbf{e}}_1\right) e^{-jk_0 x_i} \, dS \, ,$$

$$\gamma_{22} = -j\frac{\sqrt{\pi}}{\lambda_0} Z_0 \iint_S \left(\vec{\mathbf{J}} \cdot \vec{\mathbf{e}}_2\right) e^{-jk_0 x_i} \, dS \, , \tag{11.21}$$

with $\vec{\mathbf{J}}$ for $E_{i1} = 0$ and $E_{i2} = 1$.

For metallized scatterers with dimensions that are large in relation to the wavelength, the proportion originating from the front of the scatterer dominates. This can be determined approximately with the help of "physical optics". For this purpose, the surface S is first divided into "illuminated areas" S_i and "shadow areas" S_s (see Fig. 11.4).

In the shadow areas, the equivalent current is set approximately to zero:

$$J \approx 0 \qquad \text{für} \qquad \vec{\rho} \subset S_s \, . \tag{11.22}$$

In the points of the illuminated areas, the surface current $\vec{\mathbf{J}}$ is set equal to the surface current that would flow in the respective point P if it were located on an infinitely extended flat plate that has $\vec{\mathbf{n}}(P)$ as a normal vector. For such a flat plate, the tangential component of the total magnetic field is just twice as large as the tangential component of the incident field:

$$\vec{\mathbf{n}} \times \vec{\mathbf{H}} = 2\,\vec{\mathbf{n}} \times \vec{\mathbf{H}}_i \qquad (\text{for flat plate}) \, . \tag{11.23}$$

This results in the approximate assumption made in the context of "physical optics" for the equivalent currents on the illuminated surface of the metallic scattering body

$$\vec{\mathbf{J}} = \vec{\mathbf{n}} \times \vec{\mathbf{H}} \approx 2\,\vec{\mathbf{n}} \times \vec{\mathbf{H}}_i \qquad \text{for} \qquad \vec{\rho} \in S_i \, . \tag{11.24}$$

The incident plane homogeneous wave according to Eq. (11.17) includes the magnetic field

$$\vec{\mathbf{H}}_i = \frac{1}{Z_0}\left(\vec{\mathbf{e}}_1\, E_{i2} - \vec{\mathbf{e}}_2\, E_{i1}\right) e^{-jk_0 x_i} \tag{11.25}$$

and thus the approximate solution for the equivalent area flow according to Eq. (11.24)

$$\vec{\mathbf{J}} \approx \frac{2}{Z_0}\left[E_{i2}\,(\vec{\mathbf{n}} \times \vec{\mathbf{e}}_1) - E_{i1}\,(\vec{\mathbf{n}} \times \vec{\mathbf{e}}_2)\right] e^{-jk_0 x_i} \qquad (\text{for } S_i)\,. \tag{11.26}$$

If this current is inserted into the Eq. (11.21) for the scattering amplitudes, then the following follows due to

$$\vec{\mathbf{J}} \cdot \vec{\mathbf{e}}_1 = -2\frac{E_{i1}}{Z_0}\;\vec{\mathbf{n}} \cdot \vec{\mathbf{u}}_i\, e^{-jk_0 x_i}$$

and

$$\vec{\mathbf{J}} \cdot \vec{\mathbf{e}}_2 = -\,2\,\frac{E_{i2}}{Z_0}\;\vec{\mathbf{n}} \cdot \vec{\mathbf{u}}_i\;\, e^{-jk_0 x_i} \tag{11.27}$$

as the result for the scattering amplitudes:

$$\gamma_{11} = \gamma_{22} = \gamma_0 = j\,2\,\frac{\sqrt{\pi}}{\lambda_0} \iint\limits_{S_i} \left(\vec{\mathbf{n}} \cdot \vec{\mathbf{u}}_i\right) e^{(-j2k_0 x_i)}\, dS\,, \tag{11.28}$$

$$\gamma_{12} = \gamma_{21} = 0\,. \tag{11.29}$$

The integral is only to be extended over the illuminated area.

This results in a scattering matrix of the form:

$$\left[\gamma\right] = \begin{bmatrix} \gamma_0 & 0 \\ 0 & \gamma_0 \end{bmatrix} \tag{11.30}$$

with γ_0 according to Eq. (11.28). This approximation result means that in the case of a linearly polarized incident wave, the backscattered wave has the same polarization. Furthermore, the amplitude of the backscattered wave is independent of the direction of polarization.

This approximate result for single reflection at the front is confirmed by the results of more precise calculations and by measurement results as long as the front side is free of "edges", i.e. as long as the radii of curvature are large against the wavelengths under consideration. The approximation under consideration will also be quite good if the area of the scattering body is large or if the amplitude of the surface current decreases towards the edges.

In the approximation considered, $\gamma_{12} = \gamma_{21} = 0$. This means that when a linearly polarized wave is incident, there is no cross-polar component in the reflection. Furthermore, $\gamma_{11} = \gamma_{22} = \gamma_0$. This means that for the backscattering cross-section

$\sigma = \sigma_{zir}$ of the circular polarization according to Eq. (11.15), both for the left and right circular polarization:

$$\sigma = \sigma_{zir} = \frac{1}{4} \left| \gamma_{11} - \gamma_{22} \pm j\, 2\, \gamma_{12} \right|^2 = 0\,. \tag{11.31}$$

Eq. (11.31) states that the backscattering cross section for circular polarization, σ_{zir}, is zero. However, it should be noted that Eq. (11.15) assumes that the transmitting and receiving antennas are identical. If, on the other hand, we use a receiving antenna that can receive the cross-polarization, i.e. the circular polarization with the opposite direction of rotation, then

$$\frac{C_2}{C_1} = \pm\, j\,,$$

$$\text{aber} \qquad \frac{D_1}{D_2} = \mp\, j\,. \tag{11.32}$$

This gives the scattering cross-section of the cross-polar component σ_{krz}:

$$\sigma_{krz} = \frac{\left|\gamma_{11} + \gamma_{22}\right|^2}{4} = \left|\ \gamma_0\ \right|^2 . \tag{11.33}$$

The backscattering cross-section for the cross-polar component is therefore just as large as for a linear polarization.

It should also be mentioned that a scattering body that does not meet the requirements of this section, namely metallic conductivity and smoothly varying geometry, will generally generate a cross-polar polarization. In particular, cross polarization can occur at mixed dielectric-metallic objects and edges. On the other hand, any object with rotational symmetry will not cause a transformation into cross-polarized components.

Eq. (11.28) can be evaluated analytically for some simple scattering body geometries.

11.2.1 Reflection at Incidence of a Plane Homogeneous Wave on a Conducting Plane Plate

We first consider a circular flat plate (Fig. 11.5). The integrand in Eq. (11.28) is the same for every point on the front of the plate. With $x_i = 0$ and $\vec{\mathbf{n}} \cdot \vec{\mathbf{u}}_i = -1$ the following applies to the scattering amplitude γ_0:

$$\gamma_0 = \frac{j\, 2\, \sqrt{\pi}}{\lambda_0} \iint\limits_{S_i} \left(\vec{\mathbf{n}} \cdot \vec{\mathbf{u}}_i\right)\, e^{(-j2k_0 x_i)}\, ds\,, \tag{11.34}$$

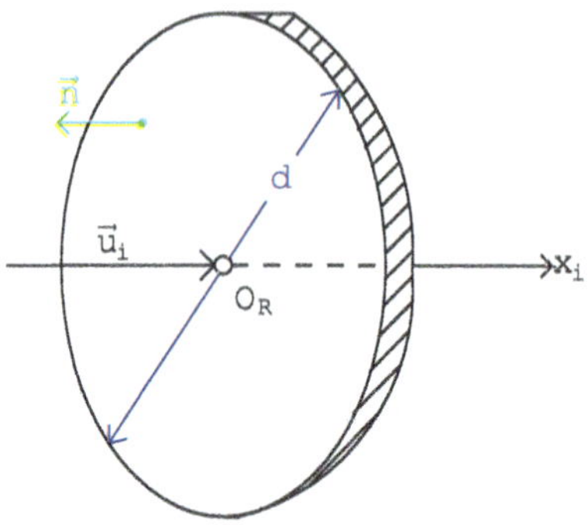

Fig. 11.5 Vertical incidence on a flat plate

$$\gamma_0 = \frac{j\,2\,\sqrt{\pi}}{\lambda_0} \int_0^{2\pi}\int_0^{r} (-1)\; e^0\; r'\, dr'\, d\varphi \tag{11.35}$$

and finally

$$\gamma_0 = -\frac{j\,2\,\sqrt{\pi}}{\lambda_0}\;\frac{\pi}{4}\; d^2\,. \tag{11.36}$$

For $\lambda_0 \ll d$, this scattering contribution dominates over all other scattering contributions, and the radar cross-section is obtained if A is the area of the disk:

$$\sigma_{lin} = |\gamma_0|^2 = \frac{4\,\pi}{\lambda_0^2}\left(d^2\,\frac{\pi}{4}\right)^2 = \frac{4\,\pi}{\lambda_0^2}\,A^2, \quad \sigma_{zir} = 0\,. \tag{11.37}$$

The relationship (11.37) also applies in general if A is the area of an arbitrarily shaped flat plate. If the flat plate is tilted by an angle $\Delta\beta$, the scattering amplitude changes in a similar way to the directional characteristic of a flat antenna (Fig. 11.6).

We had seen that the directional characteristic results from the Fourier transform of the occupancy function. The contributions resulting from the electric and magnetic surface current density were almost the same (Eq. 9.48). For the plane conductive scattering body that we consider in this section, there is only one electrical surface current density, but with doubled amplitude. Note, however, that the inclination of the flat plate by an angle $\Delta\beta$ already leads to a linearly increasing phase of the equivalent surface current density. This phase factor in the occupancy function causes the main backscattering direction to be tilted by another $\Delta\beta$ relative to the normal direction (Eq. 9.70). This is the reflection known from optics mirror reflection with angle of incidence equal to angle of reflection. If the plate is tilted by $\Delta\beta$, the angle between the direction of incidence and the main backscattering direction is already $2\Delta\beta$. A rectangular plate, for example, has a constant surface current

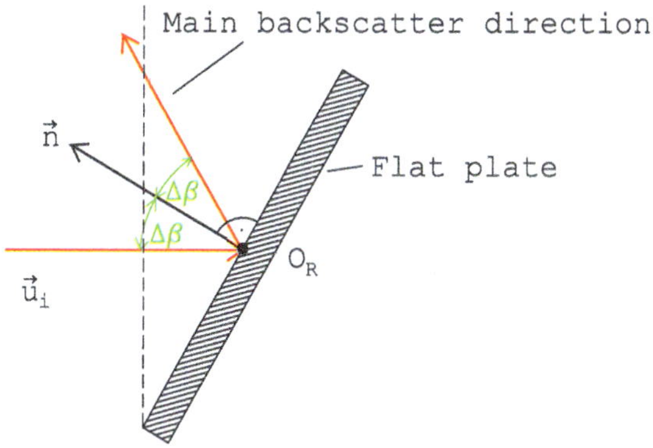

Fig. 11.6 Reflection on an inclined flat plate of ideal conductivity

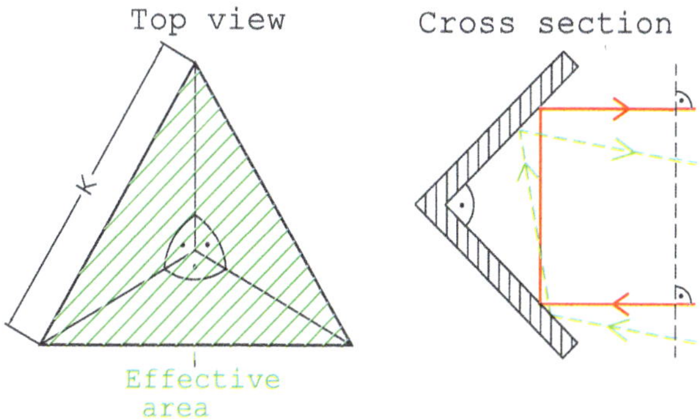

Fig. 11.7 Top view of and lateral section through the triple mirror

distribution, and thus the backscatter cross-section σ over the angle of rotation has the course of a si^2 function.

A so-called corner-reflector is popular for metrological applications. Like the corner of a room, this is made up of three metallic triangular surfaces (Fig. 11.7).

This triple mirror has the geometrical-optical property that an incident beam is reflected parallel to itself itself, regardless of the angle of incidence $\Delta\beta$ (principle of the cat's eye).

For an angle of incidence of approximately $\Delta\beta \leq 35^\circ$, a triple mirror of this type has the almost angle-independent area of the shaded area in the image 11.7 with the size of

$$A = \frac{\sqrt{3}}{4}\ k^2\,. \tag{11.38}$$

Taking Eq. (11.37) into account, the backscattering area is obtained for an angled mirror

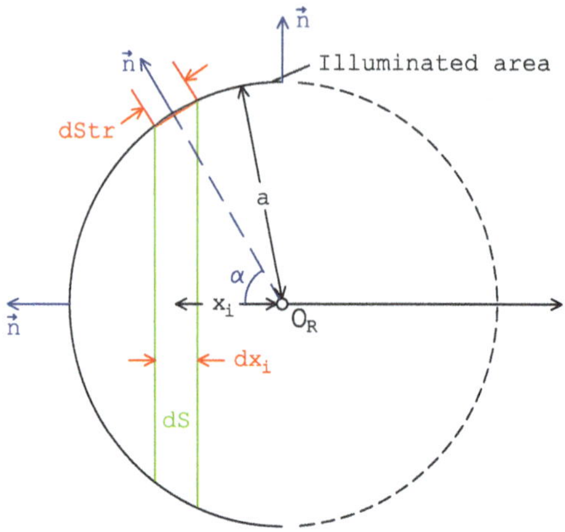

Fig. 11.8 Section through a sphere to calculate the backscatter cross-section

$$\sigma = \frac{\pi}{\lambda_0^2} \; \frac{3}{4} \; k^4 \tag{11.39}$$

for a linearly polarized incident wave.

11.2.2 Reflection on the Front of a Metallic Sphere

The integrand in Eq. (11.28) is also constant within a ring with $x_i = \text{const.}$ (Fig. 11.8).

The area dS of a small ring on the surface of the sphere is calculated from the mean circumference of the ring and its width $dStr$:

$$dS = 2\,\pi\,\sqrt{a^2 - x_i^2}\; dStr\,. \tag{11.40}$$

The width of the ring is obtained from the ratio:

$$\frac{dStr}{dx_i} = \frac{a}{\sqrt{a^2 - x_i^2}}\,. \tag{11.41}$$

This results in the following for the ring area dS:

$$dS = 2\,\pi\,a\;\,dx_i\,. \tag{11.42}$$

Since x_i is less than zero in the integration range, the following still applies

$$\vec{\mathbf{n}} \cdot \vec{\mathbf{u}}_i = \frac{x_i}{a}\,. \tag{11.43}$$

Tab. 11.1 Backscatter cross-sections of different geometries

Scattering body	Scattering cross-section
Sphere	$\sigma = \pi\, a^2$
Circular cylinder	$\sigma = 2\,\pi\, \frac{a\,L}{\lambda_0}$
Flat plate	$\sigma = 4\,\pi\, \frac{A^2}{\lambda_0^2}$

To determine the scattering amplitude of the sphere, the area in Eq. (11.28) must be integrated in the range of $-a \ll x_i \ll 0$:

$$\gamma_0 = \frac{j\,2\,\sqrt{\pi}}{\lambda_0} \int_{-a}^{0} \frac{x_i}{a}\, e^{(-j2k_0x_i)}\; 2\,\pi\, a\, dx_i\,, \tag{11.44}$$

$$\gamma_0 = \frac{j\,2\,\sqrt{\pi}}{\lambda_0}\; 2\,\pi\, \left\{ \frac{j\,a}{2\,k_0}\, e^{j2k_0a} + \frac{1}{4\,k_0^2}\, \left(1 - e^{j2k_0a}\right) \right\}. \tag{11.45}$$

If the diameter of the sphere is large compared to the wavelength ($a \gg \lambda_0$), the first summand in the curly bracket in Eq. (11.45) predominates and the result is obtained:

$$\gamma_0 = -a\sqrt{\pi}\; e^{j\,2\,k_0\,a}\,. \tag{11.46}$$

This results in a backscatter cross-section σ of the sphere:

$$\sigma_{lin} = |\,\gamma_0\,|^2 \pi\, a^2;\; \sigma_{zir} = 0\,. \tag{11.47}$$

This means that the backscattering cross-section is equal to the maximum cross-sectional area of the sphere and is independent of the frequency.

If, on the other hand, the sphere diameter d is much smaller than the wavelength ($a \ll \lambda_0$), the following applies for $\varepsilon_r \to \infty$ derived from Eq. (11.59):

$$\sigma_{lin} = frac\; \pi^5\, \lambda_0^4\; d^6; \qquad \sigma_{zir} = 0\,, \tag{11.48}$$

as will be shown in Tab. 11.1.

Tab. 11.1 summarizes the backscatter cross-sections of a flat plate, a cylinder (see Fig. 11.9) and a sphere for the case where the geometries are much larger than the wavelength.

The different λ_0 dependence of the radar cross-section is particularly noteworthy in these results. While for the sphere the value of σ is constant at high frequencies, for the cylinder it increases proportionally to the frequency f ($\sim 1/\lambda_0$) and for the flat plate and vertical incidence even proportionally f^2.

This different behavior can be understood if one considers the size of the scattering body not only in the direction of the emitter, but in all spatial directions ("scattering diagram"). Whereas with the metal sphere the scattering pattern becomes independent of the frequency at high frequencies, the focusing property of the plate increases steadily with frequency.

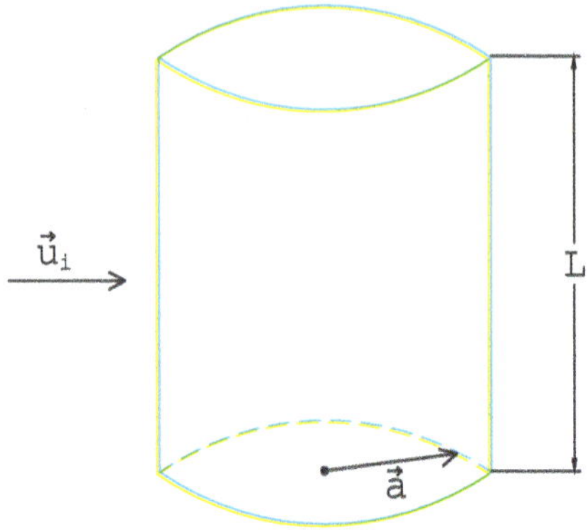

Fig. 11.9 Circular cylinder and its sizes for calculation of the backscatter cross-section

11.3 Scattering by Dielectric Bodies that are Small Against the Wavelength (Rayleigh Scattering)

In radar technology backscattering from targets composed of many small, spatially separated dielectric particles plays an important role. An example of this is a precipitation field that is composed of many individual raindrops . In preparation for such scattering processes, this section will deal with scattering from a single dielectric particle whose dimensions are small compared to the wavelength.

In this case, the scattering body can be thought of as being replaced by an electric elementary dipole in terms of its scattering behavior. Its dipole moment is obtained as a solution to an electrostatic boundary value problem. The results are particularly simple if the scattering body has a spherical shape. As this case is also of practical importance, it will be dealt with in more detail here.

When a non-magnetic sphere with the relative dielectric constant ε_r and the diameter d is placed in an electrostatic field, electrostatic charges are influenced at the boundary surface, which lead to a dipole moment p.

Before the introduction of a sphere (Fig. 11.10), the homogeneous incident field may lie in the z-direction. In spherical coordinates, we can write for this E_z component:

$$\vec{\mathbf{E}}_i = E_z\,\vec{\mathbf{u}}_z = E_z\left(\cos\vartheta\,\vec{\mathbf{u}}_r - \sin\vartheta\,\vec{\mathbf{u}}_\vartheta\right)\,. \tag{11.49}$$

It is known that after placing a dielectric sphere in a homogeneous electric field, the field inside the sphere still only has a z component. However, the field is smaller by a factor k than in the outer space. In the outer space, a dipole field with the dipole moment p (Eq. 9.11), whose value is obtained from the boundary conditions to be fulfilled, is superimposed on the homogeneous incident field.

For the interior with $r \leq d/2$ the following applies

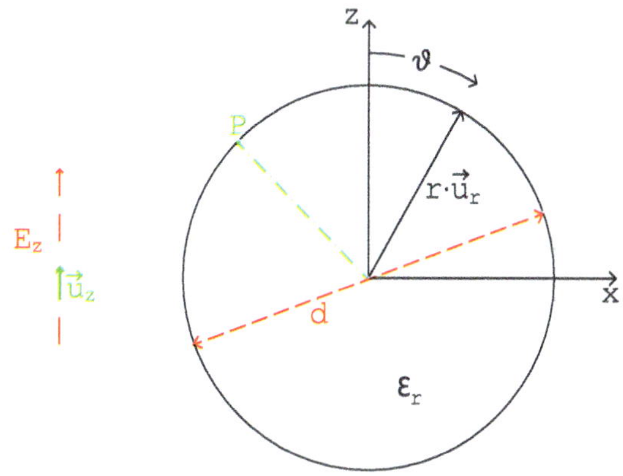

Fig. 11.10 Dielectric sphere in a homogeneous electromagnetic field

$$\vec{\mathbf{E}} = k\,E_z\,\vec{\mathbf{u}}_z = k\,E_z\,\left(\cos\vartheta\,\vec{\mathbf{u}}_r - \sin\vartheta\,\vec{\mathbf{u}}_\vartheta\right)\;,$$

and for the outer space with $r \geq d/2$ applies

$$\vec{\mathbf{E}} = E_z\,\vec{\mathbf{u}}_z + \frac{p}{4\,\pi\,\varepsilon_0\,r^3}\,\left(\sin\vartheta\,\vec{\mathbf{u}}_\vartheta + 2\,\cos\vartheta\,\vec{\mathbf{u}}_r\right)\;. \tag{11.50}$$

The continuity of the tangential component of the electric field on the surface of the sphere requires

$$E_\vartheta\left(\frac{d}{2}-0\right) = E_\vartheta\left(\frac{d}{2}+0\right) \tag{11.51}$$

and thus

$$-k\,E_z\,\sin\vartheta = \frac{p}{4\,\pi\,\varepsilon_0\,(d/2)^3}\,\sin\vartheta\;-\;E_z\,\sin\vartheta$$

or

$$E_z\,(1-k) = \frac{p}{4\,\pi\,\varepsilon_0\,(d/2)^3}\;. \tag{11.52}$$

In addition, the normal component of the electrical displacement on the surface of the sphere must be continuous:

$$\varepsilon_r\,k\,E_z\,\cos\vartheta = \frac{p}{4\,\pi\,\varepsilon_0\,(d/2)^3}\,2\,\cos\vartheta + E_z\,\cos\vartheta\;,$$

$$E_z\,(\varepsilon_r\,k-1) = \frac{2\,p}{4\,\pi\,\varepsilon_0\,(d/2)^3}\;. \tag{11.53}$$

The Eqs. (11.51) and (11.52) can be solved for k and p with the result:

$$k = \frac{3}{\varepsilon_r + 2} \,,$$

and

$$p = \varepsilon_0 \pi \frac{d^3}{2} \; \frac{(\varepsilon_r \, - \, 1)}{(\varepsilon_r \, + \, 2)} \; E_z \, . \tag{11.54}$$

In the sense of a "quasi-stationary" consideration, the result for the case $D/\sqrt{\varepsilon_r} \ll \lambda_0$ can also be transferred to alternating fields. We then obtain the following for the phasor of the induced time-harmonic dipole moment

$$\vec{\mathbf{p}} = \varepsilon_0 \; \frac{(\varepsilon_r \, - \, 1)}{(\varepsilon_r \, + \, 2)} \; \pi \; \frac{d^3}{2} \; \left(E_{i1} \, \vec{\mathbf{e}}_1 \, + \, E_{i2} \, \vec{\mathbf{e}}_2 \right) \, . \tag{11.55}$$

In Eq. (11.55), E_{i1} and E_{i2} are the components of the incident field at the location of the scattering body. The result Eq. (11.55) is also valid if the relative dielectric constant ε_r is complex due to dielectric losses. The result can be extended to a metallic sphere.

The dipole moment $\vec{\mathbf{p}}$ is associated with

$$I \, \vec{\ell} = j \, \omega \, \vec{\mathbf{p}} = j \, \frac{2 \, \pi \, c}{\lambda_0} \, \vec{\mathbf{p}} \tag{11.56}$$

a scattered field which is given at the location of the transmitter according to Eq. (9.19) with $r = R$ and $\vartheta = \pi/2$ by the following expression:

$$\vec{\mathbf{E}}_s = - \, \frac{\pi^2}{2} \; \frac{d^3}{R \, \lambda_0^2} \; \frac{(\varepsilon_r \, - \, 1)}{(\varepsilon_r \, + \, 2)} \; e^{-jk_0R} \; (E_{i1} \, \vec{\mathbf{e}}_1 \, + \, E_{i2} \, \vec{\mathbf{e}}_2) \, . \tag{11.57}$$

A comparison with Eq. (11.7) provides the following results for the scattering amplitudes

$$\gamma_{11} = \gamma_{22} = - \, \pi^{\left(\frac{5}{2}\right)} \; \frac{(\varepsilon_r \, - \, 1)}{(\varepsilon_r \, + \, 2)} \; \frac{d^3}{\lambda_0^2}$$

and

$$\gamma_{12} = \gamma_{21} = 0 \, . \tag{11.58}$$

For the radar cross-section of linear polarization, the following thus follows

$$\sigma = \left| \, \gamma_{11} \, \right|^2 = \left| \frac{(\varepsilon_r \, - \, 1)}{(\varepsilon_r \, + \, 2)} \right|^2 \; \pi^5 \; \frac{d^6}{\lambda_0^4} \, . \tag{11.59}$$

The scattering cross-section for the circular polarization is again zero.

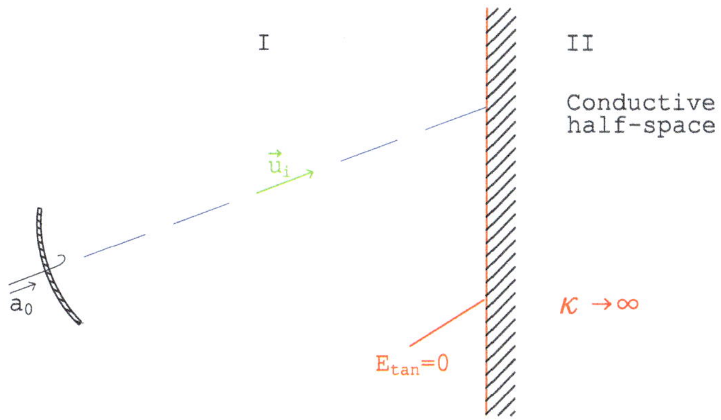

Fig. 11.11 Reflection of an antenna field on a conductive half-plane

The result according to Eq. (11.59) states that the radar cross section in the range $D/\sqrt{\varepsilon_r} \ll \lambda_0$ increases with the square of the scattering body volume and for a frequency-independent ε_r with the fourth power of the frequency.

These two statements still apply even if the scattering body deviates from the spherical shape. For any geometric shape, you also get

$$\sigma_{lin} = h_1 \frac{V^2}{\lambda_0^4}$$

and

$$\sigma_{krz} = h_2 \frac{V^2}{\lambda_0^4} . \tag{11.60}$$

Here, h_1 and h_2 are functions of ε_r, the scattering body shape and orientation as well as polarization.

The dependencies according to Eq. (11.60) are also valid for sufficiently small metallic scattering bodies.

For small metallic scattering bodies with any shape, the statement of Eq. (11.60) also applies.

It should also be noted that simple closed solutions are also obtained for rotational ellipsoids in addition to the spherical shape, provided that the dimensions are small compared to the wavelength.

11.4 Reflection on a Conducting Half-Plane

The electromagnetic field of a transmitting antenna is reflected by an infinitely extended, ideally conducting metallic half-plane as shown in Fig. 11.11.

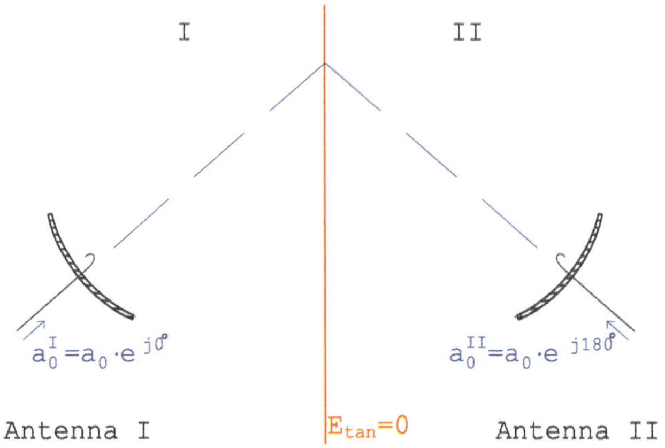

Fig. 11.12 Mirror-inverted second antenna

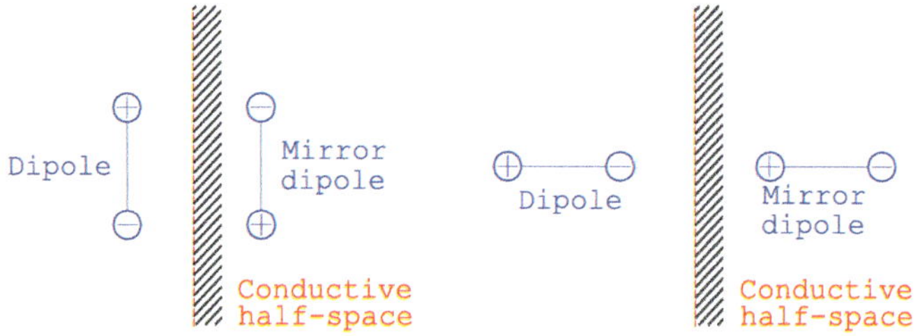

Fig. 11.13 Dipoles in front of a conductive half-space

The boundary condition that the tangential component of the electric field on the surface of the half-plane disappears can be fulfilled if a second identical transmitting antenna is arranged as a mirror image, as shown in Fig. 11.12, but is controlled in phase opposition (180°).

The conducting half-plane must be thought of as distant. The superposition of the fields of antenna I and antenna II then results in the true total field in half-space I, i.e. the incident field and reflection of the arrangement of Fig. 11.11.

For dipoles in front of a conducting half-plane, the mirroring principle can also be applied, as shown in Fig. 11.13. It should be noted that a positive (negative) charge in the mirror point results in a negative (positive) charge.

The aperture assignment of a surface antenna can also be replaced by many individual dipoles, which would then be mirrored as shown in Fig. 11.13.

An example from practice is the Yagi antenna (Fig. 11.14). If the reflecting half-space consists of a dielectric ε_r, the input wave a_0^{II} of the mirror antenna can be reduced approximately according to the reflection factor of the half-plane:

$$a_0^{\mathrm{II}} = \frac{\sqrt{\varepsilon_r} - 1}{\sqrt{\varepsilon_r} + 1}\, a_0 \,. \tag{11.61}$$

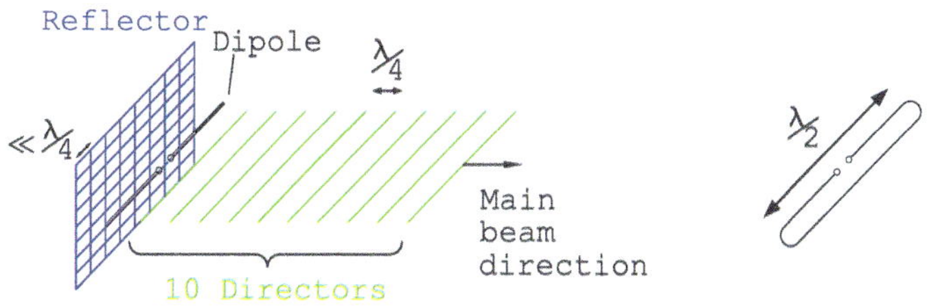

Fig. 11.14 Left: Yagi antenna, right: folding dipole

Note the distance dependence of the received power for the different reflectors. For the infinite half-plane, $|b_0|^2 \sim 1/R^2$.

11.5 Dependence of the Radar Cross-Section on the Frequency and the Aspect Angle

The principle of the interference effect can be explained using a model with two identical isotropic frequency-independent point scatterers. Fictitious isotropic point scatterers are scattering objects that have no spatial extent and whose scattering amplitude γ_0 is independent of the viewing direction or the aspect angle. The model of such a point scatterer provides a clearer representation. According to Fig. 11.15, it applies to the scattering cross-section σ (Fig. 11.16) as a function of the aspect angle θ and the angular frequency ω:

$$\sigma(\theta,\omega) = \left|\sum_i \gamma_i(\theta,\omega)\right|^2 = |\gamma_0|^2 \left| e^{j\frac{\omega L\cos\theta}{c}} + e^{-j\frac{\omega L\cos\theta}{c}} \right|^2 . \tag{11.62}$$

$$\sigma(\theta,\omega) = 4\,\sigma_0\,\cos^2\left(\frac{\omega L\cos\theta}{c}\right) . \tag{11.63}$$

The upper part of the Fig. 11.16 shows the dependence of the scattering cross-section σ on the angular frequency and in the lower part of the Fig. 11.16 on the aspect angle.

For a larger real object, there will be many more mostly unequal scattering centers, so the scattering diagram will look more complicated. Fig. 11.17 shows a measurement result on an airplane.

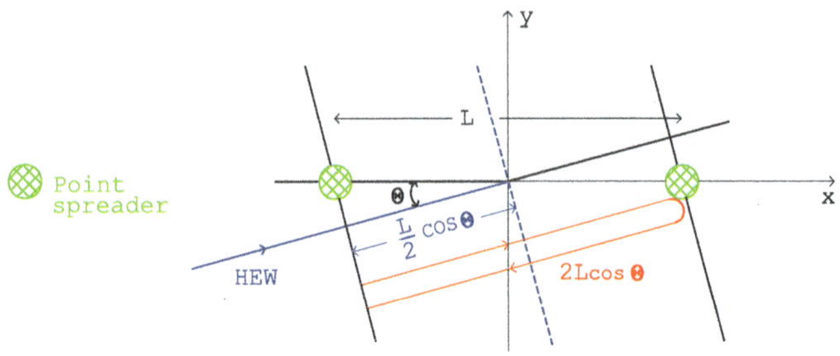

Fig. 11.15 Coordinates for two point scatterers at the incidence of a homogeneous plane wave (HEW)

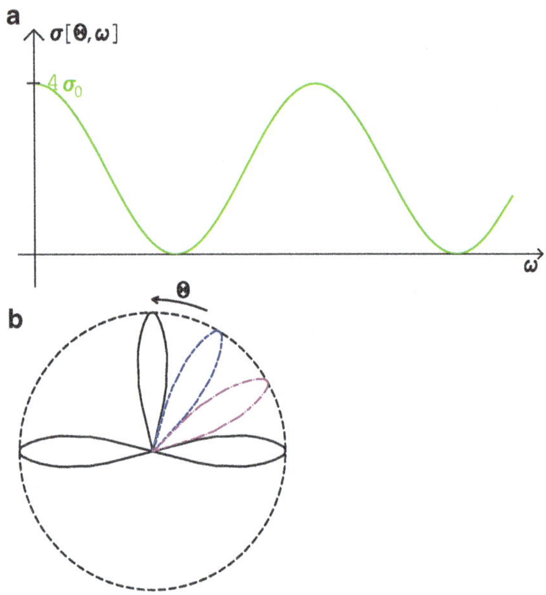

Fig. 11.16 Scattering cross-section σ as a function of the frequency ω and the aspect angle θ, where applies: $L = 4\lambda$

11.6 Volumetric Meteorological Radar Targets

In the presence of precipitation, the totality of all raindrops (or hailstones, snowflakes, etc.) represents a volumetric target whose totaldimensions are usually large compared to the beam dimensions. The detection of such a meteorological target is desirable for a weather radar. On the other hand, it is undesirable in a radar for airspace surveillance, for example, because the echoes originating from the precipitation can obscure the echoes belonging to the aircraft (the precipitation echoes are then **Clutter**).

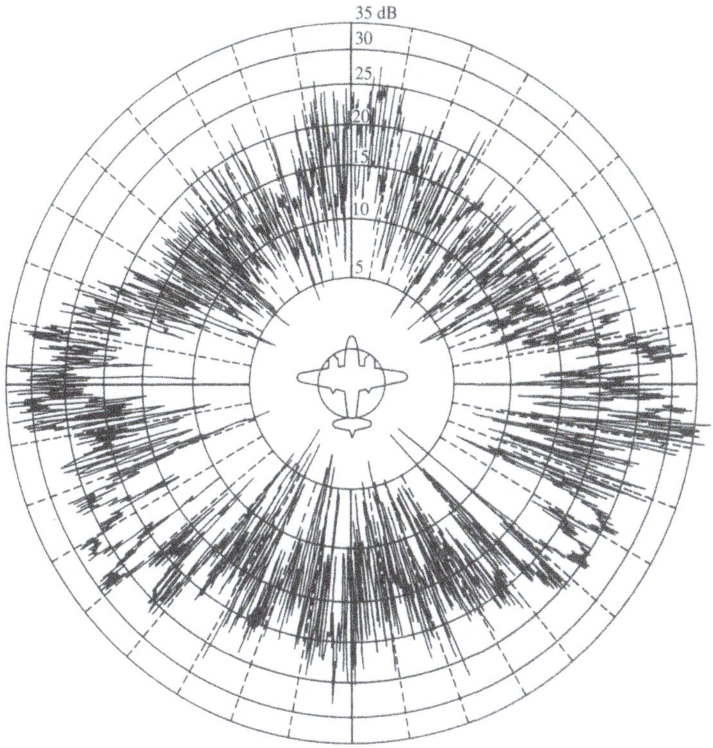

Fig. 11.17 Backscatter diagram of an airplane, [97]

For a simple calculation, the gain function of the antenna is roughly approximated. It is assumed that the total radiated power is transported in a beam with an elliptical cross-section. The diameters of this beam are thus $\Delta\theta\ R$ and $\Delta\varphi\ R$. Within this beam, the radiation density is assumed to be constant and the gain function is set equal to the gain in the main beam direction. Furthermore, a rectangular modulated sine pulse of length Δt is assumed. Due to the doubling of the travel distance due to the path being traversed twice, the echo signal is transmitted through all particles in a volume range of depth $c\ \Delta t/2$ and the cross-section

$$A = \pi\ \frac{\Delta\theta\ \Delta\varphi}{4}\ R^2 \tag{11.64}$$

influenced (Fig. 11.18).

This volume area

$$V = \pi\ c\ \frac{\Delta t}{2}\ \frac{\Delta\theta\ \Delta\varphi}{4}\ R^2 \tag{11.65}$$

is referred to as the resolution cell. The backscattered echo signal results from the interference of all individual echoes of the N particles (i. e. raindrops) located in

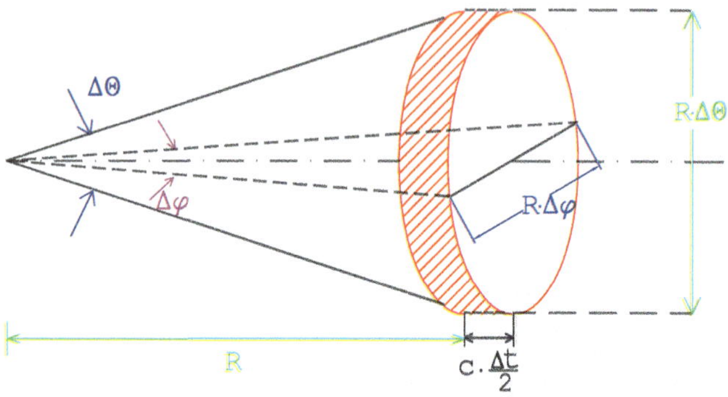

Fig. 11.18 To define a resolution cell

the resolution cell with the radar cross sections $\sigma_{0,\nu}(\nu = 1,\ldots,N)$. The radar cross-section depends on the respective polarization and frequency. If we assume randomly distributed particles with a uniformly distributed phase, we can show that the total radar cross-section σ is equal to the sum of the individual cross-sections:

$$\sigma = \sum_{\nu=1}^{N} \sigma_{0,\nu} \,. \tag{11.66}$$

If the square of the magnitude of the sum of all scattering amplitudes is formed as in Eq. (11.63), the mixed terms stand out on average, similar to uncorrelated signals.

Because the volume of the resolution cell depends on the antenna parameters $\Delta\theta$ and $\Delta\varphi$ as well as on the distance R and the pulse duration t, the radar cross-section used in Eq. (11.66) does not only depend on the properties of the target. depends on the properties of the target. Therefore, σ is conveniently related to the volume of the resolution cell, and the specific radar cross-section per unit volume Σ is obtained. For this can also be written as follows after introducing the particle number n per unit volume (with $\sigma_{0,\nu} \approx \sigma_0$ for all ν):

$$\Sigma = \frac{\sigma}{V} = \frac{1}{V} \sum_{\nu=1}^{N} \sigma_{0,\nu} = \sum_{\nu=1}^{n} \sigma_0 \,. \tag{11.67}$$

This is the `radar equation` for volumetrically distributed targets:

$$\frac{\left| b_0 \right|^2}{\left| a_0 \right|^2} = \frac{\lambda_0^2}{(4\,\pi)^3\, R^4}\, G^2\, V\, \Sigma = \frac{\lambda_0^2\, G^2}{(4\,\pi)^3}\, \frac{\Delta\theta\, \Delta\varphi}{4}\, \frac{\pi}{R^2}\, \frac{c\, \Delta t}{2}\, \Sigma \,. \tag{11.68}$$

Note that here the insertion loss has a dependence with $1/R^2$, in contrast to individual targets small in relation to the beam width, where the dependence can be described with $1/R^4$. This is due to the fact that the resolution cell becomes larger with increasing distance and thus more and more particles are detected.

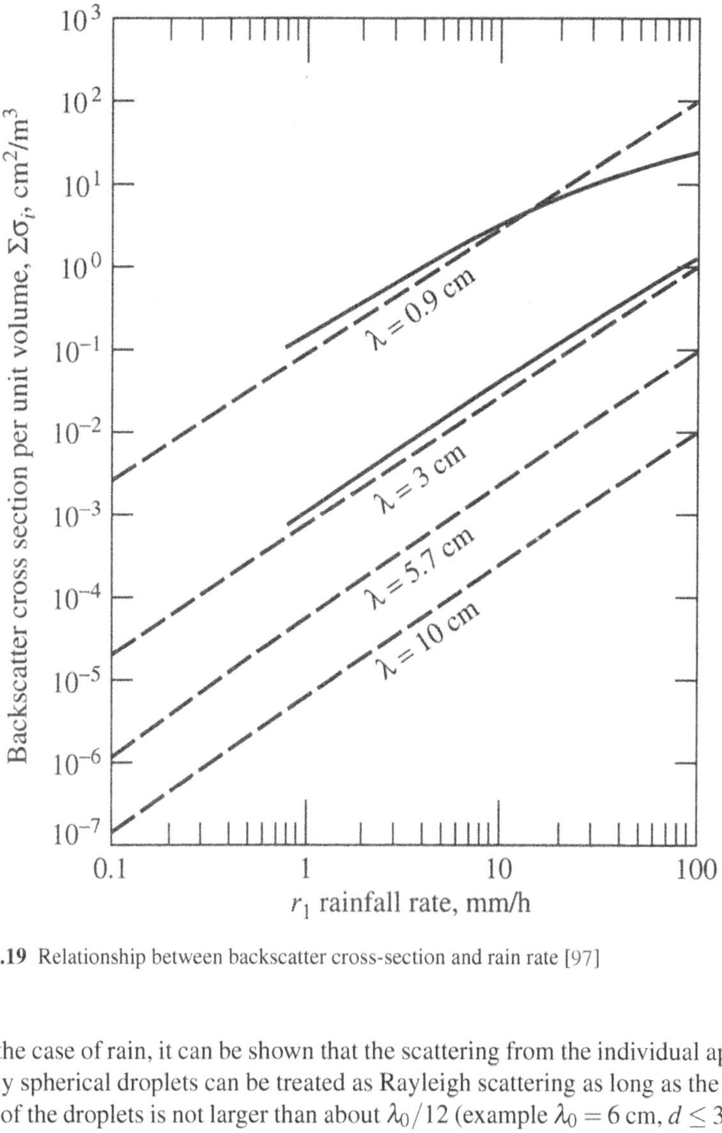

Fig. 11.19 Relationship between backscatter cross-section and rain rate [97]

In the case of rain, it can be shown that the scattering from the individual approximately spherical droplets can be treated as Rayleigh scattering as long as the diameter d of the droplets is not larger than about $\lambda_0/12$ (example $\lambda_0 = 6$ cm, $d \leq 3$ mm). If the Rayleigh cross-section is applied according to Eq. (11.59), then after summing over n raindrops per unit volume, the result is

$$\Sigma_{lin} = \frac{\pi^5}{\lambda_0^4} \left| \frac{\varepsilon_r - 1}{\varepsilon_r + 2} \right|^2 \sum_{\nu=1}^{n} d_\nu^6 \,,$$
$$\Sigma_{zir} = 0 \,. \tag{11.69}$$

Fig. 11.19 shows results on the radar cross-section per unit volume as a function of the rain rate per hour with the wavelength λ_0 as additional parameter.

List of Frequently used Formula Symbols and Abbreviations

A Area; amplifier mode of operation
$[\mathbf{A}]$ A-chain matrix of two-port elements
AB Amplifier operating mode
$[\mathbf{ABCD}]$ ABCD chain matrix of two-elements
AC Alternative current
AD Amplitude Discriminator
ADC Analog-to-digital converter
ADS Advanced Design System (from Agilent)
AF Audio Frequency
AGC Switchable amplifier with regulation, Automatic Gain Control
ALL Amplitude-Locked Loop
AM Amplitude Modulation
A_{ij} Elements of the matrix $[\mathbf{A}]$
A_w Antenna effective area
a_i, b_i Complex wave parameters (mono-mode case)
a_i^+, b_i^+ Complex wave parameters of a common mode
a_i^-, b_i^- Complex wave sizes of a push-pull mode
B Susceptance, reactive conductance, amplifier operating mode
$[\mathbf{B}]$ Inverse chain matrix of two-port elements
B_{ij} Elements of the matrix $[\mathbf{B}]$
b_0 Complex generator wave
C, C, C' Amplifier operating mode, capacitanceät, capacitanceät per length unit
$C(\theta, \varphi)$ Directional characteristic
CP Charge pump
c_0, c Speed of light
D Amplifier operating mode
DC Dircct Currcnt
DDS Direct Digital Synthesizer
DUT Device Under Test
dB Decibel
dBc Power ratio of a signal to a carrier signal, expressed in decibels

H. Heuermann, *Microwave Technology*,
https://doi.org/10.1007/978-3-658-45686-3

det() Determinant of a matrix
$\delta(\omega)$ Dirac's delta function as a function of frequency
E Amplifier operating mode
E Electric field strength
ECD Equivalent Circuit Diagram
EVM Error Vector Magnitude
F Amplifier operating mode
F Noise figure
FD Frequency domain
$F(j\omega)$ General function in the frequency domain
FOM Figure Of Merit
FTR Frequency Tuning Range
$F(s)$ Control filter
f Frequency
f_c Cut-off frequency
f_{HF} Frequency in the MHz or GHz range
f_{IF} Frequency in the DC, Hz, kHz or lower MHz range
f_{LO} Frequency of the oscillator in the MHz or GHz range
f_{RF} Frequency in the RF range
f_r Resonant frequency
f_{ZF} Frequency in the DC, Hz, kHz or lower MHz range
$f(t)$ General function in the time domain
G, G' Active conductance, active conductance per unit length, gain
GaN Gallium Nitride
G_{av} Available gain
H Magnetic field strength, amplifier operating mode
HB Harmonic Balance
HBT Heterojunction-Bipolar-Transistor
HEMT High Electron Mobility Transistors
HR Harmonic Radar
h Planck's constant
$h(t)$ Impulse response as a function of time
I Complex current
$[\mathbf{I}]$ Unit matrix
Im{} Imaginary operator
IF Intermediate Frequency
IM Intermodulation
IMP Institute for Microwave and Plasma technology, Univ. FH Aachen, director: Prof. H. Heuermann
IMT Data format from Agilent
ISM bands Bands for Industrial, Scientific and Medical applications
InP Indium Phosphide
i Complex current related to Z_0
i, j Running indices for matrix and vector elements, ranging from n=1 to the respective dimension

$i(t)$ Current in the time domain
J Current density, amplifier operating mode
k Wave number, stability number (k-factor), coupling factor of resonators
L, L' Inductance, inductance per unit length
$L(\Delta\omega)$ Single sideband noise power
LDMOS Laterally Diffused Metal Oxide Semiconductor
LO Local Oscillator
l Distance
ℓ Line length
lg () Logarithm on the base 10
M Match (termination of a wave)
M Magnetization
M_{ij} M-parameters or mixed-mode parameters
$[\mathbf{M}]$ M-parameter matrix or mixed-mode parameter matrix
MAG Maximum Available Gain
MF Mixed-Frequency
MF-S Mixed-Frequency-S-parameters
MESFET Metaloxid Semiconductor Field Effect Transistor
MSG Maximum Stable Gain
m Mass
1/N Frequency divider (value)
O Open circuit, (Open)
OP Operation amplifier
P Electrical power
PA Power Amplifier
PAE Power Added Efficiency
PD Phase Discriminator
PFD Phase Frequency Discriminator
PM Phase Modulation
P_w Active electrical power
$P(\omega)$ Sampling function in the frequency domain
$[\mathbf{P}]$ Propagation matrix
p Electrical length, dipole moment
$p(t)$ Pseudo impulse response as a function of time
Q Quality of components and resonators, reactive power, charge
Q_0 Unloaded quality factor of a resonator
Q_L Loaded quality factor of a resonator
R, R' Effective resistance, effective resistance per unit length
Re{} Real part operator
rect (t) Generalized rectangular function as a function of time
$\mathrm{rep}_\mathrm{T}\, h(t)$ Periodic repetition of the impulse response
RF Radio Frequency
Rg () Rank of a matrix
RX Receiving unit
r_x Reflection parameter of single-port and terminal elements

S Short circuit, (short), amplifier operating mode
S Poynting vector, power density, radiation density
SDR Software Defined Radio
SFCW Stepped-Frequency Continuous Wave
S_{ij} Scattering parameters, S-parameters
S_{ij}^{kl} Mixed-frequency S-parameter from the frequency l to the frequency k
$[\mathbf{S}]$ Scattering parameter matrix
$[\mathbf{S}]^{-1}$ Inverse (scattering parameter) matrix
$[\mathbf{S}]^{T}$ Transposed (scattering) matrix
$[\mathbf{S}]^{*}$ Conjugate complex (S-) matrix
$\mathrm{s_a}$ Low-frequency output signal
$\mathrm{s_r}$ Received signal
$\mathrm{s_t}$ Transmit signal
T, T Time interval, through connection (Thru or Through), temperature
T_e Effective noise temperature
TD Time domain
TEM Transversal Electro-magnetical Mode
TX Transmission unit
t Time variable, transformation ratio of an impedance transformer
track() Track of a matrix
U Complex voltage, node potentials
u Complex voltage related to Z_0, normalized complex voltage
$u(t)$ Voltage in the time domain
V Volume
VCO Voltage controlled oscillator
VNA Vectorial measuring network analyzer, vector network analyzer
v Propagation velocity
v_{gr} Group velocity
v_p Power amplification
v_{ph} Phase velocity
$W_{e,m}$ Electric or magnetic field energy
W Directional characteristic
w Width, occupancy function
X Reactance, reactance
XCO Crystal oscillator
X_{ij}^{kl} Mixed-frequency X-parameter from the frequency l to the frequency k
Y Admittance, apparent conductance
$[\mathbf{Y}]$ Y conductance matrix of two-port elements
y_x Admittance parameters of one-port and terminal elements
Z Impedance
Z_0 Characteristic impedance, reference impedance (usually $50\,\Omega$)
Z_E, Z_e Input resistance or impedance
Z_{even}, Z_+ Common mode impedance
Z_L Characteristic impedance of a line (usually $50\,\Omega$)
Z_{odd}, Z_- Differential mode impedance

$[\mathbf{Z}]$ Z-resistance matrix of two-port elements
z_x Impedance parameters of single-port and terminal elements
α, β Coefficients with alternating meaning
α Attenuation constant or coating of a cable
β Phase constant or coating of a line
γ Complex propagation or propagation constant of a line
$\gamma \ell$ Complex electrical length
$\Delta()$ Determinant of a matrix
δ Penetration depth, skin depth
$\delta(\omega)$ Dirac's delta function as a function of frequency
ε Dielectric constant
$\varepsilon_0, \varepsilon_r$ Absolute and relative dielectricät constant
η Efficiency factor
κ (Specific) conductivity
λ Wavelength
λ_D Debye length
λ_{MFP} Mean free path length
μ Permeability constant

References

1. Agilent Technologies, Schottky Barrier Diode Video Detektor, Application Note 923, 1999
2. Aref, A. F., Askar, A., Nafe A. A., Tarar, M. M., Negra, R., Efficient Amplification of Signals with high PAPR using a Novel Multilevel LINC Transmitter Architecture, European Microwave Conference 2012, pp. 1035–1038
3. Bächthold, W., Mikrowellentechnik, Vieweg, Braunschweig/Wiesbaden, 1999
4. Bächthold, W., Mikrowellenelektronik, Vieweg, Braunschweig/Wiesbaden, 2002
5. Best, R., Theorie und Anwendungen des Phase-Locked Loops, 1982
6. Blume, S., Theorie elektromagnetischer Felder, Hüthig, Heidelberg, 1988
7. Bronstein, I., Semendjajew, K., Taschenbuch der Mathematik, Harry Deutsch Verlag, Thun, Frankfurt/Main, 1987
8. Cage, O. Electronic Measurements and Instrumentation, McGraw-Hill, Tokyo, 1971
9. Chen, Z. N., Lui, D., Nakano, H., Qing, X., Zwick, T., Handbook of Antenna Technologies, Springer, Singapore, 2016
10. J. Choi, F. Iza, H. J. Do, J. K. Lee and M. H. Cho, Microwave-excited atmospheric-pressure microplasmas based on a coaxial transmission line resonator, Plasma Sources Sci. Technol., Vol. 18 025029, NO. 6, 2009
11. Cripps, S.C.: RF Power Amplifiers for Wireless Communications, Artech House, Boston, 1. Auflage, 1999
12. Peter Wright, Jonathan Lees, Johannes Benedikt, Paul J. Tasker,and Steve C. Cripps, A Methodology for Realizing High Efficiency Class-J in a Linear and Broadband PA, IEEE Transactions on Microwave Theory and Techniques, IEEE Trans. Microwave Theory Tech., MTT-57, Dec. 2009, pp. 3196–3204
13. Detlefsen, J., Siart, U., Grundlagen der Hochfrequenztechnik, Oldenbourg Verlag, München, 2003
14. Dunsmore, J.P., Handbook of Microwave Component Measurements with Advanced VNA Techniques; John Wiley and Sons: Hoboken, NJ, USA, 2020; ISBN 978-1-119-47713-6
15. Engen, G.F., Hoer, C.A., Thru-Reflect-Line: An Improved Technique for Calibrating the Dual Six Port Automatic Network Analyzer, IEEE Trans. Microwave Theory Tech., MTT-27, Dec. 1979, pp. 987–993
16. Erkens, H., Heuermann, H., Blocking Structures for Mixed-Mode-Systems, European Microwave Conf., Amsterdam, Oct. 2004, pp. 297–300
17. Essaadali, R., Jarndal, A., Kouki, A. B., Ghannouchi, F. M., Conversion Rule Between X-Parameters and Linearized Two-Port Network Parameters for Large-Signal Operating Conditions, IEEE Trans. Microwave Theory Tech., MTT-11, Nov. 2018, pp. 4745–4756
18. Eul, H.-J., Methoden zur Kalibrierung von heterodynen und homodynen Netzwerkanalysatoren, Dissertationsschrift, Institut für Hoch- und Höchstfrequenztechnik, Ruhr-Universität Bochum, 1990

H. Heuermann, *Microwave Technology*,
https://doi.org/10.1007/978-3-658-45686-3

19. EUL, H.J., SCHIEK, B., A Generalized Theory and New Calibration Procedures for Network Analyzer Self-Calibration, IEEE Trans. Microwave Theory Tech., MTT-39, Apr. 1991, pp. 724–731
20. FERRERO, A., PISANI, U., KERWIN, K.J., A New Implementation of a Multiport Automatic Network Analyzer, IEEE Trans. Microwave Theory Tech., vol. 40, 1992, pp. 2078–2085
21. FRANZ, G., Niederdruckplasmen in Mikrostrukturtechnik, Springer Verlag, 2004, 3. Auflage
22. GÄRTNER, U., Ein homodynes Streuparametermessverfahren mit digitaler Phasenmodulation für den Mikrowellenbereich, Dissertationsschrift, Institut für Hoch- und Höchstfrequenztechnik, Ruhr-Universität Bochum, 1986
23. GALLAGHER, K. A., MAZZARO, G.J.; MARTONE, A.F., SHERBONDY, K.D., NARAYANAN, R.M., Derivation and Validation of the Nonlinear Radar Range Equation. In Proceedings of the SPIE Volume 9829, Radar Sensor Technology XX, SPIE Defense + Security, Baltimore, MD, USA, 2016
24. GALLAGHER, K. A., Harmonic Radar: Theory and Applications to Nonlinear Target Detection, Tracking, Imaging and Classification. PhD thesis, Pennsylvania State University, 2015
25. GOLDBERG, B.-G., Digital Techniques in Frequency Synthesis, 1998
26. GONZALEZ G., Microwave Transistor Amplifiers, Prentice-Hall, Upper Saddle River, NJ, 2. Auflage, 1997
27. GONZALEZ G., Foundations of Oscillator Circuit Design, Artech House, Boston, 2007
28. VON GRONOW, M., HEUERMANN, H., KLING, R., Neue ultrakompakte Leuchtstofflampen mit integriertem HF-Transformator, Licht2010, Wien, Oct. 2010
29. GRONAU, G., Rauschparameter- und Streuparameter-Messtechnik, Fortschritte der Hochfrequenztechnik, Band 4, Verlagsbuchhandlung Nellissen-Wolff, Aachen, 1992
30. GRONAU, G., Höchstfrequenztechnik, Springer, Berlin, 2001
31. HARZHEIM, TH., Mixed Frequency Single Receiver Architectures and Calibration Procedures for Linear and Non-Linear Vector Network Analysis, Promotionsarbeit (FH Aachen und Univ. Luxemburg), http://hdl.handle.net/10993/39176, 2019
32. HARZHEIM, TH., HEUERMANN, H., Phase Repeatable Synthesizers as a New Harmonic Phase Standard for Nonlinear Network Analysis, IEEE Transactions on Microwave Theory and Techniques, Volume: 66, Issue:6, June 2018
33. HARZHEIM, TH., MUEMEL, M., HEUERMANN, H., A SFCW Harmonic Radar System for Maritime Search and Rescue using Passive and Active Tags. Int. J. Microw. Wirel. Technol., https://doi.org/10.1017/S1759078721000520, 2021
34. HOMEPAGE, HEUERMANN HF-TECHNIK GMBH, http://www.hhft.de
35. HOMEPAGE: HF-LEHRGEBIET DER FH-AACHEN, https://www.fh-aachen.de/fachbereiche/elektrotechnik-und-informationstechnik/labore-und-lehrgebiete/lehrgebiet-hoch-und-hoechstfrequenztechnik-prof-dr-ing-holger-heuermann/
36. HOMEPAGE: INSTITUT FÜR MIKROWELLEN- UND PLASMATECHNIK, IMP DER FH-AACHEN, https://www.fh-aachen.de/forschung/imp/
37. HEUERMANN, H., Sichere Verfahren zur Kalibrierung von Netzwerkanalysatoren für koaxiale und planare Leitungssysteme, Dissertationsschrift, Institut für Hochfrequenztechnik, Ruhr-Universität Bochum, 1995, ISBN 3-8265-1495-5
38. HEUERMANN, H., Hochfrequenztechnik. Komponenten und Mess-, Funk- RFID- sowie Lokalisierungssysteme, Springer Vieweg, ISBN 978-3-658-37825-7, 4. edition, 2023
39. HEUERMANN, H., GSOLT: The Calibration Procedure for all Multi-Port Vector Network Analyzers, MTT-S International Microwave Symposium, Philadelphia, 2003
40. HEUERMANN, H., Calibration of a VNA Without a Thru Connection for Non-Linear and Multi-Port Measurements, IEEE Trans. Microwave Theory Tech., Nov. 2008
41. HEUERMANN, H., Wahlpflichtvorlesung: Antennen und Ausbreitung, FH-Aachen, Homepage Lehrgebiet Hoch- und Höchstfrequenztechnik
42. HEUERMANN, H., Verfahren zur Kalibrierung von Netzwerkanalysatoren ohne Durchverbindungen zur Messung von linearen und nichtlinearen elektrischen Parameter, Rohde&Schwarz-Patent DE102007027142.7 and PCT/EP2007/005215 from 13.06.2007

43. HEUERMANN, H., HARZHEIM, CRONENBROECK, Th., First SIMO Harmonic Radar Based on the SFCW Concept and the HR Transfer Function, Remote Sensing 13(24), DOI: 10.3390/rs13245088, Dec. 2021
44. HEUERMANN, H., Vorrichtung zum Erzeugen eines abgekoppelten Mikrowellenplasmas, DE10 2024 102 296.5, 26.01.2024
45. HEUERMANN, FINGER, T., Microwave Spark Plug for Very High Pressure Conditions, 3th conference: Advanced Ignition Systems for Gasoline Engines, Berlin, 2014
46. HEUERMANN, H., SADEGHFAM, A., Analog Amplitude-Locked Loop Circuit to Support RF Energy Solutions, 64th International Microwave Symposium, San Francisco, May 2016
47. HOFFMANN, M.H.W., Hochfrequenztechnik, Springer-Verlag, Berlin-Heidelberg, 1997
48. HOPWOOD J., F. IZA, S. COY AND D. B. FENNER, A microfabricated atmospheric-pressure microplasma source operating in air, J. Phys. D: Appl. Phys., 38, 2005
49. IBRAHIM, I., HEUERMANN, H., Improvements in the Flicker Noise Reduction Technique for Oscillator Designs, European Microwave Conf., Rom, 2009
50. IBRAHIM, I., HEUERMANN, H., Novel Theory and Architecture of a Vector Signal Generator Implemented With Two PLLs, European Microwave Conf., Amsterdam, 2008
51. INFINEON, Halbleiter, Publicis Corporate Publishing, Erlangen, 2004
52. INAC, O., UZUMKOL, M., REBEIZ, G. M., 45-nm CMOS SOI Technology Characterization for Millmeter-Wave Applications, IEEE Trans. on Microw. Theory and Tech., Vol. 62, No. 6, June 2014
53. ITOH T., HADDAD G., HARVEY J.: RF Technologies for Low Power Wireless Communications, Wiley-Interscience, New York, 1. Auflage, 2001
54. JAFFE, R., RECHTIN, E.: Design and Performance of Phase-Lock Circuits Capable of Near-Optimum Performance over a Wide Range of Input Signal and Noise Levels, IRE Trans. on Information Theory, vol. 1, no. 1, pp. 66–76, 1955
55. JANSSEN, R., Streifenleiter und Hohlleiter, Hüthig-Verlag, Heidelberg, 1992
56. KARK, K., Antennen und Strahlungsfelder, Vieweg-Verlag, Wiesbaden, 2004
57. KAJFEZ, D., GUILLON, P., Dielectric Resonators, Artech House, Dedham, 1986
58. KAWASAKI, K.,ET. AL., An Octa-Push Oscillator at V-Band, IEEE Trans. Microwave Theory Tech., MTT-58, July 2010
59. KERSTING, R., SCHOPP, C., HEUERMANN, H., LANGER, L., SCHOMER, S., LUCKAS, B., Additive Manufacturing With Metallic Wire Using a Coaxial Fed Microwave Plasma Jet, in IEEE Transactions on Plasma Science, vol. 52, no. 2, pp. 336-343, doi: 10.1109/TPS.2023.3309231, Feb. 2024
60. J. Kim and K. Terashima, 2.45 GHz Microwave-excited Atmospheric Pressure Air Microplasmas based on Microstrip Technology, Appl. Phys. Lett. 86, 2005
61. KRAMME, R., Medizintechnik (Verfahren, Systeme, Informationsverarbeitung), 2. Auflage, Springer Verlag, S. 397
62. KUMMER, M., Grundlagen der Mikrowellentechnik, VEB Verlag Technik, Berlin, 1989
63. KUROKAWA, Power Waves and the Scattering Matrix, IEEE Trans. Microwave Theory Tech., MTT, Mar. 1965, pp. 194–202
64. LEE, T., HAJIMIRI, A., Oscillator Phase Noise: A Tutorial, IEEE Journal, Solid-State Circuits, Vol. 35, No. 3, pp. 326–336, 2000
65. LEESON, D., A Simple Model of Feedback Oscillator Noise Spectrum, IEEE Conference, Vol. 54, No. 2, pp. 329–330, 1966
66. K. LINKENHEIL, H. O. RUOSS, T. GRAU, J. SEIDEL AND W. HEINRICH, A Novel Spark Plug for Improved Ignition in Engines with Gasoline Direct Injection (GDI),*IEEE Trans. Plasma Sci.*, Vol. 33, No. 5, Oct. 2005
67. MATTHEI, G., YOUNG, L., JONES, E., Microwave Filters, Impedance-Matching Networks, and Coupling Structures, McGraw-Hill Verlag, New York, 1980
68. MAAS, S. A., Nonlinear Microwave Cirtuits, IEEE Press, New York, 1996
69. MAAS, S. A., Microwave Mixers, Artech House, London, 1993
70. MAZZARO, G. J., MARTONE, A. F., RANNEY, K. I., NARAYANAN, R. M., Nonlinear Radar for Finding RF Electronics: System Design and Recent Advancements, IEEE Trans. on Microw. Theory and Tech., 2017, pp. 1716–1726

71. MILLER, Technique Enhances the Performance of PLL Synthesizers, Microwave & RF, Jan. 1993
72. MÖLLER, M., Entwurf und Optimierung monolithisch integrierter Breitbandverstärker in Si-Bipolartechnologie für optische Übertragungssysteme, Dissertationsschrift, Ruhr-Universität Bochum, 1999
73. MAXIM, Data Sheet of MAX4003, 0.1–2.5 GHz Power detector, 19-2620; Rev 1, 03.2003
74. PARISI, S. J., 180° Lumped Element Hybrid, IEEE Microw. Theory and Tech. – Sym. Digest, 1989, pp. 1243–1246
75. PETERS, N., SCHMITZ, TH., SADEGHFAM, A., HEUERMANN, H. Concept of Balanced Antennas with Load-Invariant Base Impedance Using a Two Element LC-Coupler, Proceedings of the European Microwave Association, 2005
76. PHILIPPOW, E., Grundlagen der Elektrotechnik, VEB Verlag Technik, Berlin, 1988
77. RANDALL, W., Oscillator Design and Computer Simulation, 2nd ed. SciTech Publishing Inc., Raleigh NC., 2000
78. RAICU, D., Multiterminal Distributed Resistors as Microwave Attenuators, IEEE Trans. Microwave Theory Tech., vol. 42, No. 7, July 1994, pp. 1140–1148
79. RAZAVI, B., Design of Analog Intregrated Circiuts, McGraw-Hill-Verlag, ISBN 0-07-118839-8, 2001
80. RHODES, J.D., Theory of Electrical Filters, Wiley Verlag, New York, 1985
81. RILEY, COPELAND AND KWASNIEWSKI, Digital PLL Frequency Synthesizers, 1983 Delta-Sigma Modulation in Fractional-N Frequency Synthesis, IEEE Journal of Solid State Circuits, Vol. 28, No. 5, May 1993
82. ROHDE, U., NEWKIRK, D., RF/Microwave Circuit Design for Wireless Applications, Wiley Verlag, New York, 2000
83. ROOT, D. E., VERSPECHT, J., HORN, J., MMARCU, M., X-Parameters, Cambridge University Press, New York, 2013
84. ROHDE, U.L., Digital PLL Frequency Synthesizers: Theorie and Design, Prentice Hall, 1982
85. ROHDE, U.L., Microwave and Wireless Synthesizers, Wiley-Interscience, 1997
86. SADEGHFAM, A., SADEGHI-AHANGAR, A., ELGAMAL, A.AND HEUERMANN, H., Design and Development of a Novel Self-Igniting Microwave Plasma Jet for Industrial Applications, 67th International Microwave Symposium, Boston, June 2019
87. SCHAAF, M., Theoretische Beschreibung und Validierung einer neuartigen S11-Regelschleife für den GHz-Bereich, Masterarbeit, IMP, Aachen, 2011
88. SCHIEK, B., Grundlagen der Hochfrequenz-Messtechnik, Springer Verlag, Berlin Heidelberg, 1999
89. SCHIEK, B., SIWERIS, H.-J., Rauschen in Hochfrequenzschaltungen, Hüthig Verlag, Heidelberg, 1990
90. SCHIEL, J.-CH., TATU, S. O., WU, K., BOSISIO R. G., Six-Port Direct Digital Receiver (SPDR) and Standard Direct Receiver (SDR) Results for QPSK Modulation at High Speeds, MTT-S International Microwave Symposium, Philadelphia, 2003, pp. 931–935
91. SCHNEIDER, M.V., Microstrip Lines for Microwave Integrated Circuits, Bell Syst. Tech. J. 48, 1969, pp. 1421–1444
92. SCHNEIDER, J., Entwicklung eines homodynen Netzwerkanalysators für den Mikrowellenbereich 26,5–40 GHz, Dissertationsschrift, Institut für Hoch- und Höchstfrequenztechnik, Ruhr-Universität Bochum, 1987
93. SCHOPP, Ch., ROHRBACH, F. LANGER, L., HEUERMANN, H., Detection of Welding Wire Length by Active S11 Measurement, IEEE Transactions on Plasma Science, doi: 10.1109/TPS.2024.3356659, Jan. 2024
94. SIMONYI, K., Theoretische Elektrotechnik, VEB Deutscher Verlag der Wissenschaften, Berlin, 1989
95. SINNESBICHLER, F., Hybrid Millimeter-Wave Push-Push Oscillators Using Silicon-Germanium HBTs, IEEE Trans. on Microw. Theory and Tech., Vol. 51, No. 2, Feb. 2003
96. SCHOPP, CH., DOLL, T., GRAESER, U., HARZHEIM, TH., HEUERMANN, H., KLING, R.,AND MARSO, M. Capacitively Coupled High-Pressure Lamp Using Coaxial Line Networks, IEEE Transactions on Microwave Theory and Techniques, Volume: 64, Issue: 10, Oct. 2016

97. SKOLNIK, MERRILL I., Introduction to RADAR systems, Third Edition, ISBN 0-07-288138-0, 2001
98. RECCO AB. Homepage. http://www.recco.com/
99. STEPHAN, A., PRANTNER, A., HEUERMANN, H., Cutting Human Tissue with Novel Atmospheric-Pressure Microwave Plasma Jet, European Microwave Conf., London, Oct. 2016
100. SOKAL, N. O., SOKAL, A. D., Class-E - A New Class of High Efficiency Tuned Single-Ended Power Amplifier, IEEE J. Solid State Circuits, SC-10, No. 3, June 1975, pp. 168–176
101. SPRINGER, A., MAURER, L., WEIGEL, R., RF System Concepts for Highly Integrated RFICs for W-CDMA Mobile Radio Terminals, IEEE Trans. on Microw. Theory and Tech., Vol. 50, No. 1, Jan. 2002
102. STENGEL, B., THOMPSON, B., Neutralized Differential Amplifiers using Mixed-Mode S-Parameters, IEEE MTT-S Digest, 2003
103. STOLLE, R., HEUERMANN, H., SCHIEK, B., Auswertemethoden zur Präzisions-Entfernungsmessung mit FMCW-Systemen und deren Anwendung im Mikrowellenbereich, tm – Technisches Messen, Heft 2/95, R. Oldenbourg Verlag, München, Feb. 1995, pp. 66–73
104. SÜSSE, R., Theoretische Grundlagen der Elektrotechnik 1, Teubner Verlag, Wiesbaden, 2005
105. SWANSON, D.G., HOEFER, W., Microwave Circuit Modeling Using Electromagnetic Field Simulation, Artech House-Verlag, New-York, 2004
106. TIEBOUT, M., Low Power VCO Design in CMOS, Springer, Berlin, 2006
107. TIETZE, U., SCHENK, Ch., Halbleiter-Schaltungstechnik, 13. Auflage, Springer Verlag, Heidelberg, 2010
108. TURDUMAMATOV, S., BELDA A., HEUERMANN, H., Shaping a Decoupled Atmospheric Pressure Microwave Plasma With Antenna Structures, Maxwell's Equations, and Boundary Conditions, in IEEE Transactions on Plasma Science, doi: 10.1109/TPS.2024.3383589, April 2024
makeatletter
109. VAN DER HEIJDEN, M. P., SPIRITO, M.,DE VREEDE, L. C. N.,VAN STRATEN, F., BURGHARTZ, J. N., A 2 GHz High-Gain Differential InGaP HBT Driver Amplifier Matched for High IP3, IEEE MTT-S Digest, 2003
110. VERSPECHT, J., ROOT, D. E., Polyharmonic Distortion Modeling, IEEE microwave magazine, June 2006
111. VOGES, E., Hochfrequenztechnik, Hüthig Verlag, Heidelberg, 2003
112. WADELL, B. C., Transmission Line Design Handbook, Artech House, Boston, 1991
113. WILLIAMS, A. B.,*Electronic Filter Design Handbook,* New York, McGraw-Hill
114. WILTRON, Microstrip Measurements with the Wiltron 360 Vector Network Analyzer, Application Note AN360-7, Jan. 1990.
115. ZINKE O., BRUNSWIG H., Lehrbuch der Hochfrequenztechnik, Band 1 + Band 2, Springer Verlag, Berlin, 1999; new English version: HARTNAGEL H. L., QUAY R., ROHDE U. L., RUDOLPH M., Fundamentals of the RF and Microwave Techniques and Technologies, Springer, Cham, 2023
116. ZINKE, O., BRUNSWIG, H., Lehrbuch der Hochfrequenztechnik, Band 1 + Band 2, Springer Verlag, Berlin, 1999; new English version: HARTNAGEL, H. L., QUAY, R., ROHDE, U. L., RUDOLPH, M., Fundamentals of the RF and Microwave Techniques and Technologies, Springer, Cham, 2023

Index

N

O

P

SPRINGER NATURE

GPSR Compliance

The European Union's (EU) General Product Safety Regulation (GPSR) is a set of rules that requires consumer products to be safe and our obligations to ensure this.

If you have any concerns about our products, you can contact us on ProductSafety@springernature.com

In case Publisher is established outside the EU, the EU authorized representative is:

Springer Nature Customer Service Center GmbH
Europaplatz 3
69115 Heidelberg, Germany

The manufacturer's authorised representative in the EU is Springer Nature Customer Service Centre GmbH, Europaplatz 3, 69115 Heidelberg, Germany. If you have any concerns regarding our products, please contact ProductSafety@springernature.com

Printed and bound by CPI Group (UK) Ltd, Croydon, CR0 4YY

30/07/2026

02181976-0003